Studies in Computational Intelligence, Volume 70

Editor-in-chief
Prof. Janusz Kacprzyk
Systems Research Institute
Polish Academy of Sciences
ul. Newelska 6
01-447 Warsaw
Poland
E-mail: kacprzyk@ibspan.waw.pl

Further volumes of this series
can be found on our homepage:
springer.com

Vol. 49. Keshav P. Dahal, Kay Chen Tan, Peter I. Cowling (Eds.)
Evolutionary Scheduling, 2007
ISBN 978-3-540-48582-7

Vol. 50. Nadia Nedjah, Leandro dos Santos Coelho, Luiza de Macedo Mourelle (Eds.)
Mobile Robots: The Evolutionary Approach, 2007
ISBN 978-3-540-49719-6

Vol. 51. Shengxiang Yang, Yew Soon Ong, Yaochu Jin Honda (Eds.)
Evolutionary Computation in Dynamic and Uncertain Environment, 2007
ISBN 978-3-540-49772-1

Vol. 52. Abraham Kandel, Horst Bunke, Mark Last (Eds.)
Applied Graph Theory in Computer Vision and Pattern Recognition, 2007
ISBN 978-3-540-68019-2

Vol. 53. Huajin Tang, Kay Chen Tan, Zhang Yi
Neural Networks: Computational Models and Applications, 2007
ISBN 978-3-540-69225-6

Vol. 54. Fernando G. Lobo, Cláudio F. Lima and Zbigniew Michalewicz (Eds.)
Parameter Setting in Evolutionary Algorithms, 2007
ISBN 978-3-540-69431-1

Vol. 55. Xianyi Zeng, Yi Li, Da Ruan and Ludovic Koehl (Eds.)
Computational Textile, 2007
ISBN 978-3-540-70656-4

Vol. 56. Akira Namatame, Satoshi Kurihara and Hideyuki Nakashima (Eds.)
Emergent Intelligence of Networked Agents, 2007
ISBN 978-3-540-71073-8

Vol. 57. Nadia Nedjah, Ajith Abraham and Luiza de Macedo Mourella (Eds.)
Computational Intelligence in Information Assurance and Security, 2007
ISBN 978-3-540-71077-6

Vol. 58. Jeng-Shyang Pan, Hsiang-Cheh Huang, Lakhmi C. Jain and Wai-Chi Fang (Eds.)
Intelligent Multimedia Data Hiding, 2007
ISBN 978-3-540-71168-1

Vol. 59. Andrzej P. Wierzbicki and Yoshiteru Nakamori (Eds.)
Creative Environments, 2007
ISBN 978-3-540-71466-8

Vol. 60. Vladimir G. Ivancevic and Tijana T. Ivacevic
Computational Mind: A Complex Dynamics Perspective, 2007
ISBN 978-3-540-71465-1

Vol. 61. Jacques Teller, John R. Lee and Catherine Roussey (Eds.)
Ontologies for Urban Development, 2007
ISBN 978-3-540-71975-5

Vol. 62. Lakhmi C. Jain, Raymond A. Tedman and Debra K. Tedman (Eds.)
Evolution of Teaching and Learning Paradigms in Intelligent Environment, 2007
ISBN 978-3-540-71973-1

Vol. 63. Wlodzislaw Duch and Jacek Mańdziuk (Eds.)
Challenges for Computational Intelligence, 2007
ISBN 978-3-540-71983-0

Vol. 64. Lorenzo Magnani and Ping Li (Eds.)
Model-Based Reasoning in Science, Technology, and Medicine, 2007
ISBN 978-3-540-71985-4

Vol. 65. S. Vaidya, L. C. Jain and H. Yoshida (Eds.)
Advanced Computational Intelligence Paradigms in Healthcare-2, 2007
ISBN 978-3-540-72374-5

Vol. 66. Lakhmi C. Jain, Vasile Palade and Dipti Srinivasan (Eds.)
Advances in Evolutionary Computing for System Design, 2007
ISBN 978-3-540-72376-9

Vol. 67. Vassilis G. Kaburlasos and Gerhard X. Ritter (Eds.)
Computational Intelligence Based on Lattice Theory, 2007
ISBN 978-3-540-72686-9

Vol. 68. Cipriano Galindo, Juan-Antonio Fernández-Madrigal and Javier Gonzalez
A Multi-Hierarchical Symbolic Model of the Environment for Improving Mobile Robot Operation, 2007
ISBN 978-3-540-72688-3

Vol. 69. Falko Dressler and Iacopo Carreras (Eds.)
Advances in Biologically Inspired Information Systems: Models, Methods, and Tools, 2007
ISBN 978-3-540-72692-0

Vol. 70. Javaan Singh Chahl, Lakhmi C. Jain, Akiko Mizutani and Mika Sato-Ilic (Eds.)
Innovations in Intelligent Machines-1, 2007
ISBN 978-3-540-72695-1

Javaan Singh Chahl
Lakhmi C. Jain
Akiko Mizutani
Mika Sato-Ilic
(Eds.)

Innovations in Intelligent Machines - 1

With 146 Figures and 10 Tables

Springer

Dr. Javaan Singh Chahl
Defence Science and Technology Organisation
Edinburgh
South Australia
Australia

Dr. Akiko Mizutani
Odonatrix Pty Ltd
Adelaide
South Australia
Australia

Prof. Lakhmi C. Jain
University of South Australia
Mawson Lakes Campus
Adelaide, South Australia
Australia
E-mail:- Lakhmi.jain@unisa.edu.au

Prof. Mika Sato-Ilic
Faculty of Systems and Information Engineering
University of Tsukuba
Japan

Library of Congress Control Number: 2007927247

ISSN print edition: 1860-949X
ISSN electronic edition: 1860-9503
ISBN 978-3-540-72695-1 Springer Berlin Heidelberg New York

Springer is a part of Springer Science+Business Media
springer.com

Cover design: deblik, Berlin
Typesetting by the SPi using a Springer LaTeX macro package
Printed on acid-free paper   SPIN: 11588450   89/SPi   5 4 3 2 1 0

Foreword

Innovations in Intelligent Machines is a very timely volume that takes a fresh look on the recent attempts of instilling human-like intelligence into computer-controlled devices. By contrast to the machine intelligence research of the last two decades, the recent work in this area recognises explicitly the fact that human intelligence is not purely computational but that it also has an element of empirical validation (interaction with the environment). Also, recent research recognises that human intelligence does not always prevent one from making errors but it equips one with the ability to learn from mistakes. The latter is the basic premise for the development of the collaborative (swarm) intelligence that demonstrates the value of the virtual experience pool assembled from cases of successful and unsuccessful execution of a particular algorithm.

The editors are to be complemented for their vision of designing a framework within which they ask some fundamental questions about the nature of intelligence in general and intelligent machines in particular and illustrate answers to these questions with specific practical system implementations in the consecutive chapters of the book.

Chapter 2 addresses the cost effectiveness of "delegating" operator's intelligence to on-board computers so as to achieve single operator control of multiple unmanned aerial vehicles (UAV). The perspective of cost effectiveness allows one to appreciate the distinction between the optimal (algorithmic) and the intelligent (non-algorithmic, empirical) decision-making, which necessarily implies some costs. In this context the decision to use or not to use additional human operators can be seen as the assessment of the "value" of the human intelligence in performing a specific task.

The challenge of the development of collaborative (swarm) intelligence and its specific application to UAV path planning over the terrain with complex topology is addressed in Chapters 3 and 4. The authors of these chapters propose different technical solutions based on the application of game theory, negotiation techniques and neural networks but they reach the same conclusions that the cooperative behaviour of individual UAVs, exchanging information about their successes and failures, underpins the development of human-like intelligence. This insight is further developed in Chapter 8 where the authors look at the evolution-based dynamic path planning.

Chapter 5 emphasises the importance of physical constraints on the UAVs in accomplishing a specific task. To re-phrase it in slightly more general terms, it highlights the fact that algorithmic information processing may be numerically correct but it may not be physically very meaningful if the laws of physics are not taken fully into account. This is exactly where the importance of empirical verification comes to fore in intelligent decision-making.

The practice of processing uncertain information at various levels of abstraction (granulation) is now well recognised as a characteristic feature of human information processing. By discussing the state estimation of UAVs based on information provided by low fidelity sensors, Chapter 6 provides a reference material for dealing with uncertain data. Discussion of the continuous-discrete extended Kalman filter placed in the context of intelligent machines underlines the importance of information abstraction (granulation).

Chapters 7 and 9 share a theme of enhancement of sensory perception of intelligent machines. Given that the interaction with the environment is a key component of intelligent machines, the development of sensors providing omni directional vision is a promising way to achieving enhanced levels of intelligence. Also the ability to achieve, through appropriate sensor design, long distance (low accuracy) and short distance (high accuracy) vision correlates closely with the multi-resolution (granular) information processing by humans.

The book is an excellent compilation of leading-edge contributions in the area of intelligent machines and it is likely to be on the essential reading list of those who are keen to combine theoretical insights with practical applications.

Andrzej Bargiela
Professor of Computer Science
University of Nottingham, UK

Preface

Advanced computational techniques for decision making on unmanned systems are starting to be factored into major policy directives such as the United States Department of Defence UAS Roadmap. Despite the expressed need for the elusive characteristic of "autonomy", there are no existing systems that are autonomous by any rigorous definition. Through the use of sophisticated algorithms, residing in every software subsystem (state estimation, navigation, control and so on) it is conceivable that a degree of true autonomy might emerge. The science required to achieve robust behavioural modules for autonomous systems is sampled in this book. There are a host of technologies that could be implemented on current operational systems. Many of the behaviours described are present in fielded systems albeit in an extremely primitive form. For example, waypoint navigation as opposed to path planning, so the prospects of upgrading current implementations are good if hurdles such as airworthiness can be overcome. We can confidently predict that within a few years the types of behaviour described herein will be commonplace on both large and small unmanned systems.

This research book includes a collection of chapters on the state of art in the area of intelligent machines. We believe that this research will provide a sound basis to make autonomous systems human-like.

We are grateful to the authors and reviewers for their vision and contribution. The editorial assistance provided by Springer-Verlag is acknowledged.

Editors

Contents

Foreword V

Preface VII

Intelligent Machines: An Introduction
Lakhmi C. Jain, Anas Quteishat, and Chee Peng Lim 1
1 Introduction 1
2 Learning in Intelligent Machines 2
3 Application of Intelligent Machines 3
3.1 Unmanned Aerial Vehicle (UAV) 3
3.2 Underwater Robot 4
3.3 Space Vehicle 4
3.4 Humanoid Robot 5
3.5 Other Attempts in Intelligent Machines 6
4 Chapters Included in this Book 7
5 Summary 7
References 8

Predicting Operator Capacity for Supervisory Control of Multiple UAVs
M.L. Cummings, Carl E. Nehme, Jacob Crandall, and Paul Mitchell ... 11
1 Introduction 11
2 Previous Experimental Multiple UAV studies 12
3 Predicting Operator Capacity through Temporal Constraints 14
3.1 Wait Times 15
3.2 Experimental Analysis of the Fan-out Equations 16
3.3 Linking Fan-out to Operator Performance 24
3.4 The Overall Cost Function 25
3.5 The Human Model 27
3.6 Optimization through Simulated Annealing 28
3.7 Results of Simulation 29

4 Meta-Analysis of the Experimental and Modeling Prediction methods 33
5 Conclusions 36
References 36

Team, Game, and Negotiation based Intelligent Autonomous UAV Task Allocation for Wide Area Applications
P.B. Sujit, A. Sinha, and D. Ghose 39
1 Introduction 39
2 Existing Literature 41
3 Task Allocation Using Team Theory 42
3.1 Basics of Team Theory 42
3.2 Problem Formulation 43
3.3 Team Theoretic Solution 45
3.4 Simulation Results 47
4 Task Allocation using Negotiation 50
4.1 Problem Formulation 50
4.2 Decision-making 53
4.3 Simulation Results 58
5 Search using Game Theoretic Strategies 61
5.1 N-person Game Model 62
5.2 Solution Concepts 63
5.3 Simulation Results 69
6 Conclusions 72
References 72

UAV Path Planning Using Evolutionary Algorithms
Ioannis K. Nikolos, Eleftherios S. Zografos, and Athina N. Brintaki 77
1 Introduction 77
1.1 Basic Definitions 77
1.2 Cooperative Robotics 79
1.3 Path Planning for Single and Multiple UAVs 80
1.4 Outline of the Current Work 85
2 B-Spline and Evolutionary Algorithms Fundamentals 86
2.1 B-Spline Curves 86
2.2 Fundamentals of Evolutionary Algorithms (EAs) 88
2.3 The Solid Boundary Representation 89
3 Off-line Path Planner for a Single UAV 90
4 Coordinated UAV Path Planning 92
4.1 Constraints and Objectives 92
4.2 Path Modeling Using B-Spline Curves 93
4.3 Objective Function Formulation 94
5 The Optimization Procedure 97
5.1 Differential Evolution Algorithm 97
5.2 Radial Basis Function Network for DE Assistance 99

5.3 Using RBFN for Accelerating DE Algorithm 102
6 Simulation Results 102
7 Conclusions 107
7.1 Trends and challenges 108
References 109

Evolution-based Dynamic Path Planning for Autonomous Vehicles
Anawat Pongpunwattana and Rolf Rysdyk 113
1 Introduction 113
2 Dynamic Path Planning 116
3 Probability of Intersection 122
4 Planning Algorithm 125
4.1 Algorithm for Static Planning 125
4.2 Algorithm for Dynamic Planning 134
5 Planning with Timing Constraints 135
6 Planning in Changing Environment 138
7 Conclusion 142
8 Acknowledgments 143
References 144

Algorithms for Routing Problems Involving UAVs
Sivakumar Rathinam and Raja Sengupta 147
1 Introduction 147
2 Single Vehicle Resource Allocation Problem in the Absence of Kinematic Constraints 148
2.1 Problem Formulation 148
2.2 Relevant Literature 149
2.3 Algorithms 150
3 Multiple Vehicle Resource Allocation Problems in the Absence of Kinematic Constraints 155
3.1 Literature Review 155
3.2 Single Depot, Multiple TSP(SDTSP) 156
3.3 Multiple Depot, Multiple TSP (MDMTSP) 158
3.4 Generalized Multiple Depot Multiple TSP (GMTSP) 159
4 Resource Allocation Problems in the Presence of Kinematic Constraints 162
4.1 Problem Formulation 162
4.2 Literature Review 163
4.3 Alternating Algorithm for the Single UAV Case 164
4.4 Approximation Algorithm for the Multiple UAV Case 165
5 Summary and Open Problems 169
References 170

State Estimation for Micro Air Vehicles
Randal W. Beard ... 173
1 UAV State Variables ... 174
2 Sensor Models ... 176
2.1 Rate Gyros ... 176
2.2 Accelerometers ... 177
2.3 Pressure Sensors ... 177
2.4 GPS ... 179
3 Simulation Environment ... 180
4 State Estimation via Model Inversion ... 182
4.1 Low Pass Filters ... 182
4.2 State Estimation by Inverting the Sensor Model ... 183
5 The Continuous-Discrete Kalman Filter ... 188
5.1 Dynamic Observer Theory ... 189
5.2 Essentials from Probability Theory ... 189
5.3 Continuous-Discrete Kalman Filter ... 191
6 Application of the EKF to UAV State Estimation ... 195
6.1 Roll and Pitch Estimation ... 195
6.2 Position and Course Estimation ... 197
7 Summary ... 198
References ... 198

Evolutionary Design of a Control Architecture for Soccer-Playing Robots
Steffen Prüter, Hagen Burchardt, and Ralf Salomon ... 201
1 Introduction ... 201
2 The Slip Problem ... 204
2.1 Slip and Friction ... 204
2.2 Experimental Analysis ... 205
2.3 Self-Organizing Kohonen Feature Maps and Methods ... 206
2.4 Results ... 207
3 Improved Position Prediction ... 209
3.1 Latency Time ... 209
3.2 Experimental Analysis ... 210
3.3 Back-Propagation Networks and Methods ... 211
4 Local Position Correction ... 213
4.1 Increased Position Accuracy by Local Sensors ... 213
4.2 Embedded Back-Propagation Networks ... 213
4.3 Methods ... 214
4.4 Results ... 215
5 Path Planning using Genetic Algorithms ... 217
5.1 Gene Encoding ... 218
5.2 Fitness Function ... 218
5.3 Evolutionary operations ... 219
5.4 Continous calculation ... 219
5.5 Calculation Time ... 220

5.6 Finding a Path in Dynamic Environments ... 220
6 Discussion ... 221
References ... 222

Toward Robot Perception through Omnidirectional Vision
José Gaspar, Niall Winters, Etienne Grossmann, and José Santos-Victor ... 223
1 Introduction ... 223
1.1 State of the Art ... 225
2 Omnidirectional Vision Sensors: Modelling and Design ... 226
2.1 A Unifying Theory for Single Centre of Projection Systems ... 228
2.2 Model for Non-Single Projection Centre Systems ... 229
2.3 Design of Standard Mirror Profiles ... 230
2.4 Design of Constant Resolution Cameras ... 233
2.5 The Single Centre of Projection Revisited ... 237
3 Environmental Perception for Navigation ... 238
3.1 Geometric Representations for Precise Self-Localisation ... 239
3.2 Topological Representations ... 246
4 Complementing Human and Robot Perceptions for HR Interaction ... 255
4.1 Interactive Scene Reconstruction ... 257
4.2 Human Robot Interface based on 3D World Models ... 262
5 Conclusion ... 263
References ... 265

Intelligent Machines: An Introduction

Lakhmi C. Jain*, Anas Quteishat**, and Chee Peng Lim**

School of Electrical & Information Engineering*
University of South Australia
School of Electrical & Electronic Engineering**
University of Science Malaysia

Abstract. In this chapter, an introduction to intelligent machine is presented. An explanation on intelligent behavior, and the difference between intelligent and repetitive natural or programmed behavior is provided. Some learning techniques in the field of Artificial Intelligence in constructing intelligent machines are then discussed. In addition, applications of intelligent machines to a number of areas including aerial navigation, ocean and space exploration, and humanoid robots are presented.

1 Introduction

"Intelligence" is an expression commonly used for humans and animals, and only until recently for machines. But what is intelligence? How can we say that this creature or machine is intelligent? Indeed, a lot of explanations and definitions for intelligence exist in the literature. Among them, a comprehensible excerpt from [1] with respect to intelligence is as follows.

> *"A very general mental capability that, among other things, involves the ability to reason, plan, solve problems, think abstractly, comprehend complex ideas, learn quickly and learn from experience"*

In general, it is believed that the main factors involved in "intelligence" are the capabilities of autonomously learning and adapting to the environment. So, unless the creature or machine learns from its environment, it may not be considered as intelligent. An interesting example is the behavior of the digger wasp, a *Sphex ichneumoneus* insect [2]. When the female wasp returns to its hole with food, she will first leave the food at the threshold and go inside the hole to check for intruders. If there is no intruder, she will take the food inside. However, if the food is moved, say a few inches, from the original position, she will put the food back on the threshold, go inside, and check for intruders again. The same procedure is repeated again and again if she found the food is displaced. This shows that the element of intelligence, i.e. ability to adapt to new circumstances, is missing in this behavior of the Sphex insect.

L.C. Jain et al.: *Intelligent Machines: An Introduction*, Studies in Computational Intelligence (SCI) **70**, 1–9 (2007)
www.springerlink.com

When we talk about intelligent machines, the first thing that normally appears in our mind is robots. Indeed, robots have been invented to substitute humans in performing a lot of tasks involving repetitive and laborious functions, for examples pick-and-place operations in manufacturing plants. However, robots that are operated based on a programmed manner and in a fully controlled environment are not considered as intelligent machines. Such robots will easily fail when the application and/or the environment contain some uncertain condition. As an example, in applications that involve hazardous and uncertain environments such as handling of radioactive and explosive materials, exploration into space and ocean, robots that can react to changes in their surrounding are very much needed. As a result, robots have to be equipped with "intelligence" so that they can be more useful and usable when operating in uncertain environments.

To be considered as an intelligent machine, the machine has to be able to interact with its environment autonomously. Interacting with the environment involves both learning from it and adapting to its changes. This characteristic differentiates normal machines from intelligent ones. In other words, a normal machine has a specific programmed set of tasks in which it will execute accordingly. On the other hand, an intelligent machine has a goal to achieve, and it is equipped with a learning mechanism to help realize the desired goal [3].

The organization of this chapter is as follows. In section 2, some learning methodologies for intelligent machines are discussed. In section 3, applications of intelligent machines to a number of areas including unmanned aerial vehicles, robots for space and ocean exploration, humanoid robots are presented. A description of each chapter included in this book is presented in section 4, and a summary of this chapter is included in section 5.

2 Learning in Intelligent Machines

When tackling learning from the machine perspective, Artificial Intelligence (AI) has become one of the main fields of interest. The definition of AI can be considered from three viewpoints [4]: (i) computational psychology–mimicking and understanding human intelligence by the generation of a computer program that behaves in the same way; (ii) computational philosophy–formulating a model that is implementable in a computer for understanding intelligent behaviors at the human level; and (iii) machine intelligence–attempting to program a computer to carry out tasks, until recently, only people could do.

In general, the learning process in intelligent machines involves acquiring information about its environment, and deploying the information to establish knowledge about the environment, and, subsequently, generalizing the knowledge base so that it can handle uncertainty in the environment. A number of machine intelligence techniques have been developed to introduce learning in machines, e.g. imitation learning [5] and reinforcement learning [6]. For robot learning, researchers have proposed a multi-learning method that makes use of more than one learning techniques [3]. Besides, different aspects of research in robotics have been conducted, which include robot mobility and control [7], robot perception [8], as well as the use of soft computing techniques for intelligent robotic systems [9]. On the other hand, the divide and conquer principle is applied to the learning tasks [10]. Each algorithm is given a specific task to handle. The learning algorithms are chosen carefully after considering the characteristics of the specific task. Another potential solution to learning is intelligent agents. Agents collect data and learn about the surrounding environment, and adapt to it [11]. The learning process in agents also requires a self-organizing mechanism to control a society of autonomous agents [12]. It should be noted that the task of imparting learning into intelligent machines is not an easy one; however the learning capability is what makes a machine intelligent.

3 Application of Intelligent Machines

The applications of intelligent machines are widespread nowadays, extending, for example, from Mars rover invented by NASA to intelligent vacuum cleaners found in our homes. Some examples of intelligent machines are as follows.

3.1 Unmanned Aerial Vehicle (UAV)

There are some aerial missions and tasks that are not suitable for human pilots either because it is too dangerous like military operations, or it takes a long time in the air like mapping tasks. Yet, these tasks are important. UAVs have been invented to carry out such mission-critical tasks [13]. Typically, an UAV comprises onboard processing capabilities, vision, GPS (Global Positioning System) navigation, and wireless communication. One of the main functions of an UAV is to navigate in an uncontrolled environment, which also is often an unknown environment, safely, and, at the same time, to perform its required task [14]. What makes an UAV intelligent is the ability to fly to its target under varying conditions. As it is not possible to predict all possible navigation scenarios in one program, the UAV has to learn from its environment, and adapt to the changes as they occur in order to reach the destination.

An UAV used to collect data in the atmosphere between satellite and the ground base is created by National Oceanic and Atmospheric Administration (NOAA), USA. The UAV is able to fill the gap where land-based and satellite-based observations fall short, thus giving a view of the planet never seen before [15]. Another UAV, a version of the military MQ9 Predator B, is used by the Department of Homeland Security, USA to monitor remote and inaccessible regions of the border. The UAV is equipped with special cameras and other sensors, and is able to stay in the air for up to 30 hours [16].

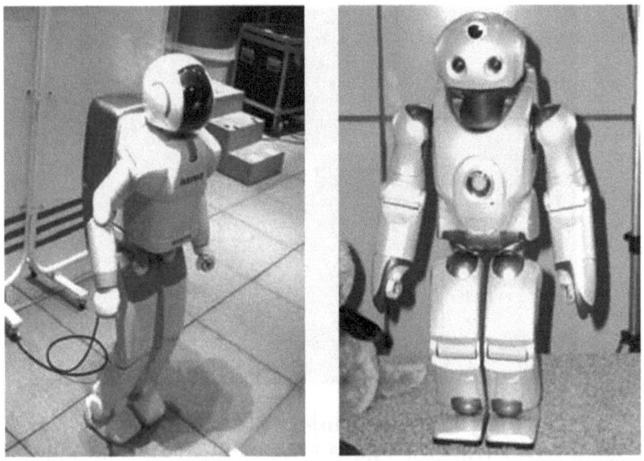

(a) ASIMO (b) QRIO

Fig. 4. Humanoid robots
(public domain images, courtesy of wikimedia commons)

(a) The Actroid robot (b) Face of Actroid

Fig. 5. Snapshots of the Actroid robot
(copyright of Aleksandar Lazinica, used by permission)

people when it hears "Hello". This office reception robot is also able to control its motions expressively within the context of a conversation, e.g., facial expressions, lip movements, and behaviour.

3.5 Other Attempts in Intelligent Machines

a. Unmanned Combat Air Vehicle (UCAV) project [24]: the objective of this project is to demonstrate the effectiveness of using UCAV to effectively and affordably prosecute twenty-first century lethal strike missions within the emerging global command and control architecture.

b. Micromechanical Flying Insect (MFI) project [25]: the objective of this project is to create an insect-like device that is capable of flying autonomously.
c. Medical micro-robot project [26]: this projects aims to create the world's smallest micro-robot as wide as human hair at about 250 micron. This micro-robot will be used to transmit images and deliver microscopic payloads to parts of the body outside the reach of existing catheter technology.

4 Chapters Included in this Book

This book includes nine chapters. Chapter one introduces intelligent machines and presents the chapters included in this book. Chapter two by Cummings et al. is on predicting operator capacity for supervisory control of UAVs. The authors have considered a cost-performance model in this study. Chapter three by Sujit et al. is on team, game and negotiation based intelligent autonomous UAV task allocation for a number of applications. The authors have also presented a scheme of searching in an unknown environment. Chapter four by Nikolas et al. is on path planning using evolutionary algorithms. The authors have used Radial Basis Function Neural Network in evolutionary environment in the design of their off-line path planner for UAV. Chapter five by Rathinam and Sengupta is on algorithms on routing problems related to UAVs. The authors have presented a class of routing problems and including review and recent developments.

Chapter six by Beard is on state estimation for micro air vehicles. The author has presented mathematical models for the sensors for multiple air vehicles. Chapter seven by Pongpunwattana and Rysdyk is on evolution-based dynamic path panning for autonomous vehicles. The algorithms take into account the uncertain information of the environment and dynamics of the system. Chapter eight by Prüter et al. is on evolutionary design of control architecture for soccer-playing robots. Artificial intelligence techniques are used to compensate the effect of slipping wheels, changing friction values, noise and so on. The final chapter by Gasper et al. is on robot perception through omnidirectional vision. The authors have examined how robots can use images which convey only 2D information to drive its actions in 3D space. The design of a navigation system considering sensor design, environmental representations, navigation control and user interaction is presented.

5 Summary

This chapter has presented an introduction to intelligent machines. A discussion on intelligence and the difference between intelligent and natural repetitive or programmed behaviors are given. The importance of an intelligent machine to learn from its changing environment and to adapt to the new circumstances is discussed. Although there are various machine intelligence techniques to impart learning to machines, it is yet to have a universal one for this purpose. Some applications of intelligent machines are highlighted, which include unmanned aerial vehicles, underwater robots, space vehicles, and humanoid robots, as well as other projects in realizing intelligent machines. It is anticipated that intelligent machines will ultimately play a role, in one way or another, in our daily activities, and make our life comfortable in future.

References

1. "Mainstream Science on Intelligence", *Wall Street Journal*, Dec. 13, 1994, p A18.
2. "Artificial Intelligence", Encyclopædia Britannica. 2007. *Encyclopædia Britannica Online*, <http://www.britannica.com/eb/article-9009711>, access date: 10 Feb 2007
3. S. Takamuku and R.C. Arkin, "Multi-method Learning and Assimilation", *Mobile Robot Laboratory Online Publications*, Georgia Institute of Technology, 2007.
4. S.C. Shapiro, Artificial Intelligence, in A. Ralston, E.D. Reilly, and D. Hemmendigner, Eds. *Encyclopedia of Computer Science*, Fourth Edition,. New York Van Nostrand Reinhold, 1991
5. S. Schaal, "Is imitation learning the route to humanoid robots?" *Trends in Cognitive Scienes*, vol. 3, pp. 233–242, 1999.
6. J. Peters, S. Vijayakumar, and S. Schaal, "Reinforcement learning for humanoid robotics", *Proceedings of the third IEEE-RAS International Conference on Humanoid Robots*, 2003.
7. S. Patnaik, L. Jain, S. Tzafestas, G. Resconi, and A. Konar, (eds), *Innovations in Robot Mobility and Control*, Springer, 2006.
8. B. Apolloni, A. Ghosh, F. Alpaslan, L. Jain, and S. Patnaik, (eds), *Machine Learning and Robot Perception*, Springer, 2006.
9. L.C. Jain, and T. Fukuda, (editors), *Soft Computing for Intelligent Robotic Systems*, Springer-Verlag, Germany, 1998.
10. P. Langley, "Machine learning for intelligent systems," *Proceedings of Fourteenth National Conference on Artificial Intelligence*, pp. 763–769, 1997.
11. F. Sahin and J.S. Bay, "Learning from experience using a decision-theoretic intelligent agent in multi-agent systems", *Proceedings of the 2001 IEEE Mountain Workshop on Soft Computing in Industrial Applications*, pp. 109–114, 2001.
12. J. Pearl, *Probabilistic Reasoning in Intelligent Systems: Networks of Plausible Inference*, Morgan Kaufmann, 1988.
13. D.A. Schoenwald, "AUVs: In space, air, water, and on the ground", *IEEE Control Systems Magazine*, vol. 20, pp. 15–18, 2000.
14. A. Ryan, M. Zennaro, A. Howell, R. Sengupta, and J.K. Hedrick, "An overview of emerging results in cooperative UAV control", *Proceedings of 43rd IEEE Conference on Decision and Control*, vol. 1, pp. 602–607, 2004.
15. "NOAA Missions Now Use Unmanned Aircraft Systems", *NOAA Magazine Online (Story 193)*, 2006, <http://www.magazine.noaa.gov/stories/mag193.htm>, access date: 13 Feb, 2007

16. S. Waterman, "UAV Tested For US Border Security", *United Press International*, <http://www.spacewar.com/reports/UAV_Tested_For_US_Border_Security_ 999.html>, access date: 30 March 2007
17. "First Flight-True UAV Autonomy At Last" *Agent Oriented Software*, (Press Release of 6 July 2004), <http://www.agent-software.com/shared/resources/pressReleases.html>, access date: 14 Feb. 2007
18. J. Yuh, "Underwater robotics", *Proceedings of IEEE International Conference on Robotics and Automation*, vol. 1, pp. 932–937, 2000.
19. "Intelligent Machines, Micromachines, and Robotics", <http://www.ntu.edu.sg/mae/Research/Programmes/Imr/>, access date:12 Feb 2007
20. J. Chestnutt, M. Lau, G. Cheung, J. Kuffner, J. Hodgins, and T. Kanade, "Footstep Planning for the Honda ASIMO Humanoid", *Proceedings of the IEEE International Conference on Robotics and Automation*, pp. 629–634, 2005.
21. F. Tanaka, B. Fortenberry, K. Aisaka, and J. R. Movellan, "Developing dance interaction between QRIO and toddlers in a classroom environment: Plans for the first steps", *Proceedings of the IEEE International Workshop on Robot and Human Interactive Communication*, p. 223–228 2005.
22. K.F. MacDorman and H. Ishiguro, "The uncanny advantage of using androids in cognitive and social science research," *Interaction Studies*, vol. 7, pp. 297–337, 2006.
23. A. Lazinica, "Highlights of IREX 2005", <http://www.ars-journal.com/ars/Free_Articles/IREX-2005.htm>, access date: 20 March, 2007
24. "X-45 Unmanned Combat Air Vehicle (UCAV)", <http://www.fas.org/man/dod-101/sys/ac/ucav.htm>, access date: 14 Feb 2007
25. "Micromechanical Flying Insect (MFI) Project", <http://robotics.eecs.berkeley. edu/~ronf/MFI/>, access date: 14 Feb 2007
26. E. Cole, "Fantastic Voyage: Departure 2009", <http://www.wired.com/news/technology/medtech/0,72448-0.html?tw=wn_technology_1>, access date: 14 Feb 2007

Predicting Operator Capacity for Supervisory Control of Multiple UAVs

M.L. Cummings, Carl E. Nehme, Jacob Crandall, and Paul Mitchell

Humans and Automation Laboratory,
Massachusetts Institute of Technology,
Cambridge, Massachusetts

Abstract. With reduced radar signatures, increased endurance, and the removal of humans from immediate threat, uninhabited (also known as unmanned) aerial vehicles (UAVs) have become indispensable assets to militarized forces. UAVs require human guidance to varying degrees and often through several operators. However, with current military focus on streamlining operations, increasing automation, and reducing manning, there has been an increasing effort to design systems such that the current many-to-one ratio of operators to vehicles can be inverted. An increasing body of literature has examined the effectiveness of a single operator controlling multiple uninhabited aerial vehicles. While there have been numerous experimental studies that have examined contextually how many UAVs a single operator could control, there is a distinct gap in developing predictive models for operator capacity. In this chapter, we will discuss previous experimental research for multiple UAV control, as well as previous attempts to develop predictive models for operator capacity based on temporal measures. We extend this previous research by explicitly considering a cost-performance model that relates operator performance to mission costs and complexity. We conclude with a meta-analysis of the temporal methods outlined and provide recommendation for future applications.

1 Introduction

With reduced radar signatures, increased endurance and the removal of humans from immediate threat, uninhabited (also known as unmanned) aerial vehicles (UAVs) have become indispensable assets to militarized forces around the world, as proven by the extensive use of the Shadow and the Predator in recent conflicts.

Current UAVs require human guidance to varying degrees and often through several operators. For example, the Predator requires a crew of two to be fully operational. However, with current military focus on streamlining operations and reducing manning, there has been an increasing effort to design systems such that the current many-to-one ratio of operators to vehicles can be inverted (e.g., [1]). An increasing body of literature has examined the

M.L. Cummings et al.: *Predicting Operator Capacity for Supervisory Control of Multiple UAVs*, Studies in Computational Intelligence (SCI) **70**, 11–37 (2007)
www.springerlink.com

effectiveness of a single operator controlling multiple UAVs. However, most studies have investigated this issue from an experimental standpoint, and thus they generally lack any predictive capability beyond the limited conditions and specific interfaces used in the experiments.

In order to address this gap, this chapter first analyzes past literature to examine potential trends in supervisory control research of multiple uninhabited aerial vehicles (MUAVs). Specific attention is paid to automation strategies for operator decision-making and action. After the experimental research is reviewed for important "lessons learned", an extension of a ground unmanned vehicle operator capacity model will be presented that provides predictive capability, first at a very general level and then at a more detailed cost-benefit analysis level. While experimental models are important to understand what variables are important to consider in MUAV control from the human perspective, the use of predictive models that leverage the results from these experiments is critical for understanding what system architectures are possible in the future. Moreover, as will be illustrated, predictive models that clearly link operator capacity to system effectiveness in terms of a cost-benefit analysis will also demonstrate where design changes could be made to have the greatest impact.

2 Previous Experimental Multiple UAV studies

Operating a US Army Hunter or Shadow UAV currently requires the full attention of two operators: an AVO (Aerial Vehicle Operator) and a MPO (Mission Payload Operator), who are in charge respectively of the navigation of the UAV, and of its strategic control (searching for targets and monitoring the system). Current research is aimed at finding ways to reduce workload and merge both operator functions, so that only one operator is required to manage one UAV. One solution investigated by Dixon et al. consisted of adding auditory and automation aids to support the potential single operator [2]. Experimentally, they showed that a single operator could theoretically fully control a single UAV (both navigation and payload) if appropriate automated offloading strategies were provided. For example, aural alerts improved performance in the tasks related to the alerts, but not others. Conversely, it was also shown that adding automation benefited both tasks related to automation (e.g. navigation, path planning, or target recognition) as well as non-related tasks. However, their results demonstrate that human operators may be limited in their ability to control multiple vehicles which need navigation and payload assistance, especially with unreliable automation. These results are concordant with the single-channel theory, stating that humans alone cannot perform high speed tasks concurrently [3, 4]. However, Dixon et al. propose that reliable automation could allow a single operator to fully control two UAVs.

Reliability and the related component of trust is a significant issue in the control of multiple uninhabited vehicles. In another experiment, Ruff et al. [5]

found that if system reliability decreased in the control of multiple UAVs, trust declined with increasing numbers of vehicles but improved when the human was actively involved in planning and executing decisions. These results are similar to those experimentally found by Dixon et al. in that systems that cause distrust reduce operator capacity [6]. Moreover, cultural components of trust cannot be ignored. Tactical pilots have expressed inherent distrust of UAVs as wingmen, and in general do not want UAVs operating near friendly forces [7].

Reliability of the automation is only one of many variables that will determine operator capacity in MUAV control. The level of control and the context of the operator's tasks are also critical factors in determining operator capacity. Control of multiple UAVs as wingmen assigned to a single seat fighter has been found to be "unfeasible" when the operator's task was primarily navigating the UAVs and identifying targets [8]. In this experimental study, the level of autonomy of the vehicles was judged insufficient to allow the operator to handle the team of UAVs. When UAVs were given more automatic functions such as target recognition and path planning, overall workload was reduced.

In contrast to the previous UAVs-as-wingmen experimental study [6] that determined that high levels of autonomy promotes overall performance, Ruff et al. [5] experimentally determined that higher levels of automation can actually degrade performance when operators attempted to control up to four UAVs. Results showed that management-by-consent (in which a human must approve an automated solution before execution) was superior to management-by-exception (where the automation gives the operator a period of time to reject the solution). In their scenarios, their implementation of management-by-consent provided the best situation awareness ratings and the best performance scores for controlling up to four UAVs.

These previous studies experimentally examined a small subset of UAVs and beyond showing how an increasing number of vehicles impacted operator performance, they were not attempting to predict any maximum capacity. In terms of actually predicting how many UAVs a single operator control, there is very little research. Cummings and Guerlain [9] showed that operators could experimentally control up to 12 Tactical Tomahawk missiles given significant missile autonomy. However, these predictions are experimentally-based which limits their generalizability. Given the rapid acquisition of UAVs in the military, which will soon follow in the commercial section, predictive modeling for operator capacity will be critical for determining an overall system architecture. Moreover, given the range of vehicles with an even larger subset of functionalities, it is critical to develop a more generalizable predictive modeling methodology that is not solely based on expensive human-in-the-loop experiments, which are particularly limited for application to revolutionary systems.

In an attempt to address this gap, in the next section of this paper, we will extend a predictive model for operator capacity in the control of unmanned ground vehicles to a UAV domain [10], such that it could be used to predict operator capacity, regardless of vehicle dynamics, communication latency, decision support, and display designs.

3 Predicting Operator Capacity through Temporal Constraints

While little research has been published concerning the development of a predictive operator capacity model for UAVs, there has been some previous work in the unmanned ground vehicle (robot) domain. Coining the term "fan-out" to mean the number of robots a human can effectively control, Olsen et al. [10, 11] propose that the number of homogeneous robots or vehicles a single individual can control is given by:

$$FO = \frac{NT + IT}{IT} = \frac{NT}{IT} + 1 \tag{1}$$

In this equation, FO (fan-out) is dependent on NT (Neglect Time), the expected amount of time that a robot can be ignored before its performance drops below some acceptable threshold, and IT (Interaction Time) which is the average time it takes for a human to interact with the robot to ensure it is still working towards mission accomplishment. Figure 1 demonstrates the relationship of IT and NT.

While originally intended for ground-based robots, this work has direct relevance to more general human supervisory control (HSC) tasks where operators are attempting to simultaneously manage multiple entities, such as in the case of UAVs. Because the fan-out adheres to Occam's Razor, it provides a generalizable methodology that could be used regardless of the domain, the human-computer interface, and even communication latency problems. However, as appealing as it is due to its simplicity, in terms of human-automation interaction, the fan-out approach lacks two critical considerations: 1) The important of including wait times caused by human-vehicle interaction, and 2) How to link fan-out to measurable "effective" performance. These issues will be discussed in the subsequent section.

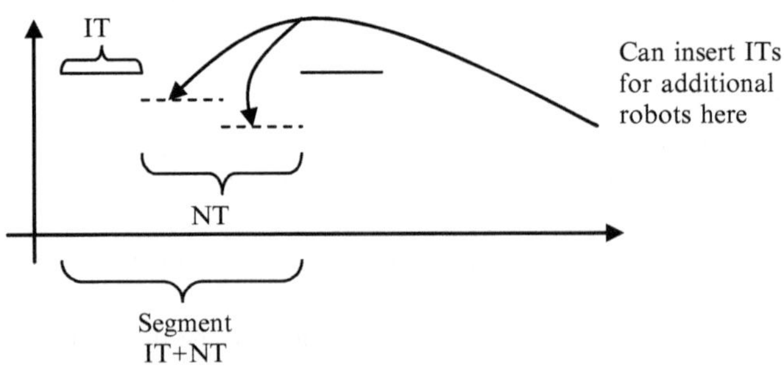

Fig. 1. The relationship of NT and IT for a Single Vehicle

3.1 Wait Times

Modeling interaction and neglect times are critical for understanding human workload in terms of overall management capacity. However, there remains an additional critical variable that must be considered when modeling human control of multiple robots, regardless of whether they are on the ground or in the air, and that is the concept of Wait Time (WT). In HSC tasks, humans are serial processors in that they can only solve a single complex task at a time [3, 4], and while they can rapidly switch between cognitive tasks, any sequence of tasks requiring complex cognition will form a queue and consequently wait times will build. Wait time occurs when a vehicle is operating in a degraded state and requires human intervention in order to achieve an acceptable level of performance. In the context of a system of multiple vehicles or robots, wait times are significant in that as they increase, the actual number of vehicles that can be effectively controlled decreases, with potential negative consequences on overall mission success.

Equation 2 provides a formal definition of wait time. It categorizes total system wait time as the sum of the interaction wait times, which are the portions of IT that occur while a vehicle is operating in a degraded state (WTI), wait times that result from queues due to near-simultaneous arrival of problems (WTQ), plus wait times due to operator loss of situation awareness (WTSA). An example of WTI is the time that an unmanned ground vehicle (UGV) idly waits while a human replans a new route. WTQ occurs when a second UGV sits idle, and WTSA accumulates when the operator doesn't even realize a UGV is waiting. In (2), X equals the number of times an operator interacts with a vehicle while the vehicle is in a degraded state, Y indicates the number of interaction queues that build, and Z indicates the number of time periods in which a loss of situation awareness causes a wait time. Figure 2 further illustrates the relationship of wait times to interaction and neglect times.

Increased wait times, as defined above, will reduce operator capacity, and Equation 3 demonstrates one possible way to capture this relationship. Since

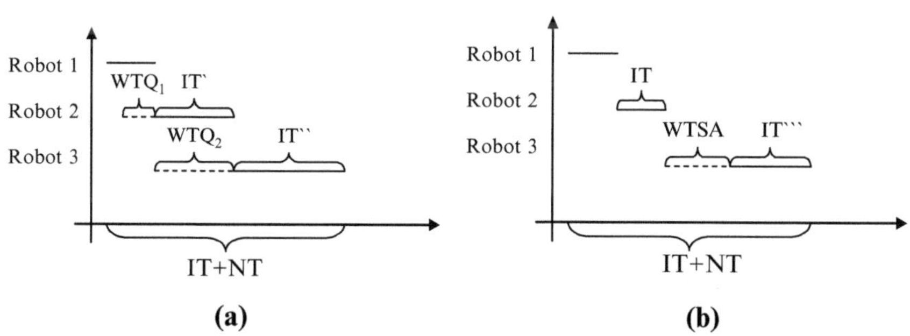

Fig. 2. Queuing wait times (a) versus situational awareness wait times (b)

WTI is a subset of IT, it is not explicitly included (although the measurement technique of IT will determine whether or not WTI should be included in the denominator.)

$$WT = \sum\nolimits_{i=1}^{X} WTI_i + \sum\nolimits_{j=1}^{Y} WTQ_j + \sum\nolimits_{k=1}^{Z} WTSA_k \tag{2}$$

$$FO = \frac{NT}{IT + \sum_{j=1}^{Y} WTQ + \sum_{k=1}^{Z} WTSA_k} + 1 \tag{3}$$

While the revised fan-out (3) includes more variables than the original version, the issue could be raised that the additional elements may not provide any meaningful or measurable improvement over the original equation which is simpler and easier to model. Thus to determine how this modification affects the fan-out estimate, we conducted an experiment with a UAV simulation test bed, holding constant the number of vehicles a person controlled. We then measured all times associated with equations 1 and 3 to demonstrate the predictions made by each equation. The next section will describe the experiment and results from this effort.

3.2 Experimental Analysis of the Fan-out Equations

In order to study operator control of multiple UAVs, a dual screen simulation test bed named the Multi-Aerial Unmanned Vehicle Experiment (MAUVE) interface was developed (Fig. 3). This interface allows an operator to effectively supervise four independent homogeneous UAVs simultaneously, and intervene as the situation requires. In this simulation, users take on the role of an operator responsible for supervising four UAVs tasked with destroying a set of time-sensitive targets in a suppression of enemy air defenses (SEAD) mission. The left side of the display provides geo-spatial information as well as a command panel to redirect individual UAVs. The right side of the display provides temporal scheduling decision support in addition to data link "chat windows" commonly in use in the military today [12]. Details of the display design such as color mappings and icon design are discussed elsewhere [13].

The four UAVs launched with a pre-determined mission plan, so initial target assignments and routes were already completed. The operator's primary job in the MAUVE simulation was to monitor each UAV's progress, replan aspects of the mission in reaction to unexpected events and in some cases manually execute mission critical actions such as arming and firing of payloads. The UAVs supervised by participants in MAUVE were capable of 6 high-level actions: traveling en route to targets, loitering at specific locations, arming payloads, firing payloads, performing battle damage assessment, and returning to base, generally in this order.

In the MAUVE simulations, flight control was fully automated as was the basic navigation control loop in terms of heading control. Operators were occasionally required to replan route segments due to pop-up threat areas so the

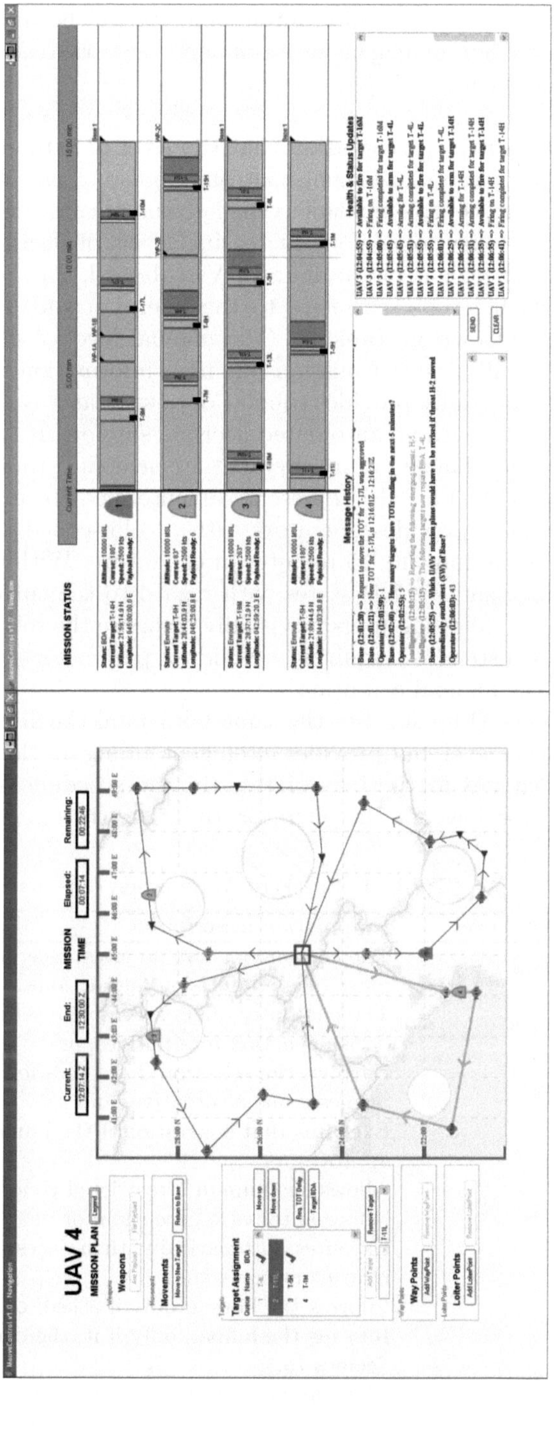

Fig. 3. The MAUVE Dual Screen Interface

navigation loop was only partially automated. As will be discussed in more detail next, the mission management autonomy was varied as an independent facto in the experiment.

Levels of Autonomy. Recognizing that the level of autonomy introduced in the mission/payload management control loop can significantly impact an operator's ability to control multiple vehicles, and thus neglect, interaction, and wait times, we developed four increasing levels of decision support for the temporal management of the four UAVs: Manual, Passive, Active, and Super-active, which loosely correlate to the Sheridan and Verplank Levels [14] of 1, 2, 4, 6 (shown in Table 1). The manual level of decision support (Fig. 1a) presents all required mission planning information in a text-based table format. It essentially provides tabular data such as waypoints, expected time on targets, etc., with no automated decision support. It is representative of air tasking orders that are in use by military personnel today.

The passive LOA (Fig. 4b) represents an intermediate mission management LOA in that it provides operators with a color-coded timeline for the expected mission assignments 15 minutes in the future. With this visual representation, recognizing vehicle states with regard to the current schedule is perceptually-based, allowing users to visually compare the relative location of display elements instead of requiring individual parameter searches such as what occurs in the manual condition.

The active LOA (Fig. 4c) uses the same horizontal timeline format as the passive automation level, but provides intelligent aiding. In the active version, an algorithm searches for periods of time in the schedule that it predicts will cause high workload for the operator, directing the operator's attention

Table 1. Levels of Automation

Automation Level	Automation Description
1	The computer offers no assistance: human must take all decision and actions.
2	The computer offers a complete set of decision/action alternatives, or
3	Narrows the selection down to a few, or
4	Suggests one alternative, and
5	Executes that suggestion if the human approves, or
6	Allows the human a restricted time to veto before automatic execution, or
7	Executes automatically, then necessarily informs humans, and
8	Informs the human only if asked, or
9	Informs the human only if it, the computer, decides to.
10	The computer decides everything and acts autonomously, ignoring the human.

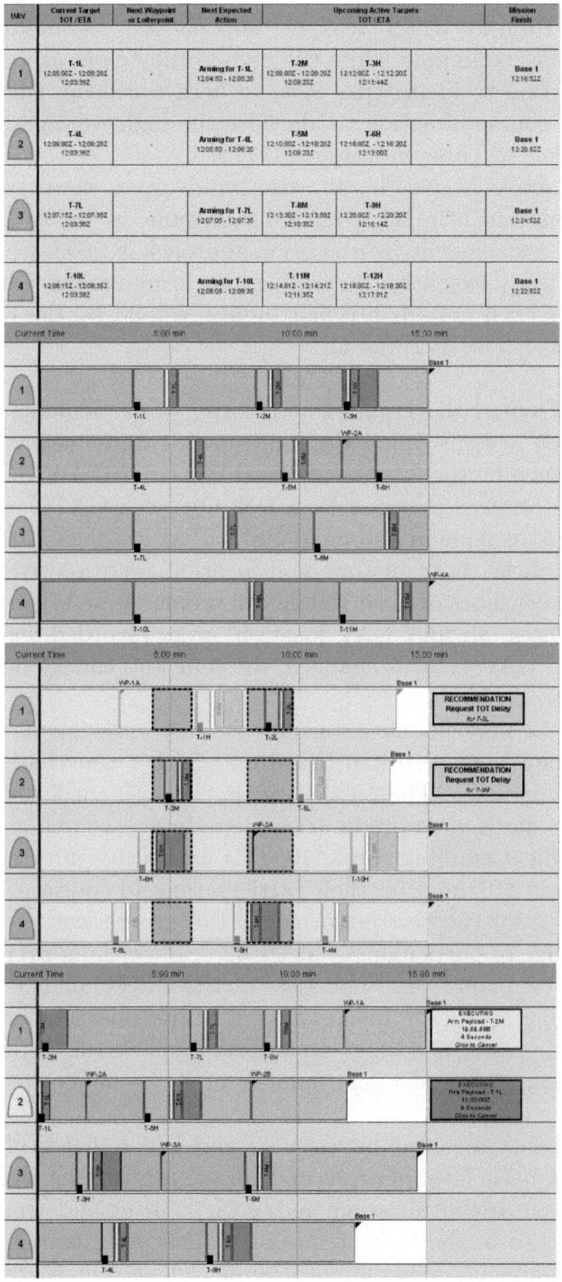

Fig. 4 a-d. Increasing levels of decision and action support in Mauve towards them. High workload areas, or "bottlenecks," are highlighted through a reverse shading technique while the rest of the colors are muted, but still visible. In addition to identifying areas of high workload, the computer also recommends a course of action to alleviate the high workload areas, such as moving a particular Time on Target (TOT).

The super-active LOA (Fig. 4d) also builds upon the passive level visual timeline, but instead of making recommendations to the operator as in the active LOA, a management-by-exception approach is taken whereby the computer automatically executes the arming and firing actions when the rules of engagement for such actions are met, unless vetoed by the operator in less than 30 seconds (LOA 6, Table 1).

Experiment Protocol. Training and testing of participants was conducted on a four screen system called the multi-modal workstation (MMWS) [15], originally designed by the Space and Naval Warfare (SPAWAR) Systems Center. The workstation is powered by a Dell Optiplex GX280 with a Pentium 4 processor and an Appian Jeronimo Pro 4-Port graphics card. During testing, all mouse clicks, both in time and location, were recorded by software. In addition, screenshots of both simulation screens were taken approximately every two minutes, all four UAV locations were recorded every 10 seconds, and whenever a UAV's status changed, the time and change made were noted in the data file.

A total of 12 participants took part in this experiment, 10 men and 2 women, and they were recruited based on whether they had UAV, military and/or pilot experience. The participant population consisted of a combination of students, both undergraduates and graduates, as well as those from the local reserve officer training corps (ROTC) and active duty military personnel. All were paid $10/hour for their participation. In addition, a $50 incentive prize was offered for the best performer in the experiment.

The age range of participants was 20–42 years with an average age of 26.3 years. Nine participants were members of the ROTC or active duty USAF officers, including seven 2nd Lieutenants, a Major and a Lieutenant Colonel. While no participants had large-scale UAV experience, 9 participants had piloting experience. The average number of flight hours among this group was 120.

All participants received between 90 and 120 minutes of training until they achieved a basic level of proficiency in monitoring the UAVs, redirecting them as necessary, executing commands such as firing and arming of payload, and responding to online instant messages. Following training, participants tested on two consecutive 30 minute sessions, which represented low and high workload scenarios. These were randomized and counter-balanced to prevent a possible learning effect. The low replanning condition contained 7 replanning events, while the high replanning condition contained 13. Each simulation was run several times faster than real time so an entire strike could take place over 30 minutes (instead of several hours).

Results and Discussion. In order to determine whether or not the revised fan-out prediction in (3) provided a more realistic estimate than the original fan-out (1), the number of vehicles controlled in the experiment was held constant (four) across all levels of automation. Thus if our proposed prediction was accurate, we should be able to predict the actual number of vehicles the operators were controlling. As previously discussed, all times were measured through interactions with the interface which generally included mouse movements, selection of objects such as vehicles and targets for more information, commanding vehicles to change states, and the generation of communication messages.

Neglect time was counted as the time when operators were not needed by any single vehicle, and thus were monitoring the system and engaging in secondary tasks such as responding to communications. Because loiter paths were part of the preplanned missions, oftentimes to provide for buffer periods, loiter times were generally counted as neglect times. Loitering was only counted as a wait time when a vehicle was left in a loiter pattern past a planned event due to operator oversight. Interaction time was counted as any time an operator recognized that a vehicle required intervention and specifically worked towards resolving that task. This was measured by mouse movements, clicks, and message generations. The method of measuring NT and IT, while not exactly the same as [11], was driven by experimental complexity in representing a more realistic environment. However, the same general concepts apply in that neglect time is that time when each vehicle operated independently and interaction time is that time one or more vehicles required operator attention.

As discussed previously, wait times were only calculated when one or more vehicle required attention. Wait time due to interactions (e.g., the time it took an operator to replan a new route once a UAV penetrated a threat area) was subsumed in interaction time. Wait time due to queuing occurred when, for example, a second UAV also required replanning to avoid an emergent threat and the operator had to attend to the first vehicle's problem before immediately moving to the second. Wait time due to the loss of situation awareness was measured when one or more vehicles required attention but was not noticed by the operator. This was the most difficult wait time to capture since operators had to show clear evidence that they did not recognize a UAV required intervention. Examples of wait time due to loss of situation awareness include the time UAVs spend flying into threat areas with no path correction, and leaving UAVs in loiter patterns when they should be redirected.

Figures 5 and 6 demonstrate how the wait times varied both between the two fan-out equations as well the increasing levels of automation under low and high workload conditions respectively. Using the interaction, neglect, and wait times calculated from the actual experiment, the solid line represents the predictions using (1), the dashed line represents the predictions of (2), and the dotted line shows how many UAVs the operators were actually controlling, which was held constant at four.

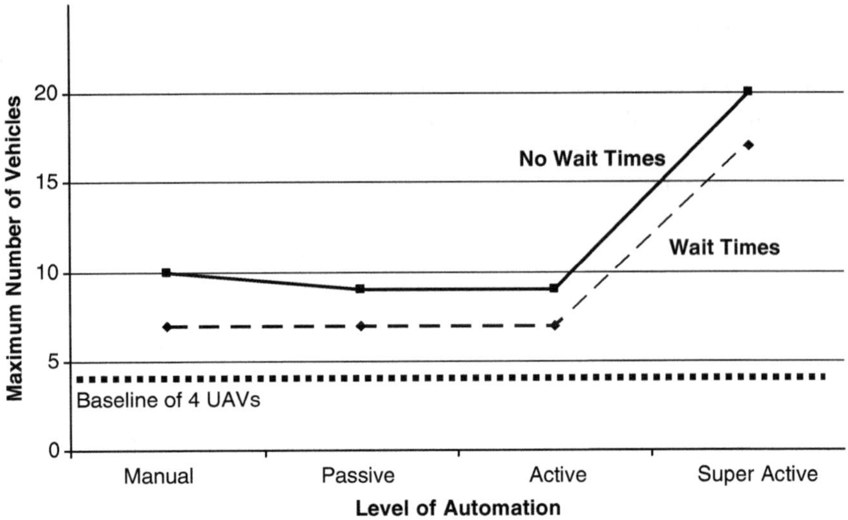

Fig. 5. Low Workload Operator Capacity Prediction

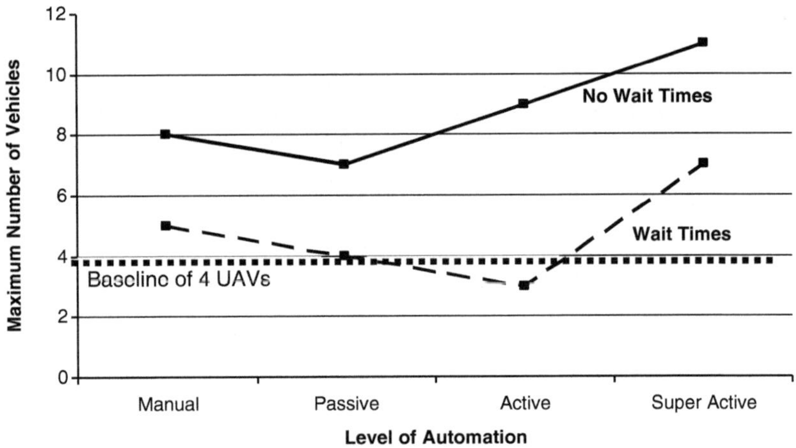

Fig. 6. High Workload Operator Capacity Predictions

Low Workload Predictions. Under the low workload condition, three important trends should be noted. Under the lower levels of automation for both the original and revised fan-out equations, operator capacity was essentially flat, and a significant increase was not seen until the use of a higher automation strategy, management-by-exception, was implemented. It is important to remember that the metric is time and not overall decision quality or performance. However, independent performance measures indicated that at the low workload level, operators were able to effectively control all four vehicles [16].

The second trend of note is the fact that for the low workload condition, the revised fan-out model (3) provides a more conservative estimate of approximately 20% under that of the model that does not consider wait times (1). However, the third important trend in this graph demonstrates that for both (1) and (3) the predictions were much higher than the actual number of UAVs controlled. This spare capacity under the low workload condition was empirically observed, in that subjective workload measures (NASA-TLX) and performance scores were statistically the same when compared across all four levels of autonomy (lowest pair wise comparison pvalue = .111 (t = 1.79, DOF = 8), and p = .494 (t = .72, DOF = 8) respectively).

Thus for the low workload condition across all levels of automation, operators were underutilized and performing well. Thus they theoretically could have controlled more vehicles. Using the revised FO model (3), under the manual, passive, and active conditions, operators' theoretical capacity could have increased by ~75% (up to 7 vehicles). Under the highest autonomy for mission management, predictions estimate operators could theoretically control (as an upper limit) four times as many, ~17 vehicles. Previous air traffic control (ATC) studies have indicated that 16-17 aircraft are the upper limit for en route air traffic controllers [17]. Since controllers are only providing navigation assistance and not interacting with flight controls and mission sensors (such as imagery), the agreement between ATC en route controller capacities and low workload for UAV operators is not surprising.

High Workload Predictions. While the low workload results and predictions suggest that operators are capable of controlling more than four vehicles in MAUVE, the results from the high workload scenarios paint an entirely different picture. The high workload scenarios were approximately double the workload over the low workload scenarios, and represent a worst case scenario. Performance results indicate that those operators with the active level of automation were not able to control their four UAVs effectively, but all other operators were with varying degrees of success. As in the low workload condition, the revised fan-out model (3) is the more conservative and as demonstrated in Figure 6, more closely predicts the actual number of four vehicles assigned to each operator. Moreover, while under the low workload condition, the estimates of controller capacity dropped almost uniformly across automation levels by 20% for the original fan-out model. However, under high workload, they dropped 36–67% for the model that includes wait times. The largest difference between conditions occurred for the active level of automation. In addition to the lower number of predicted vehicles, the active condition produced statistically lower performance scores (e.g., t = 2.26, DOF = 8, p = 0.054 for the passive-active comparison). This was attributed to the inability of subjects in the active condition to correctly weight uncertainty parameters and is discussed in detail elsewhere [16].

As in the low workload results, subjects performed the best (in terms of time management) under the highest level of automation for mission

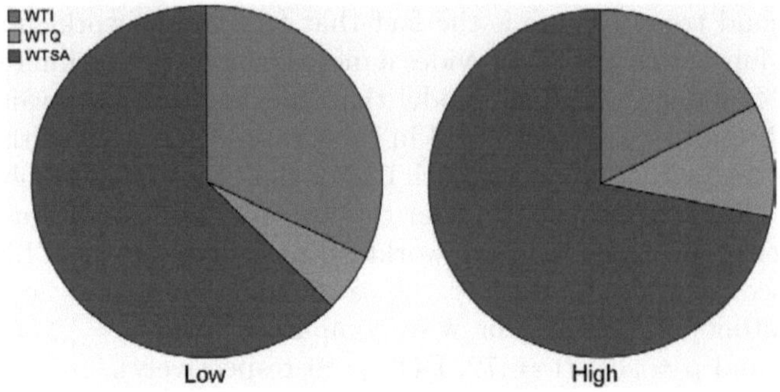

Fig. 7. Wait Time Proportions

management (super-active), with a theoretical maximum of seven vehicles. However, under this condition in the experiment, subjects exhibited automation bias and approved the release of more weapons on incorrect targets than for the passive and active levels. Automation bias, the propensity for operators to take automated recommendations without searching for disconfirming evidence, has been shown to be a significant problem in command and control environments and also operationally for the Patriot missile [18]. Thus increased operator capacity for management-by-exception systems must be weighed against the risk of incorrect decisions, by either the humans or the automation.

Wait Time Proportions. Figures 5 and 6 demonstrate that the inclusion of wait times in a predictive model for operator capacity in the control of MUAVs can radically reduce the theoretical maximum limit. Figure 7 demonstrates the actual proportions of wait time that drove those results. Strikingly, under both low and high workload conditions, the wait times due to the loss of situation awareness dominated overall wait times.

This partitioning of wait time components is important because it demonstrates where and to what degree interventions could potentially improve both human and system performance. In the case of the experiment detailed in this chapter, clearly more design intervention, form both and automation and HCI perspective, is needed that aids operators in recognizing that vehicles need attention. As previously demonstrated, some of the issues are directly tied to workload, i.e., operators who have high workloads have more loss of situation awareness. However, often loss of situation awareness occurred because operators did not recognize a problem which could mitigated through better decision aiding and visualization.

3.3 Linking Fan-out to Operator Performance

Results from the experiment conducted to compare the original fan-out (1) and the revised fan-out estimate which includes wait times (3), demonstrate the revised model is both more conservative and closer to the actual number of vehicles under successful control. While under low workload, both the experiment and prediction indication that operators could have controlled more vehicles than four, the only high workload scenario in which operators demonstrated any spare capacity was with the super-active (management-by-exception) decision support. Moreover, wait time caused by the lack of situation awareness dominated overall wait time. In addition, this research demonstrates that both workload and automated decision support can dramatically affect wait times and thus, operator capacity.

While more pessimistic than the original fan-out equation (1), the revised fan-out equation can really only be helpful for broad "ballpark" predictions of operator capacity. This methodology could provide system engineers with a system feasibility metric for early manning estimations, but what primarily limits either version of the fan-out equation is the inability to represent any kind of cost trade space. Theoretically fan-out, revised or otherwise, will predict the maximum number of vehicles an operator can *effectively* control, but what is effective is often a dynamic constraint. Moreover, the current equations for calculating fan-out do not take into account explicit performance constraints. In light of the need to link fan-out to some measure of performance, as well as the inevitability of wait times introduced by human interaction, we propose that instead of a simple maximum limit prediction, we should instead find the optimal number of UAVs such that the mission performance is maximized.

3.4 The Overall Cost Function

Maximizing UAV mission performance is achieved when the overall performance of all of the vehicles, or the team performance, is maximized. Consider multiple UAVs that need to visit multiple targets, either for destruction (SEAD missions as discussed previously) or imaging (typical of Intelligence, Search, and Reconnaissance (ISR) missions). A possible cost function is expressed in (4):

$$\begin{aligned} C = &\text{Total\_Fuel\_Cost} + \text{Total\_Cost\_of\_Missed\_Targets} \\ &+\text{Total\_Operational\_Cost} \end{aligned} \tag{4}$$

Total_Fuel_Cost is the amount of fuel spent by all the vehicles for the duration of the mission multiplied by the cost of consuming that fuel. The Total_Cost_of_Missed_Targets is the number of targets not eliminated by any of the UAVs multiplied by the cost of missing a single target. The Total_Operational_Cost is the total operation time for the mission multiplied by some operational cost per time unit, which would include costs such as maintenance and ground station operation costs. This more detailed cost function is given in (5).

$$C = \text{cost\_of\_fuel}^{*}\text{total\_UAV\_distance} + \text{cost\_per\_missed\_target}^{*}\#\text{\_of\_missed\_targets} + \text{operation\_cost\_per\_time}^{*}\text{total\_time} \quad (5)$$

In order to maximize performance, the cost function should be minimized by finding the optimal values for the variables in the cost equation. However, the variables in the cost equation are themselves dependent on the number of UAVs and the specific paths planned for those UAVs. One way to minimize the cost function is to hold the number of UAVs variable constant at some initial value and to vary the mission routes (individual routes for all the UAVs) until a mission plan with minimum cost is found. We then select a new setting for the number of UAVs variable and repeat the process of varying the mission plan in order to minimize the cost. After iterating through all the possible values for the number of UAVs, the number of UAVs with the least cost and the corresponding optimized mission plan are then the settings that minimize the cost equation. As the number of UAVs is increased, new routing will be required to minimize the cost function. Thus, the paths, which determine time of flight, are a function of number of UAVs.

Moreover, if a target is missed, then there is an additional, more significant cost. When the number of UAVs planned is too low, the number of missed targets increases and hence the cost is high. When the number of UAVs is excessive, more UAVs are used than required and thus additional, unnecessary costs are incurred. We therefore expect the lowest cost to be somewhere in between those two extremities, and that the shape of the cost curve is therefore concave upwards[1] (Figure 8). The profile in Figure 8 does not include the effect of wait times, and it does not take into account the interaction between the vehicles and the human operator.

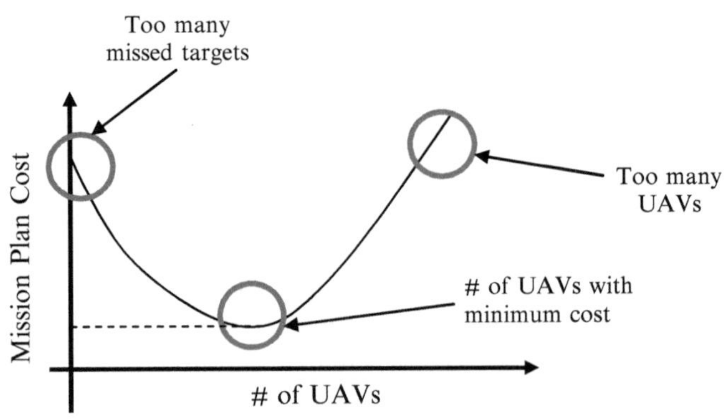

Fig. 8. Mission Plan Costs as a Function of Number of UAVs

[1] Note that this claim is dependent on the assumption that the UAVs independently perform tasks.

In terms of wait times, any additional time a vehicle spends in a degraded state will add to the overall cost expressed in (5). Wait times that could increase mission cost can be attributed to 1) Missing a target which could either mean physically not sending a UAV to the required target or sending it outside its established TOT window, and 2) Adding flight time through route mismanagement, which in turn increases fuel and operational costs. Thus, wait times will shift the cost curve upwards. However, because wait times will likely be greater in a system with more events, and hence more UAVs, we expect the curve to shift upwards to a greater extent as the number of UAVs is increased.

In order to account for wait times in a cost-performance model, which as previously demonstrated is critical in obtaining a more accurate operator capacity prediction, we need a model of the human in our MUAV system, which we detail in the next section.

3.5 The Human Model

Since the human operator's job is essentially to "service" vehicles, one way to model the human operator is through queuing theory. The simplest example of a queuing network is the single-server network shown in Figure 9.

Modeling the human as a single server in a queuing network allows us to model the queuing wait times, which can occur when events wait in the queue for service either as a function of a backlog of events or the loss of situation awareness. For our model, we model the inter-arrival times of the events with an exponential distribution, and thus the arrivals of the events will have a Poisson distribution. In terms of our model, the events that arrive are vehicles that require intervention to bring them above some performance threshold. Thus neglect time for a vehicle is the time between the arrival of events from that particular vehicle and interaction time is the same as the service time.

The arrival rate of events from each vehicle is on average one event per each (NT + IT) segment. The total arrival rate of events to the server (the operator) is the average arrival rate of events from each vehicle multiplied by the number of vehicles.

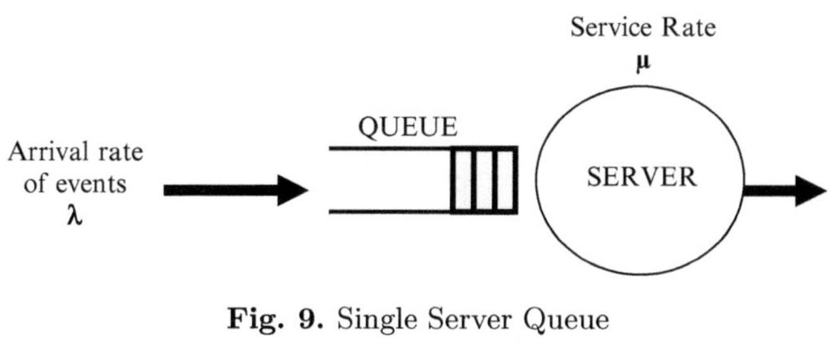

Fig. 9. Single Server Queue

$$Arrival\_rate = \lambda = \#\_of\_UAVs * \frac{1\_event}{(NT + IT)} = \frac{\#\_of\_UAVs}{NT + IT}\frac{events}{time} \quad (6)$$

In terms of the service rate, by definition, the operator takes, on average, an IT length of time to process each event. Therefore assuming that the operator can constantly service events (i.e., does not take a break while events are in the queue):

$$Service\_rate = \mu = \frac{1}{IT}\frac{events}{time} \quad (7)$$

By using Little's theorem, we can show that the mean time an event spends in the queue is:

$$W_q = \frac{\lambda/\mu}{\mu - \lambda} \quad (8)$$

For the purposes of our predictive model, we will assume that this wait time in the queue (Wq, eqn. 8) includes both situation awareness wait times (WTSA) as well as wait times due to operator engagement in another task (referred to as WTQ in the previous section).

Now that we have established our operator model based on queuing theory, we will now show how this human model can be used to determine operator capacity predictions through simulated annealing optimization.

3.6 Optimization through Simulated Annealing

The model that captures the optimization process for predicting the number of UAVs that a single operator can control is depicted in Figure 10. The optimizer takes in as input the number_of_UAVs, the mission description (including the number of targets and their locations), parameters describing the vehicle attributes (such as UAV speed), and other parameters including the weights that are used to calculate the cost of the mission plan. The optimizer in our model (programmed in MATLAB®) iterates through the #_of_UAVs variable, applying a Simulated Annealing algorithm to find the optimal paths plan, as described earlier. The #_of_UAVs with the smallest cost is then selected as that corresponding to the optimal setting. As previously discussed, the human is modeled as a server in a priority queuing system that services events generated by the UAVs according to arrival priorities. The average arrival and

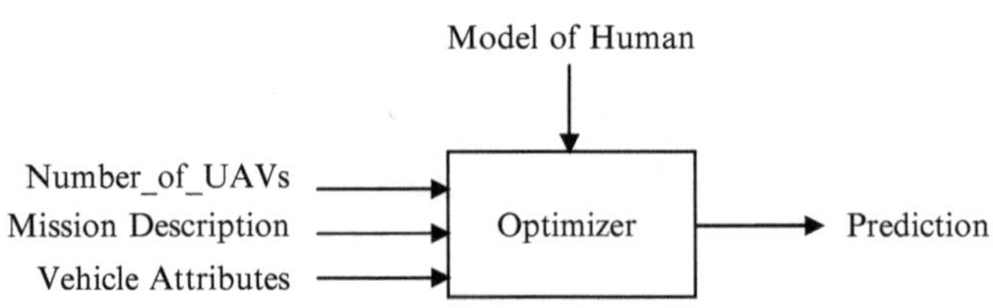

Fig. 10. Optimization Model

Table 2. Optimization Parameters

Name	Unit	Value
Mission Data (includes number of targets, time on targets, and locations)	-	5–10 targets
UAV speed	mi/hr	100
UAV Endurance	hr	5
UAVs launch location	Cartesian	0,0
Cost_per_missed_target	$/target	1500
Cost_of_fuel_per_min	$/min	10
Cost_of_operations_per_min	$/min	1
NT	min	3[2]
IT	min	0.3[1]

service rates as well as their corresponding probabilistic distributions are as assumed earlier.

We chose the simulated annealing (SA) technique for heuristic-based optimization. There were several benefits to selecting the SA technique over other optimization techniques. First, SA is a technique that is well suited to avoiding local minima, a property that is necessary when sub-optimal solutions can exist while searching for the global optimum as is the case in evaluating different mission plans. Also, SA introduces randomness such that the technique generates alternative acceptable solutions on different runs, hence allowing the system designer to seek alternative optimal designs when initial solutions are not feasible. Two limitations of SA are that problems with many constraints can be difficult to implement and that run times can be long. Our problem, however, is one of few constraints and hence their implementation was not an issue. Also, since optimization takes place in mission planning stages and not in time-critical mission replanning, the long run times have a minimal adverse effect.

Model Parameters, Constraints, and Variables. The list of parameters established for the design process is presented in Table 2. We selected generic UAV capabilities that would be exhibited by small-to-medium size UAVs engaged in an ISR mission such as the Hunter or Shadow. Our cost function was discussed previously (5) and Table 3 details the constraints used in our model.

3.7 Results of Simulation

We first investigated the cost-UAV number relationship for the theoretical best case in which the human operator is "perfect" and introduces no delays in the system. In Figure 11, the optimized cost is plotted against the number

[2] Interaction and neglect times were determined using the MAUVE interface described previously.

Table 3. Constraints for Simulated Annealing

Constraints
• A UAV cannot visit targets for which it cannot meet the times on target
• Each UAV must visit at least one target
• UAV routes must start and end at launch locations.

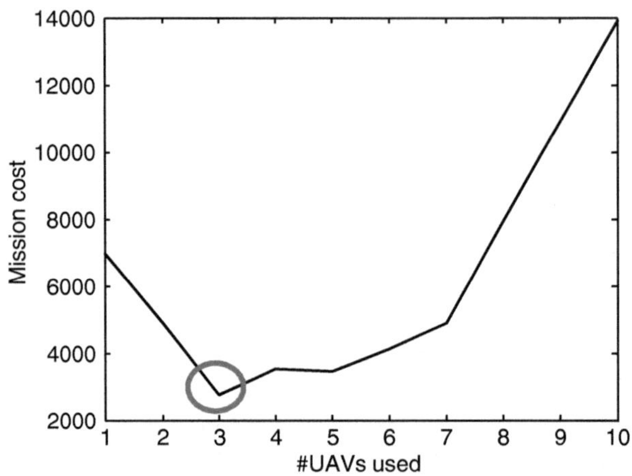

Fig. 11. Minimum Cost versus Number of UAVs

of UAVs variable, with a mission plan of 10 targets. As predicted, the curve is concave upwards and has a global minimum where the cost is minimized.

We then proceeded to include the effect of the human operator and hence, wait times. It was assumed that during periods of wait times, UAVs loitered in the same spot and therefore maintained the same geographic location on the map. In order to demonstrate how plan complexity could affect the problem, we included three scenarios in which 5, 7, and 10 targets were represented, as depicted in Figure 12.

Figure 12a shows the effect of a relatively simple mission with only 5 targets. In general, the effect of wait times on the cost curve is minimal, i.e., the minimum theoretical best case is equal to that with the operator wait time case. This is not unexpected, since simple missions have a rather small sensitivity to wait times. The simplicity of the missions does not overburden the operator who is operating in a robust cognitive state, and can accommodate the wait times without incurring increased costs.

Figure 12b demonstrates how the curves can shift as a function of increasing task complexity (7 versus 5 targets). The cost curves between the theoretical best case and the operator wait time model clearly deviate and the optimal number of UAVs that should be controlled decreases from 4 to 3 UAVs. This

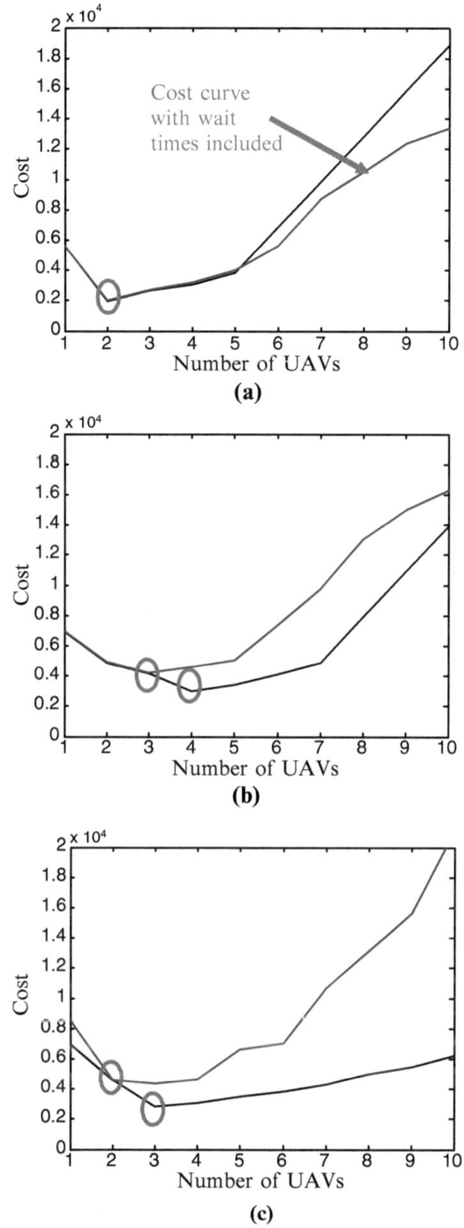

Fig. 12. Cost surves with (a) 5 targets (b) 7 targets, and (c) 10 targets

is primarily due to the fact that the wait times generally affect the longer routes where the probability of missing targets increases.

The results for 10 targets are shown in Figure 12c, which demonstrate a significant divergence from the theoretically perfect operator model and the

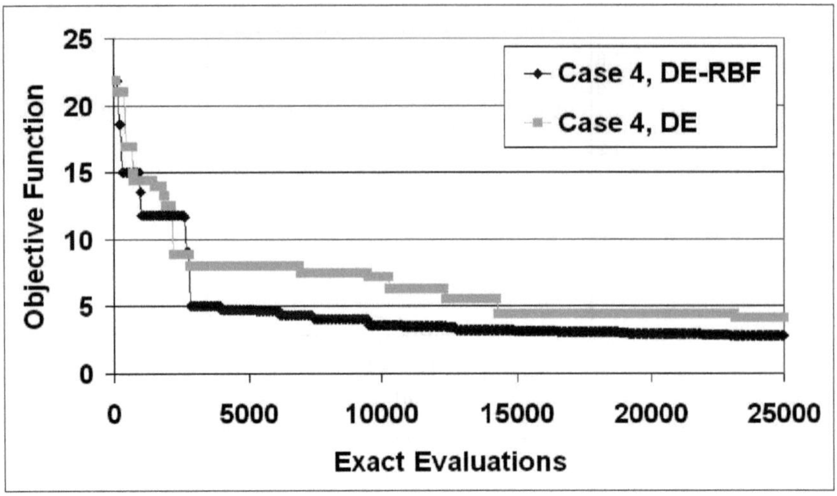

Fig. 13. Operational Demands vs. Human Limitations in Mission Planning

model with wait times, especially beyond 5 UAVs. Interestingly, as in the case with 5 targets, the minimum cost occurs at 2 UAVs for 10 targets. However, in the case of 5 targets, only 2 UAVs were needed to meet the operational requirements and the operator could meet this demand. However, with 10 targets, 2 UAVs became the minimum cost point because of operator limitations, i.e., the wait times incurred by controlling more vehicles became unacceptably high. Thus, at the inflection point in these curves, the left region is primarily constrained by operational demands, but the right region is dominated by human performance limitations, specifically wait times, as seen in Figure 13.

Another interesting trend across the predictions in Figures 12a-c is the relatively flatness of the curves in the left region. For example, there is generally a plateau in performance such that dramatic increases are not seen in cost until more than 5 UAVs are managed. In the case of 5 targets and up to 5 UAVs under control, costs increased at a rate of 33% per additional UAV. Beyond 5 UAVs, the mission increased at a rate of 74% per UAV, a much sharper increase, which suggests that operator performance is relatively robust up to 5 UAVs, at which point the operator is saturated and severely limits overall mission success. These graphs demonstrate that the more complex mission requirements added to the cognitive load of the operators, thus workload had to be reduced by reducing the number of UAVs under control.

In terms of the fan-out and revised fan-out equations (1 & 3), using the same neglect and interaction times as in the cost-based simulation model (Table 2), as well as the wait times derived from the queuing theory model (8), the predictions are seen in Figure 14. While the original fan-out estimated a constant 11 vehicles, given the wait times that would build with increasing numbers of UAVs under control, the optimal control point is 5 vehicles

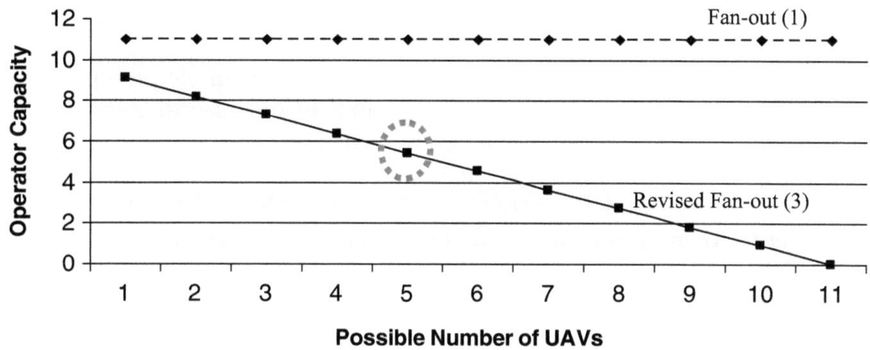

Fig. 14. Predictions Using Cost-Based Simulation Inputs

using the revised fan-out. Interestingly this number is very close to what was experimentally observed in the previously described experiment.

4 Meta-Analysis of the Experimental and Modeling Prediction methods

Two methods for determining maximum operator capacity for supervisory control of multiple UAVs have been presented, both based on operator interactions and wait times for mission tasks, as well as neglect times during which one or more vehicles operate autonomously. The strengths and weaknesses of each method will now be discussed, as well as how these methods could be used synergistically.

In the first method, the original fan-out equation that related operator interaction and vehicle neglect times (1) was revised to include operator wait times (3). An experiment was conducted to determine if the revised fan-out predictions more closely matched actual human-in-the-loop control scenarios. The results showed that the revised fan-out model produced more conservative estimates when modified to include wait times caused by human interactions, which include interaction wait time, wait time in the queue, and wait time due to the loss of situation awareness.

While this temporal-based method for computing fan-out gives more conservative general estimates, it lacks the cost-benefit analysis trade space representation that can be found through optimization methods that provide for sensitivity analysis. For example, in the experiment, it was estimated that operators could control 7–16 UAVs in a low workload scenario, but only 3–7 vehicles in high workload settings. The ranges resulted from increasing levels of automation as an experimental independent variable. Because these predictions were based on experimental data (which were discrete across four different levels of automation), there can be no post-hoc sensitivity analysis, only refining the experimental method and running more human subject trials, which is very expensive and labor intensive.

In comparison, optimization methods such as the example presented here provide not only predictions for operator capacity but also directly link the capacity to a system performance measure, which was cost in our example. By developing the estimates through the fan-out approach, there is only the consideration of a vaguely defined threshold for acceptable operator performance. Furthermore, there is no way to directly infer how this human performance affects the overall system, which is actually the more critical variable, particularly in command and control settings. Moreover, while it was very expensive in terms of experimental design for human subjects to examine mission complexity in terms of low and high workload, in the cost-based simulation method, mission complexity was represented by the number of targets, which was relatively not costly to alter. Thus, this type of prediction method allows for more specific and detailed predictions for operator capacity, as well as how the external environment (i.e., number of targets) will affect overall mission success.

However, while the simulation estimations provide for multivariate sensitivity analysis across operator and system performance metrics, one drawback is the inability to directly correlate the predictions to possible design interventions. As previously discussed, the cost-based simulation links the external environment to both operator and system performance, but it inherently lacks the ability to parse out which system parameters could and should be changed to improve operator and autonomy performance. For example, in the SA model, all wait times are included in a single measure, however the wait times (interaction, queuing, and situation awareness) fundamentally have different causes. In addition, as demonstrated in Figure 7, the different types of wait times can have dramatically different values and without the ability to model and see the separate effects of different wait time sources, it is not clear what design interventions could occur to mitigate them (such as improved decision support or increased vehicle autonomy.)

Moreover, a cost-based simulation cannot represent the impact of specific automation strategies on operator performance. It is often assumed that as autonomy levels increase (as depicted in Table 1), the need for human interaction decreases, and thus lowers system wait times. However, as can be seen in Figure 15, these assumptions are not always accurate. In the experiment previously discussed, we predicted that as system autonomy increased, wait times due to an operator workload queue (referred to as wait time in the queue in the previous section) would decrease. However, the dotted line demonstrates what queuing wait times were actually observed, and there was clearly an anomaly with the active condition that corresponds to LOA 4 in Table 1. Described more in detail in [16], what was hypothesized to be a decision support tool to mitigate operator workload actually degraded operator performance and caused increased, instead of decreased, wait times. This insight was only gained through the experimental derived interaction, neglect, and

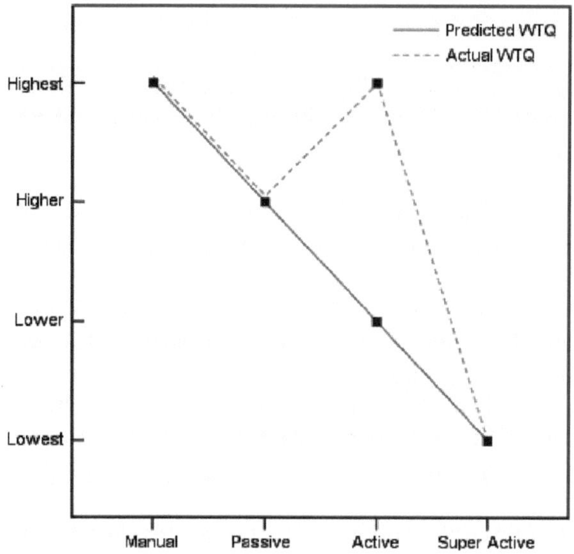

Fig. 15. Wait Times in the Queue across Levels of Automation

wait times. Because the SA optimization approach and other similar stochastic approaches assume an a priori distribution (both in arrival rates and service times), if such simulation methods are not used in conjunction with experimentally derived data, results are highly speculative and lack external validity.

This last point about the problem with assumptions highlights an inherent limitation to both methods: Estimating interaction, neglect, and wait times. As previously discussed, for the cost-based simulation, a distribution must be selected for wait times, and presently there is little theoretical or empirical basis for doing so. In addition, interaction and neglect times must be selected a priori and while these could be estimated from system design parameters, they are highly contextual and will likely dramatically change with different levels of autonomy, decision support, mission complexity, operator training, etc.

Similarly, even experimentally derived interaction, neglect, and wait times can be difficult to measure. Unfortunately the times and the associated costs (degraded performance, etc.) are very difficult to capture in performance-based simulations such as the one reported in this study. Through using software that tracked users' cursor movements and activation of control devices, we were able to determine on a gross level when a subject was actively engaged with a particular UAV, but subtle transitions are difficult to capture. The use of psychophysiologic measurement devices may be of use in addition to performance-based measures but the application of these methods needs significantly more investigation.

5 Conclusions

With the recognition that intelligent autonomy could allow a single operator to control multiple vehicles (including air, ground, and water), instead of the converse which is true today, there is increasing interest in predicting the maximum numbers of autonomous vehicles an operator can control. A critical system architecture question is then how many vehicles could one operator control? While there are other methods that could be used to predict this number (e.g., cognitive modeling which suffers from the ability to represent highly complex systems, and simulations and experiments with advanced prototypes, which suffer from exorbitant development costs), we demonstrated, through two different methods, how this number can be estimated by considering the temporal elements of supervisory control of multiple UAVs.

In the first method, we demonstrated that past equations of fan-out omitted important aspects of human interactions with multiple UAVs. We suggest an alternative equation that captures some of these aspects using wait times. However, these temporal approaches to measuring fan-out are limited since these results are not explicitly linked to performance. In comparison, we used cost-based simulation model that links operator performance to both mission costs and complexity; however, it suffers from problematic assumptions and an inability to highlight specific areas for design interventions.

While each method has strengths and weaknesses, they are not mutually exclusive. The two approaches can be synergistic in that temporal data gathered experimentally for initial rough estimates such as fan-out can provide more valid simulation models. Predictions then made through optimization simulations can be furthered refined through sensitivity analyses and appropriately focused human-in-the-loop experiments. In this way, effects of increasing UAVs and/or system autonomy can be seen on system performance as well as operator performance. In terms of application, this iterative approach to predicting operator capacity would likely provide the most benefit early in the systems engineering conceptual stages when unmanned aerial systems are still in development and uncertainty in system parameters is high.

Acknowledgments

The research was supported by grants from Boeing Phantom Works and Lincoln Laboratory.

References

1. J. Franke, V. Zaychik, T. Spura, and E. Alves, “Inverting the Operator/Vehicle Ratio: Approaches to Next Generation UAV Command and Control,” presented at Association for Unmanned Vehicle Systems International and Flight International, Unmanned Systems North America Baltimore, MD, 2005.

2. S. Dixon, C. Wickens, and D. Chang, "Mission Control of Multiple Unmanned Aerial Vehicles: A Workload Analysis," *Human Factors*, in press.
3. A.T. Welford, "The psychological refractory period and the timing of high-speed performance - a review and a theory," *British Journal of Psychology*, vol. 43, pp. 2–19, 1952.
4. D.E. Broadbent, *Perception and Communication*. Oxford: Pergamon, 1958.
5. H.A. Ruff, S. Narayanan, and M.H. Draper, "Human Interaction with Levels of Automation and Decision-Aid Fidelity in the Supervisory Control of Multiple Simulated Unmanned Air Vehicles," *Presence*, vol. 11, pp. 335–351, 2002.
6. S. Dixon, C.D. Wickens, and D. Chang, "Unmanned Aerial Vehicle Flight Control: False Alarms Versus Misses," presented at Humans Factors and Ergonomics Society 48th Annual Meeting, New Orleans, 2004.
7. M.L. Cummings and D. Morales, "UAVs as Tactical Wingmen: Control Methods and Pilots' Perceptions," in *Unmanned Systems*, vol. February, 2005.
8. S.L. Howitt and D. Richards, "The Human Machine Interface for Airborne Control of UAVs," presented at 2nd AIAA "Unmanned Unlimited" Systems, Technologies, and Operations—Aerospace, Land, and Sea Conference and Workshop, San Diego, CA, 2003.
9. M.L. Cummings and S. Guerlain, "Developing Operator Capacity Estimates for Supervisory Control of Autonomous Vehicles," *Human Factors*, in press.
10. D.R. Olsen and S.B. Wood, "Fan-out: Measuring Human Control of Multiple Robots," presented at CHI2004, Vienna, Austria, 2004.
11. D.R. Olsen and M.A. Goodrich, "Metrics for Evaluating Human-Robot Interactions," presented at Performance Metrics for Intelligent Systems, Gaithersburg, MD, 2003.
12. M.L. Cummings, "The Need for Command and Control Instant Message Adaptive Interfaces: Lessons Learned from Tactical Tomahawk Human-in-the-Loop Simulations," *CyberPsychology and Behavior* vol. 7, 2004.
13. M.L. Cummings and P.M. Mitchell, "Managing Multiple UAVs through a Timeline Display," presented at AIAA InfoTech, Arlington, VA, 2005.
14. T.B. Sheridan and W.L. Verplank, "Human and Computer Control of Undersea Teleoperators," MIT, Cambridge, Man-Machine Systems Laboratory Report 1978.
15. G. Osga, K. Van Orden, N. Campbell, D. Kellmeyer, and D. Lulue, "Design and Evaluation of WarfighterTask Support Methods in a Multi-Modal Watchstation," SPAWAR, San Diego 1874, 2002.
16. M.L. Cummings and P.J. Mitchell, "Automated Scheduling Decision Support for Supervisory Control of Multiple UAVs," *Journal of Aerospace Computing, Information, and Communication*, in press.
17. B. Hilburn, P.G. Jorna, E.A. Byrne, and R. Parasuraman, "The Effect of Adaptive Air Traffic Control (ATC) Decision Aiding on Controller Mental Workload," in *Human-automation Interaction: Research and Practice*. Mahwah, NJ: Lawrence Erlbaum, 1997, pp. 84–91.
18. M. L. Cummings, "Automation Bias in Intelligent Time Critical Decision Support Systems," presented at AIAA Intelligent Systems, Chicago, IL, 2004.

2. S. Dixon, C. Wickens, and D. Chang, "Mission Control of Multiple Unmanned Aerial Vehicles: A Workload Analysis," Human Factors, [illegible]
3. A. T. Welford, "The psychological refractory period and the timing of high-speed performance—a review and a theory," British Journal of Psychology, vol. 43, pp. [illegible], 1952.
[illegible]

Team, Game, and Negotiation based Intelligent Autonomous UAV Task Allocation for Wide Area Applications

P.B. Sujit[1], A. Sinha[2], and D. Ghose[3]

[1] Department of Electrical and Computer Engineering, Brigham Young University, Provo, Utah, USA
[2] Department of Aerospace Engineering, Indian Institute of Science, Bangalore, India
[3] Department of Aerospace Engineering, Indian Institute of Science, Bangalore, India
pbsujit@byu.edu, asinha@aero.iisc.ernet.in,dghose@aero.iisc.ernet.in

Abstract. Unmanned aerial vehicles (UAV) have the potential to be used for search and surveillance missions, and as munitions in the battlefield. The UAVs are deployed in swarms as they may not have sufficient computational, sensor, and operational capability to complete the task single-handedly. A desirable feature for these UAV swarms is the capability of intelligent autonomous decision making and coordination, with minimal or no centralized control. In this chapter, we present decentralized and distributed task allocation schemes based on concepts from team theory, game theory, and from negotiation techniques used in decision-making problems arising in economics, and apply these to design intelligent decision-making strategies for multiple UAV systems performing a wide area search and surveillance mission. We also address the task of searching an unknown environment, which is a major component in such missions, separately using game theoretical concepts.

1 Introduction

Unmanned aerial vehicles are being extensively used for military and civilian applications, like search, surveillance and as munitions in the battlefield. They play a crucial role in information gathering from hostile and unknown regions. These UAVs can also be used as munitions to search, attack and destroy targets in an unknown region. The UAVs used for these applications may have limited effectiveness and may not have the required stealth capability and munition payload to complete the task single-handedly. Hence, there is a necessity for such UAVs to be deployed in swarms. A desirable feature for these UAV swarms is the capability of intelligent autonomous decision making and coordination. The UAVs operating in an unknown region are expected to

P.B. Sujit et al.: *Team, Game, and Negotiation based Intelligent Autonomous UAV Task Allocation for Wide Area Applications*, Studies in Computational Intelligence (SCI) **70**, 39–75 (2007)
www.springerlink.com

carry out several tasks in relation to the targets or other entities of interest present in the region [1, 2]. An efficient task allocation method is necessary to assign UAVs to targets.

The classical solution for such task allocation problems is a centralized one that generates the necessary commands for the UAVs. But, centralized task allocation systems have well known limitations and do not address scalability issues too well. Hence, there is a necessity to develop decentralized task allocation algorithms. These algorithms must be suitable for implementation in a multiple agent UAV swarm, should be scalable, and also have low computational overheads. An efficient task allocation strategy should have the ultimate objective to complete the mission in minimum time by cooperating and coordinating with other UAVs. Cooperation can be achieved by explicit or implicit communication with neighbouring UAVs.

In this chapter, we will present decentralized and distributed task allocation schemes based on concepts from team theory, game theory, and from negotiation techniques used in decision-making problems arising in economics, and apply these to design intelligent decision-making strategies for multiple UAV systems performing a wide area search and surveillance mission [3]-[8]. In this context, we will explore the role that communication between UAVs plays during decision-making.

The overall problem of task allocation is modelled as a sequence of tasks that the UAVs need to carry out on a target. The allocation of tasks will depend on various factors such as the proximity of the UAV to the target, its perception of the target status, its capability to carry out the task at hand, the choice that it may have in carrying out a given task or obtain greater benefit by performing some other tasks, where the choice can be between some alternate targets or tasks, and so on. All this needs to be carried out in a decentralized and distributed manner.

In team theoretical task allocation, each UAV takes decision autonomously. The UAV senses the status of the target and evaluates the expected benefit of attacking the target. The UAV also senses the presence of other UAVs within its sensor radius, and estimates the probability of the neighbouring UAVs attacking the target. Based on these values (expected benefit of attacking a target and the probability of the other UAVs attacking the same target), a linear programming problem is formulated. The UAV decides on a task/target assignment based on the solution provided by this formulation, which is proven to be team optimal. An important feature of the decision-making process is that, there is no explicit communication between UAVs. This formulation is especially useful in a hostile environment where communication between UAVs is either minimal or just not possible.

In negotiation based task allocation we restore the communication among UAVs for decision making. Each UAV broadcasts its intentions to attack a target, along with its perceived benefit in doing so, to its neighbours. A UAV evaluates all the proposals that it receives. The evaluation is carried out by comparing the benefits proposed by other UAVs in attacking the same target.

It is shown that negotiation based task allocation can efficiently allocate tasks and targets to UAVs, and detect and resolve conflicts between neighbouring UAVs. The decision-making mechanism has very low computational overheads, and is shown to be scalable to large number of UAVs and targets.

One of the major tasks that need to be carried out in such wide area operations is that of search and it has enough facets of its own to merit a separate treatment. The search task is carried out to detect targets in an unknown region. The problem associated with search is to develop coordination algorithms for multiple UAVs to minimize search route duplication and maximize the information collected during the operation. We show that intelligent distributed algorithms, based on game theory, can be developed to perform such search tasks.

2 Existing Literature

Task allocation of UAVs is an active research area for the past few years. When a UAV detects a target, it broadcasts the information to all the other UAVs in the search space. Since, the information is common to all the UAVs, each UAV independently solves a task allocation algorithm and determines its task. The various task allocation schemes developed by researchers are based on network flow model [9] [10], mixed integer linear programming (MILP) [11], dynamic programming [12] or genetic algorithm [13]. The task allocation can also have additional objectives like minimize path lengths [14], or timing constraints [10]. Turra et al. [15] present a task allocation algorithm for multiple UAVs performing search, identification, attack, and verification tasks in an unknown region for targets that move in real time. These authors also address the problem of obstacle avoidance. Jin et al. [16] propose a probabilistic task allocation scheme for the scenario presented in [9, 1].

Recently, market based approaches have shown a considerable increase in performance for task allocation strategies to multiple agent applications. Dias and Stenz [17, 18] introduce a novel approach for coordinating robots based on the free market architecture in economics. The approach defines revenues and cost functions across the possible plans for executing a specified task. The task is accomplished by decomposing it into sub-tasks and allowing the robots to bid and negotiate to carry out these sub-tasks. Gerkey and Mataric [19] use an auction mechanism for multi-robot coordination while they analyze the communication and computational complexity involved in multi-robot task allocation in [20]. The authors categorize the task allocation methods based on the capability of the robots and the kind of tasks involved. Mataric et al. [21] develop various task allocation strategies and study their performance on a multi robot application with sensor noise. Their simulation study is compared with results obtained using experiments. Gurfil [22] also uses an auction mechanism for task allocation among multiple UAVs performing search and destroy mission. Lagoudakis et al. [23] use auction algorithm for assigning unexplored tasks to group of mobile robots. The paper also provides some theoretical bounds on the computational complexity of the proposed algorithm. Sariel and Balch [24] present an auction scheme with various objectives. The various objectives are based on traveling salesman problems.

In reality, the UAVs are subjected to limited communication constraint which most of the researchers have not addressed. Hence, the algorithm developed either cannot adhere to the limitation imposed by UAVs, or they cannot be easily extended. Hence, there is necessity to develop task allocation algorithms that are distributed and can perform efficiently. In this chapter we propose team theory, game theory, and negotiation based task allocation schemes that consider the communication constraint of the UAVs into account and show that they perform better than various other strategies.

3 Task Allocation Using Team Theory

3.1 Basics of Team Theory

A team (as defined in [25]) is a group of individuals each of whom takes decision about something different but who receive a common reward as a joint result of all those decisions. The individuals get information about the external situation (state of the environment) through observations and communication and so the information available to different individuals are different. They take decision based on their respective information. The state of the environment is a random variable, the probability distribution of which is known *a priori* to the individuals. Based on the state of the environment and the decision taken, the team incurs a common payoff. Team theory deals with finding the best communication and the best decision rules, given the payoff function, the probability distribution of the environment and the communication cost.

Let $T = \{1, 2, \ldots, N\}$ denote a team of N individuals or decision makers and S denote the set of alternative states of the environment. We consider the set S to be discrete and finite, i.e., the possible configuration/state of the environment is finite. The probability mass function defined on the set S is given by $\gamma(s)$.

When the state of the environment is $s \in S$, each decision maker receives the signal $y_i(s)$ as information through the process of observation. Let $Y_i = \{y_i\}$ denote the set of alternative signals that the decision maker i can receive as information. The function $\eta_i : S \rightarrow Y_i$ is called the information function of the ith decision maker and so

$$y_i = \eta_i(s) \tag{1}$$

The set of all information function $\eta = \{\eta_1, \eta_2, \ldots, \eta_N\}$ is called the information structure of the team.

Based on the information y_i received by the ith team member, it takes the decision x_i. Let $X_i = \{x_i\}$ be the set of alternative decisions that the ith

decision maker can take. Then, the function $\delta_i : Y_i \rightarrow X_i$ is called the decision function for the ith decision maker and we have

$$x_i = \delta_i(y_i) \tag{2}$$

Considering the decision function of all the decision makers, the vector $\delta = \{\delta_1, \delta_2, \ldots, \delta_N\}$ is called the team decision function. There can be some constraints on the team decision functions. For example, for every $s \in S$, let $k(s) \in \mathbb{R}^n$ be a close convex set. We will consider only those decision functions for which $\delta(\eta(s)) \in k(s), \forall s$. Let $x = \{x_1, x_2, \ldots, x_N\}$ denote the team decision.

The outcome of the decisions of the team members depends jointly on the state s and the team decision x and it is determined according to some function $u(s, x)$ which is pre-specified. Hence, the payoff of the team is given by

$$\omega = u(s, x) \tag{3}$$

The team decision problem is concerned with finding the maximum expected payoff with respect to the team decision function i.e.,

$$\max_{\delta} E[\omega(s, x)] = \max_{\delta} \sum_{x \in k(s)} \gamma(s) u(s, \delta(\eta(s))) \tag{4}$$

If the payoff function is linear in the decision variables, the team is called a linear team. As shown in [26], the solution of the linear team can be obtained by solving a linear programming problem in the decision function space. Let the payoff function be $\omega = \sum_i C_i x_i$, where C_i is a function of the state and so it is also a random variable. Then, the objective function is given as

$$\max_{x \in k(s)} E\left[\sum_i C_i x_i\right], \tag{5}$$

3.2 Problem Formulation

Let us consider a battlefield scenario where N UAVs are deployed to search and destroy targets within a stipulated time. Thus the team T consists of the N UAVs, which are the decision makers. The environment comprises of the targets of different strengths scattered on a plain. Assume that there are M targets, the location and the strength of which are not known *a priori* to the UAVs. We define the state of the environment as $s = (\{Z_i^u\}_{i=1}^N, \{Z_j^t\}_{j=1}^M, \{V_j\}_{j=1}^M)$ where Z_i^u is the position of the ith UAVs, Z_j^t is the position of the jth target and V_j is the strength/values of the jth target.

The UAVs can observe the environment within a given sensor radius. We assume that there is no communication among the UAVs. Thus, the information available to the UAVs about the state of the environment are the number of targets, their values and the number of other UAVs present within the sensor radius. It is assumed that UAVs know the position of both the targets and other UAVs precisely but the strength of the targets are known with certain probability which is a function of the distance between the UAV and the target.

Let us assume that at time t_s, UAV i can see m_i number of targets and n_i number of UAVs within its sensor range. It senses the value of the targets with probability $p(d_{ij}), j = 1, \cdots, m_i$ where d_{ij} is the distance between the ith UAV and the jth target. Based on these information, the ith UAV decides its action, which can be attacking any one of the j targets or opting for search. The decision taken by the ith UAV is given as $x_i = [x_{i1}, x_{i2}, \ldots, x_{im_i}, x_{i(m_i+1)}]$ where $x_{ij} \in \{0, 1\}$ denotes whether the ith UAV will perform the task j or not. Here, the task $j = m_i + 1$ is the search task.

Let the benefit of performing the task x_{ij} by the ith UAV be C_{ij}. In general, C_{ij} is a function of the states of the environment, i.e. the current position of the UAVs, the current position of the targets and the value of the targets. We will model C_{ij} in the next section. The payoff of the whole team is then given as

$$\omega = \sum_{i=1}^{n_i} \sum_{j=1}^{m_i+1} C_{ij} x_{ij} \tag{6}$$

The objective of the team is to maximise ω with respect to the action, x_{ij} taken by the UAVs at time t_s. We assume that the overall mission of the UAVs will be optimal if the decision taken at each time step is optimal. However, there may be some constraints on the combined decision taken by the team $x = [x_1, x_2, \cdots, x_n]$ due to practical limitations. At any given time, an UAV can perform only one task, so

$$\sum_{j=1}^{m_i} x_{ij} = 1, \quad i = 1, 2, \cdots, n_i \tag{7}$$

For the mission to be effective, it is necessary that only one UAV be assigned to one target at a given time. However the number of targets present may be more than the number of UAVs and so it may not be possible to assign UAVs to all the targets. Thus, we have

$$\sum_{i=1}^{n_i} x_{ij} \leq 1, \quad j = 1, 2, \cdots, m_i \tag{8}$$

Hence, the optimization problem can be posed as maximize ω in (6) with respect to x subjected to the constraints in (7)-(8) and $x_{ij} \in \{0, 1\}, \forall i, j$. Here, both the objective function as well as the constraints are linear. Thus, it can be solved using linear programming. However, linear programming give solution in $x_{ij} \in [0, 1]$. For this class of problem, it is easy to see that for every $x_{ij} = \zeta$ such that $0 < \zeta < 1$, there exits a solution $\hat{x}_{ij}$ =0 or 1 that will give a better or equal performance. Since we assume decentralized control of the UAVs, each UAV solves the optimization problem individually to decide on its action.

3.3 Team Theoretic Solution

The problem defined in Section 3.1 assumes that the optimization problem is solved globally. However, in the scenario that we consider, the UAVs do not have global information. Each UAV solves the optimization problem with only local information available to it. Moreover, the value of the target status is a random variable. Hence, we use concepts from team theory to solve this optimization problem.

Before reformulating this problem, we define the benefit C_{ij} that the ith UAV gets by performing the task j. If it is a search task then

$$C_{is} = \frac{\text{time left in the mission}}{\text{total flight time}} \tag{9}$$

If it is the task of attacking the target j then,

$$C_{ij} = V_j w_r \quad - \quad S_{ij} \tag{10}$$

where, V_j = value of target j, w_r = the weightage given to the search task over the task of attacking a target, and

$$S_{ij} = \frac{\text{time to reach the target } j \text{ by UAV } i}{\text{total flight time}} \tag{11}$$

However, the ith UAV knows the values of the target j with some probability. The probability distribution is assumed to be linear and is shown in Figure 1(a). Let $p_r(d_{ij})$ define the probability of target j to have a value r at a distance d_{ij}. Here, $r = \{0, 0.5, 1\}$ where, when $r = 1$, the target has not been attacked and is intact, when $r = 0.5$ the target is partially destroyed, and when $r = 0$ the target is fully destroyed. Thus, C_{ij}'s are random variables with probability distribution $p(d_{ij}) = [p_1(d_{ij}), p_{0.5}(d_{ij}), p_0(d_{ij})]$.

Speculation/BDA: Since speculation on the target is done at every time step, and is reflected on the value of targets, we will not attach any separate benefit to the speculative task.

Each UAV also has to estimate the benefits that its neighbouring UAV (say the kth UAV) will get from the different tasks that it can perform. It calculates the benefits as follows:

Search task: The search task is similar to that defined above, hence the search value is the same for all UAVs.

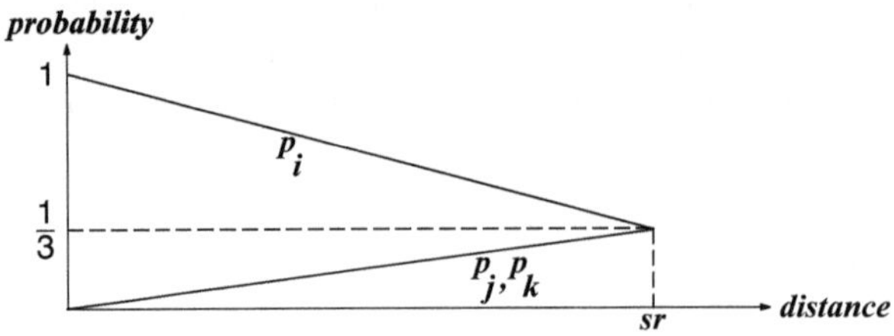

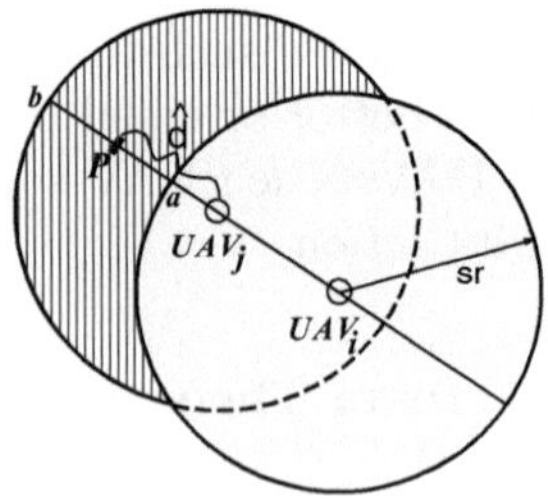

Fig. 1. (a) Probability distribution of the values of the targets as a function of distance (b) Determination of virtual targets

Attacking target j: If target j is within the sensor radius of the kth UAV then

$$C_{kj} = V_j w_r - S_{kj} \tag{12}$$

If target j is not in the sensor range of the kth UAV then $C_{kj} = 0$. Here, we have assumed that all the UAVs have the same sensor range and hence the ith UAV can estimate whether the jth target is within the sensor range of the kth UAV.

Attacking virtual target: The concept of virtual target is used to estimate the environment beyond the sensor range of the ith UAV (see Figure 1(b)). The ith UAV cannot see the shaded region which the jth UAV can see. Depending on the number of targets present in that shaded region, the behaviour of the jth UAV will vary. To estimate the number of targets that might be there, we assume that the targets are uniformly distributed. We take into consideration the combined effect of all these target, which we assume to be placed at a point p, equidistant from point (a, b). This combined target is called the virtual target for the kth UAV. The benefit that the kth UAV gets for attacking this virtual target $\hat{k}$ is

$$C_{k\hat{k}} = (\text{average value of target}) n_k w_r \quad - \quad S_{k\hat{k}} \tag{13}$$

where, n_k is the number of targets that can be present in the shaded region. Therefore, $n_k = n_i$ (area of shaded region)$/(\pi s_r^2)$ and $C_{l\hat{k}} = 0, \forall\, l = 1, \ldots, n_i$, $l \neq k$, and s_r is the sensor range. That is, for any other UAVs, the benefit of attacking the virtual target $\hat{k}$ of the kth UAV is zero. Let us denote $T_v^i = \{\hat{k}_1, \hat{k}_2, \cdots, \hat{k}_{n_i - 1}\}$ to be the set of virtual targets for the $n_i - 1$ neighbours that the ith UAV has to take into account.

Since, C_{ij} are random variables, the ith UAV will maximize the expected payoff. The expectation is calculated on the basis of the joint probability distribution $P_i(s)$ on the state. Here, we assume that the value of the targets are independent, therefore

$$P_i(s) = \prod_{j=1}^{m_i} p(d_j) \tag{14}$$

The objective is to maximize the expected payoff $E(\omega)$ with the constraints defined in Section 3.1, thus each UAV solves the following linear programming problem:

$$\max_{x} \quad E(\sum_{ij} c_{ij} x_{ij}) \tag{15}$$

$$i = 1, \ldots, n_i; \; j = 1, \ldots, m_i, m_i + 1, (m_i + 1) + 1, \ldots, (m_i + 1) + (n_i - 1)$$

subject to

$$\sum_j x_{ij} = 1, \forall i; \quad \sum_i x_{ij} \leq 1 \forall j; \quad x_{i,\hat{j}} = 0, \; \forall i, \text{ and } \hat{j} \in T_v^i; \quad x_{ij} \in [0, 1], \; \forall i, j$$

where $j = m_i + 1$ is a search task, $j = (m_i + 1) + 1, \ldots, (m_i + 1) + (n_i - 1)$ represent the virtual targets.

3.4 Simulation Results

We demonstrate the effectiveness of using team theory for a multi-UAV task allocation problem using a simulation environment. Consider a geographical search space of 100×100 with 20 targets present in the geographical region, as shown in Figure 2(a). The search and attack operation is carried out for 200 time steps, which also represents the flight time of the UAVs. The sensor range of each UAV is 20. The location of the targets are not known a *priori* to the UAVs. All the targets in the search space have the same target value for these set of simulations, however, in general, the target may have different target values depending on their threat levels. The targets are located randomly in the search space. We use 7 UAVs for the mission. The UAVs perform search, attack and speculative tasks on the target. We compare the results when UAVs use team theory based decision making with other types of task allocations, namely, greedy allocation, and limited sensor range with full communication.

Greedy Allocation

In this allocation scheme, each UAV decides to move to a target that would give maximum benefit. Since the value of the targets are random variables, we consider the expected value of the target to calculate the benefit C_{ij}. Hence, the ith UAV's decision is given by:

$$\max_j C_{ij} = \max_j [E(V_j) w_r - S_{ij}] \tag{16}$$

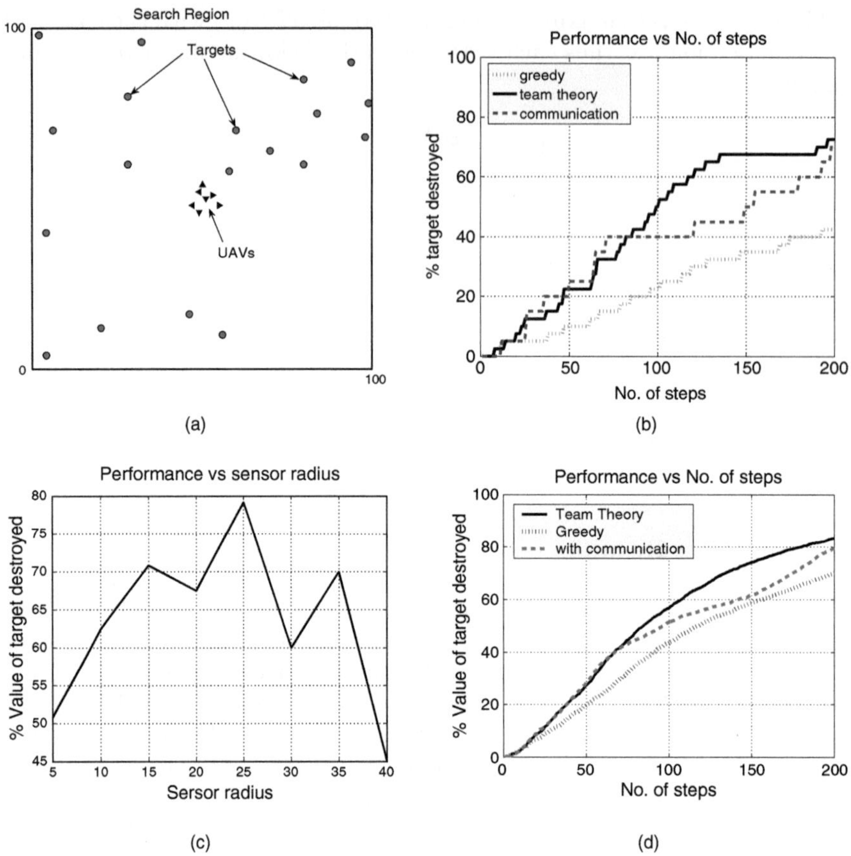

Fig. 2. (a) Search region with UAVs and targets (b) Number of targets destroyed completely (c) Performance of target value destroyed with variation in sensor range for the UAVs (d) Average target value destroyed; performance on averaging over 20 different maps

Limited Sensor Range with Full Communication

Here, each UAV has limited sensor range s_r but can communicate with all the other UAVs. Whenever new information is sensed by a UAV, the UAV broadcasts the information to all the other UAVs. We assume that there are no communication delays. Hence, all the UAVs have the same information about the state of the environment at any given time. So, all the UAVs solve the same LP problem. Moreover, the concept of virtual target does not apply here, as the ith UAV knows the number of targets present in the neighbours' sensor region through communication. Similar to the greedy strategy, the UAV would like to maximize the expected value of the target. The ith UAV solves the following problem

$$\max_{x_{ij}} \quad \sum C_{ij} x_{ij} \tag{17}$$

subject to

$$\sum_j x_{ij} = 1; \quad \sum_i x_{ij} \leq 1; \quad x_{ij} \in [0,1], \ \forall i,j$$

where $i = 1, \ldots, N$ and $j = 1, \ldots, t_a$, with t_a representing all the targets detected so far.

Figure 2(b) shows the performance curves for 7 UAVs performing search and attack tasks on a 100×100 search space shown in Figure 2(a). For evaluation of the performance by each strategy we use the percentage values of the target destroyed (T_d). For instance, at time step t_i, if, say, t_c targets are completely destroyed, t_h targets are half destroyed, and t_n targets are not attacked, then

$$T_d = t_c + 0.5t_h + 0t_n \tag{18}$$

The target value destroyed (T_d) provides an insight into how many targets are half destroyed or fully destroyed in the search space. We can see that as time passes the number of targets being destroyed increases and hence the target value destroyed (T_d) also increases. The performance of greedy strategy is found to be the worst compared to other two strategies. However, team theoretic strategy performs the best in spite of there being no communication between UAVs.

Figure 2(b) show the performance of a particular simulation. To obtain the average performance of all the strategies, we carry out the simulation for 20 different random target maps for 200 time steps, each with the same UAV positions. During the search task, it is logical that, after some time, during which search is carried out and if no targets are found, the UAV has to change its direction, so that there is a better chance of finding a target. Hence, after every 10 steps of search task, the UAVs change their direction of search by a random angle. Hence, the performance of the target destroyed sometimes depends on the random change in search direction. Hence, to average out the randomness of search we simulate search and attack operation over each target map three times and consider the average performance. Figure 2(d) shows the average performance of each strategy for 20 such randomly generated target maps. From the figure we can see that initially all the strategies perform almost at the same level but as time progresses, team theoretic strategy outperforms the other strategies. This is a significant result since the team theoretic strategy assumes no communication between UAVs and has limited sensor range. In case of full communication, there is considerable communication cost and the computational cost are also more, when compared to team theoretic strategies, as the UAV has to consider all the other UAVs information about the targets. The greedy strategy has a tendency to move in groups and thus not effectively using the resources of having multiple UAVs for the mission. Team theory perform better and is scalable to large scale systems as the information sensing is local and consequently the computational effort is less.

Figures 2(b) and 2(d) shows that the team theoretic strategy performs better than the other strategies. Another study examines the effect of sensor radius on T_d (Figure 2(c)). Here, we considered a random target map and carried out three simulations for each sensor radius. The effect of sensor radius shown is the average of the three simulations. The figure shows that for this particular case sensor radius of about 25 gives the best performance compared to any other sensor radius. The performance of team theory, greedy and full communication strategies depends on the sensor range. If the sensor radius is small, a UAV can sense very small area and the decision taken will not be effective. We expect that with increase in sensor range the performance will also improve. In the case of team theory, this is not true because if we consider a large sensor range, the estimated value of the virtual target will be incorrect. This is because the area sensed by the kth UAV can include regions beyond the search region space where there are no targets. But, the ith UAV does not consider this fact and assumes equal density of targets everywhere. This unnecessarily gives more weightage on the virtual target and the overall performance decreases. This effect can be seen in Figure 2(c). This problem can be resolved if we consider other parameters such as target density gradients or restriction to the search space.

The ratio of search value to the target value also plays a crucial role. If we give equal priority to search and attacking a target then the UAV may opt for search task even though there is a target near it. On the other hand, if we increase the value of the target then there is a possibility that the UAV may loiter in the vicinity of a target which is already destroyed. In our simulations, we considered the search value to be 25% of the target attack value and this yielded good results. But, a more focused study is necessary to examine this aspect of the problem.

4 Task Allocation using Negotiation

In this section, we present a task allocation algorithm for multiple UAVs performing search and attack tasks in an unknown region using negotiation scheme for the scenario given in Section 3. Here we assume that once a target is attacked, it is destroyed and hence battle damage assessment task on the target is not necessary to be performed. This is one of the very few applications available that exploits the use of negotiation for a network of UAVs involved in a practical problem of decision-making.

4.1 Problem Formulation

Consider N UAVs/agents performing a search and destroy operation on a bounded region consisting of M targets whose exact positions are not known *a priori*. The basic problem of task allocation is to efficiently assign agent $A_i \in N$, to target $m_i \in M$, such that the mission is completed as quickly as possible. The task allocation problem can be solved by using either a centralized controller or a decentralized controller. In the former case each agent communicates the information it has to the central controller that solves a task allocation algorithm and assigns each agent to a particular task. However, implementing this task allocation strategy in real-time requires large communication overheads and will not be scalable to large number of agents and targets. Also, these strategies are not robust to failures. Hence, a decentralized task allocation strategy, which avoids many of these problems, may be more advantageous if implemented on a multi-agent system. One way of implementing a decentralized task allocation strategy would be by making each agent broadcast its information to all the other agents so that each agent has the required information to solve the task allocation problem independently and assign a task for itself. The implementation of this task allocation strategy also requires large amount of communication among the agents. To reduce this demand one can define a neighbourhood concept for each agent so that an agent communicates its information only to those agents that are in its neighbourhood. The neighbourhood can be range dependent, in which case it is dynamic or pre-defined, in which case it is static or randomly selected. In this work, we will assume only range dependent neighbourhood for agents.

The implementation of decentralized task allocation with finite communication range poses several challenging problems. For instance, consider Case A in Figure 3 where agent A_1 and A_2 have target T_1 in their sensor range and an allocation has to take place as to which agent should be assigned to the target. The task allocation can be done using a greedy strategy, in which case both the agents would move towards the same target which is not desirable. Another task allocation mechanism used in multi-robot literature is based

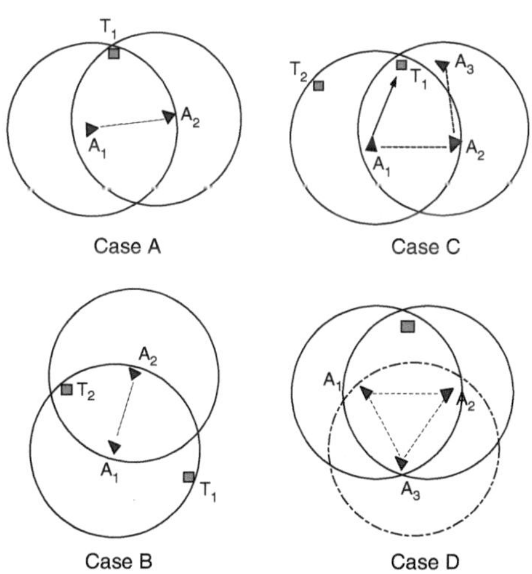

Fig. 3. Some scenarios for decision-making

on auctions [20]. But in the application under consideration since the system of UAVs is decentralized, each agent would become an auctioneer and hence both the agents would auction the same target.

Consider Case B in Figure 3, where A_1 has T_1 and T_2 in its sensor range while A_2 has only T_2. The auction mechanism requires broadcast of all the target and their associated costs. Resolving conflicts using auctions is a difficult task. In Case C, we can see that A_1 sees T_1 while A_3 is already on its way to attack T_1. So, A_1 wastes some resource in moving towards a target that is already assigned, Since the communication is limited it does not have access to the assignment of other agents. Instead of T_1 it could have attacked T_2. Here too greedy and auction algorithm would not yield good performance. In Case D, agent A_3 gets the auction information from A_1 and A_2 about T_1, now A_3 does not know to which agent it has to send the bid. A modification to the standard auction algorithm may eliminate some of the difficult issues, however this would complicate the decision-making rules for multiple agents using auction mechanism locally. These complications in using auctions for limited communication cases motivate us to use negotiation as a tool to handle these situations efficiently. In Case A, A_1 and A_2 can negotiate on which agent would be assigned to target T_1. While in Case B, A_1 and A_2 can negotiate such that one agent attacks T_1 and the other moves towards T_2. In Case C, A_2 can detect a conflict between A_1 and A_3 and send decisions such that A_1 or A_3 move towards T_1. However, in Case D, A_3 actually negotiates between A_1 and A_2, which are not neighbours, and detects possible conflict and hence provides an efficient task allocation decision.

However, the implementation of negotiation scheme involves designing of negotiation rules over which the decision-making process takes place. In the next section we describe the negotiation scheme employed for decision-making.

At every time step each agent has to perform a task. The task can be (i) searching for a target or (ii) attacking a target. Each agent senses its environment consisting of other agents and targets. An agents' assignment for a task depends on four different situations. These situations are dependent on the availability of neighbouring agents and targets. The four situations, in which agent A_i has to perform a task and play a role in the decision making process are:

1. *No targets and no neighbours*
   Task: Search
   Decision role: Continue to move in the same direction
2. *No targets but has neighbours*
   Task: Perform search or attack. The target information may be provided by the neighbouring agents.
   Decision role: Acts as a negotiator for neighbouring agents
3. *Targets are present but no neighbours*
   Task: Attack
   Decision role: Select a target that yields maximum value

4. *Target as well as neighbours are present*
   Task: Search or attack
   Decision role: Negotiate with neighbours

Once an agent A_i is present within a distance d from the target, we assume that the agent can destroy the target effectively. An agent has to negotiate with its neighbouring agents for an efficient task allocation. The agents are not subjected to any turn radius constraints and hence can move in any direction. The agents have to maximize the number of targets destroyed in the search space by coordinating with its neighbouring agents through negotiation.

4.2 Decision-making

Negotiation as a Tool to Handle Uncertainty in Agent Actions

In general, negotiation refers to the communication process that facilitates coordination and cooperation among a group of agents [27]. In multi-agent systems, its aim is to resolve problems related to resource allocation and task assignments between various agents in a decentralized setting.

Our approach is somewhat similar to Rubinstein's model of strategic negotiation [28] where agents make proposals that are either accepted or rejected by other agents; and whether an agent implements its proposal or not depends on what other agents do. However, our approach is different from Rubinstein's model on many counts due to the nature of the task allocation problem. Unlike most negotiation models we do not have a situation where each proposal is vetted by all the other agents. In fact, due to the connectivity restrictions, we have a network of agents where an agent is not necessarily directly connected to all other agents. So, each agents decision is based on the response of only those agents that are connected to it. Moreover, unlike in Rubinstein's model, agents make simultaneous offers at pre-defined decision epochs and the actions are accordingly distributed between agents. Another way in which our model differs from Rubinstein's model is that in a task allocation problem the need for negotiation arises mainly because of lack of information about the action of other agents. So, the whole process of negotiation is geared towards determining the action of an agent in a coordinated autonomous fashion without assuming any kind of hierarchy or priority among agents.

A coordinated decision by an agent would be one that is not in conflict with the decision of its neighbors. There is no conflict except that which arises due to uncertainty of agent actions. For example, it occurs when more than one agent is planning to attack the same target, thus decreasing the effectiveness of the mission. Resolution of such conflicts can be effected either by

(i) Direct communication/negotiation as in the case when an Agent A_i and another agent A_j are within communication range.
(ii) Indirect negotiation when an Agent A_i and another agent A_j, $A_j \notin \mathcal{N}(A_i)$ want to attack the same target T, and A_i and A_j are connected through

a sequence of communication links through other agents. For instance, they may be connected through a third agent A_k with $A_j \in \mathcal{N}(A_k)$ and $A_k \in \mathcal{N}(A_i)$.

In the first case, since A_j is within the communication range of A_i, it can exchange information with A_j and resolve the conflict. While in the second case, A_i does not know about the existence of A_j and so direct communication is not feasible. So, the intermediate agents are important in the negotiation process. In the negotiation scheme developed next, we will show that it is the neighboring agents who contribute to the decision-making of agent A_i.

Negotiation Scheme

Each agent A_i performs the following actions during decision-making: (i) Sends/receives proposals (ii) Processes received proposals and sends Accept/Reject decisions to proposing agents (iii) Computes own route decision (iv) Implements decision. All these actions happen within each negotiation cycle. This is shown in Figure 4. Note that an agent A_i that has no targets will have only the second segment, while the agents that have targets as well as neighbouring agents will have all the four segments of decision-making.

The different segments of the negotiation cycle are described below:

Send/receive proposals (NC1): Each agent evaluates the benefit associated with each target. Let $b_i(T_j)$ be the expected benefit that A_i gets by attacking target T_j, which is given by

$$b_i(T_j) = V_j w_r \ - \ S_{ij} \tag{19}$$

where, V_j = value of target T_j, w_r = the weight given to search task over the task of attacking a target, S_{ij} = (time to reach the target T_j by agent A_i)/(total flight time). The benefit set $\mathcal{B}_i$ of A_i consists of benefits for all the tasks an agent has. Let $\mathcal{T}_i$ be the set of all targets. The benefit set for agent A_i is represented as:

$$\mathcal{B}_i = \{b_i(T_j) \mid T_j \in \mathcal{T}_j\} \tag{20}$$

Agent A_i chooses a target T_{S_i} for which A_i gets the maximum value, as

$$S_i = \arg\max_j \{b_i(T_j) \in \mathcal{B}_i\} \tag{21}$$

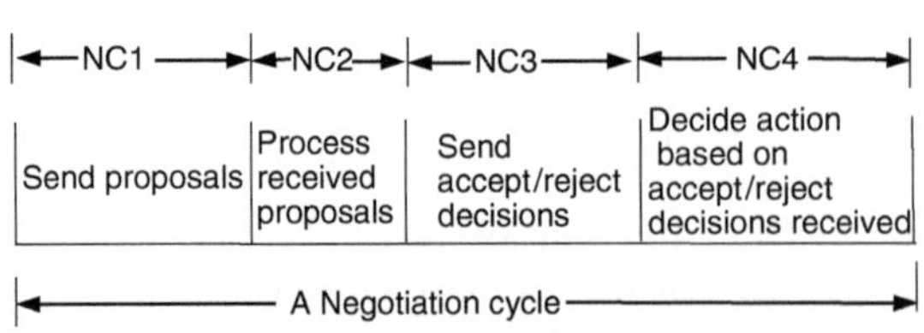

Fig. 4. Negotiation cycle

The proposal of agent A_i, sent to its neighboring agents, is of the form $Q_i = (A_i, T_{S_i}, b_i(T_{S_i}))$, containing the proposer agent's identification, proposed target, and the value associated with T_{S_i}.

Processing received proposals (NC2) and sending decisions (NC3): Let $\mathcal{Q}_i$ be the set of proposals received by agent A_i from its neighbors A_j, including its own proposal.

$$\mathcal{Q}_i = \{(A_j, C_{S_j}, \beta_j); L(A_j) \in \mathcal{N}(L(A_i), q_c)\}$$

Let T_k^i be a target that appears in at least one of the proposals received by A_i. That is, $T_k^i = T_{S_j}$ for some $Q_j \in \mathcal{Q}_i$. For each such T_k^i, define $\mathcal{A}(T_k^i)$ as the set of agents that have proposed T_k^i, and $\mathcal{B}(T_k^i)$ as the set of values associated with agents in $\mathcal{A}(T_k^i)$. So,

$$\begin{aligned} \mathcal{A}(T_k^i) &= \{A_j \mid Q_j \in \mathcal{Q}_i, T_{S_j} = T_k^i\} \\ \mathcal{B}(T_k^i) &= \{b_i(T_j) \mid A_j \in \mathcal{A}(T_k^i))\} \end{aligned} \tag{22}$$

Using the above sets ($\mathcal{A}(T_k^i)$ and $\mathcal{B}(T_k^i)$), agent A_i sends *accept* or *reject* decision to its neighbors using the following rules:

Rule 1: An agent A_i sends *accept* to agent A_j, if

$$\mathcal{A}(T_k^i) = \{A_j\} \tag{23}$$

That is, $\mathcal{A}(T_k^i)$ is a singleton containing only agent A_j (note that A_j could be A_i itself).

Rule 2: If $\mathcal{A}(T_k^i)$ is not a singleton then agent A_i sends *accept* to that agent in $\mathcal{A}(T_k^i)$ which obtain the maximum value by attacking target T_k^i and *reject* to all other agents in $\mathcal{A}(T_k^i)$. That is, *accept* is sent to $A_{j'} \in \mathcal{A}(T_k^i)$ if,

$$j' = \arg\max_j \{b_i(T_j) \in \mathcal{B}(T_k^i)\} \tag{24}$$

Note that Rule 2 subsumes Rule 1. But they are stated separately for clarity. Again $A_{j'}$ can be A_i itself.

Rule 3: An agent can send only one *accept* for one target. If there are more than one j' then the agent selects one of them.

Rule 4: For A_i to decide on its action at the current search step it has to get *accept* from all its neighboring agents to which it had sent its proposals.

Rule 1 implies that when an agents' proposal is not in conflict with other agents' proposals an *accept* can be sent without considering the other agents' decisions. When more than one agent proposes to attack T_k then there is a conflict between the proposing agents which A_i has to resolve. The conflict can be resolved by comparing the benfits' proposed by the agents. Agent A_i compares the $b_i(T_j)$ received for target T_k and sends *accept* decision to an agent A_k which has the highest $b_i(T_j)$ and *reject* decisions to the remaining agents. An agent A_i can receive a mix of *accept* and *reject* decisions from its neighbors. If we allow the agent to attack a target T_k, since it has got acceptance from some of the agents, this assignment would cause ineffective performance as multiple agents will get assigned to the same target. Hence, *Rule 4* guards against agents getting multiple assignment. Rules 1-4 are the key to the negotiation scheme. While implementing *Rule 3*, we may encounter situations where more than one agent has the same $b_i(T_j)$, in which case we use a deadlock resolution scheme that resolves such deadlocks.

Computing route decision (NC4): Agent A_i decides whether to implement or discard its proposed task based on the *accept* or *reject* decisions received from its neighbors. The agent implements its proposal if it receives *accept* decisions from all its neighbors and discards it if the agent receives a *reject* from even one of its neighbors. An agent that received a *reject* for its proposal from at least one neighbor will go on to the next negotiation cycle and this process will continue till it receives all *accept* decisions. An agent that has arrived at a decision (after receiving accept from all its neighbors) will not send any more proposals during subsequent negotiation cycles. The sequence of negotiation cycles will terminate automatically when all the agents have converged to a decision. Later we will prove that only a finite number of negotiation cycles are necessary. When an agent A_i receives reject for all its proposals, it adopts the search task.

Additional Target Information Exchange

An agent that has received acceptance to its proposal may have other targets within its sensor range. An agent A_i can send this information to its neighbouring agents who can use it. The information that an agents sends is the target location and its value as perceived by A_i. This information will be more useful for those agents that may not have decided any targets but are neighbours of A_i. The target information broadcast by A_i can also be useful if all the proposals of agent $A_j \in \mathcal{N}(A_i)$ are rejected.

Once an agent receives the available targets from agent A_i, it can make assignment to any of the targets based on random number generation, greedy strategy, or start a negotiation with its neighbouring agents for obtaining an assignment. Here, we use greedy strategy for simplicity.

Deadlock Resolution Mechanism

We define a deadlock, when an agent A_i is unable to decide to whom it has to send an acceptance. This situation can happen when more than one agent, with the same $b_i(T_j)$ value, seeks target T_j to attack. Since the $b_i(T_j)$ values are same, use of *Rule 2* is not possible and agent A_i cannot send acceptance to all the agents as that will violate *Rule 3*. There are two possible ways of resolving deadlock: loss information and token algorithm.

Loss information: In this scheme, agent A_i requests for more information from agents in $\mathcal{A}(T_k^i)$. This additional information will aid in effective decision-making. The additional information that an agent requests is the value of possible loss that each proposing agent suffers if it chooses the next best action instead of the proposed action. Let the new benefit vector for agent A_k be $\hat{\mathcal{B}_k}$ and the loss λ_k be evaluated using (25) as,

$$\hat{\mathcal{B}_k} = \{\mathcal{B}_k \setminus b_i(C_{S_K})\}; \quad \lambda_k = \max \mathcal{B}_k - \max \hat{\mathcal{B}_k} \tag{25}$$

where, $'\setminus'$ denotes set difference. When agent A_i requests for loss information, the loss λ_k is sent to agent A_i. Let Λ_i represent the set of loss information received from all the agents in $\mathcal{A}(T_k^i)$. An *accept* is sent to an agent A_j that satisfies the condition in (26) and *reject* is sent to the remaining agents.

$$A_j = \arg\max_i(\Lambda_i) \tag{26}$$

Suppose there are multiple $b_i(T_j)$'s that are at the next highest level, then the same procedure needs to be repeated. Using the loss information does not guarantee that the deadlock will be resolved. This situation can arise when multiple agents have the same loss value. In that case, we use a token algorithm as given below.

Token Algorithm : Every agent A_i carries a unique token number K_i. Whenever the above situation (of the loss being equal) occurs wherein the agent is unable to decide to whom it has to send acceptance, the agent requests for token number of the agents $A_k, A_k \in \mathcal{A}(T_k^i)$. Agent A_i compares these token numbers and chooses an agent A_j with the least token number. The token number of A_j is increased by a number $\hat{N}$, where $\hat{N}$ is an arbitrary large number greater than N. This scheme ensures that an agent that has been selected earlier in this situation, will not be selected again in a similar situation if there is at least one other agent which has not been selected before.

Some Theoretical Results

Theorem 1. *If more than one agent is proposing a target T_j, then at least one of the agents will receive all acceptances from its neighbors.*

Proof. Let $\mathcal{A}(T_j^i)$ be the set of agents proposing target T_j as their proposal. Then, by *Rule* 2, agent A_i sends an *accept* decision to agent A_j which has the maximum $b_i(T_j)$. If there are multiple agents with same $b_i(T_j)$ then A_i invokes the deadlock resolution mechanism by which one agent would receive an *accept*. □

Theorem 2. *The negotiation terminates in a finite number of negotiation cycles.*

Proof. From Theorem 1 we observe that, at each negotiation cycle, at least one of the agents gets all *accept* and so decides upon a target for its next step. Since there are a finite number of agents, in a finite number of negotiation cycles each agent would decide upon a target to attack. If the target are not available then they continue to search task. Hence, all the agents would decide upon a task in a finite number of negotiation cycles. The maximum number of negotiation cycles an agent can go through is N. □

4.3 Simulation Results

A simulation study is conducted on a battlefield scenario of size 100×100. Through these simulations we show that the negotiation scheme performs better than greedy strategy in terms of average number of targets destroyed. The simulation is carried out using 7 UAVs for 100 different sets of target positions with each set having 20 targets. The *a priori* knowledge about number of targets present in the space and their initial positions are not available to the UAVs. We also study the performance of negotiation and greedy schemes for various sensor radius.

From Figure 5 we can see that the negotiation scheme outperforms the greedy strategy. The number of targets using negotiation scheme is higher and

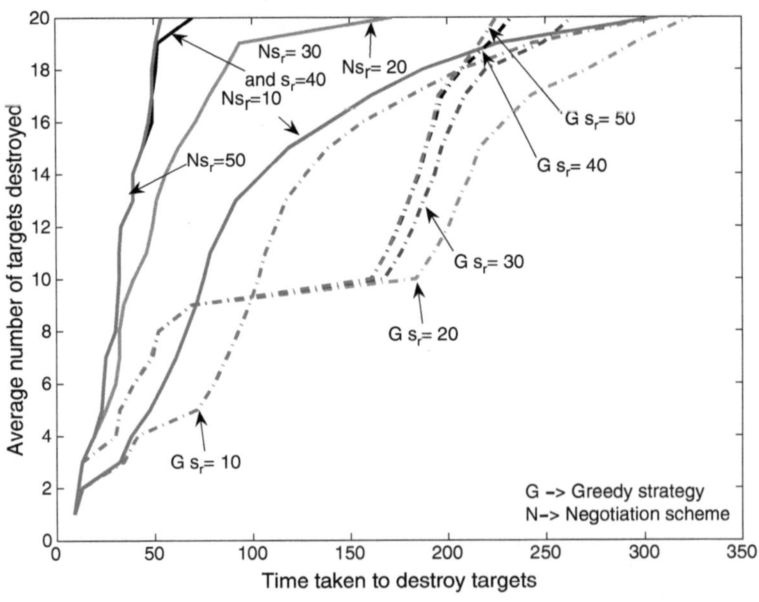

Fig. 5. Average number of target hits for 100 different target positions

the time taken to accomplish the mission is comparatively low. An expected result of increase in performance with increase in sensor range can be seen for the performance curves of negotiation scheme in the figure. However, this intuitive result is not true for greedy strategy.

The performance of greedy strategy with sensor radius $s_r = 10$ is better than higher sensor radius $s_r = 20$ to $s_r = 50$. This is due to the fact with low sensor radius, the UAVs are unable to sense the targets initially and hence move in the initial heading direction (spreading out). But, with higher sensor radius, the agents are able to sense the target from their initial positions and hence all the UAVs move in the direction of sensed target as a swarm. Hence, the performance is worse when compared to lower sensor radius.

We carried out another set of simulations to study the performance of task allocation algorithm for different target distributions on the search space. In order to conduct these experiments we define a proximity factor that determines the nature of the distribution or spread of targets in the search space. The proximity factor is defined as:

$$\rho = \frac{S_r}{\frac{1}{N}\sum_{i=1}^{N}\sqrt{(x_i - x_c)^2 + (y_i - y_c)^2}} \tag{27}$$

where N is number of targets, (x_i, y_i) represents the position of the ith target location, (x_c, y_c) the mean of all the target positions and S_r the sensor radius. Low proximity factor implies well separated targets compared to the sensor radius. While high proximity factor ensures that the targets are placed very closely. Figure 6 show different target distributions in the search space.

The simulations are carried out using 7 UAVs for a search space consisting of 50 targets, with different proximity factors. Figure 7 shows the performance of negotiation and negotiation with target information based task allocation

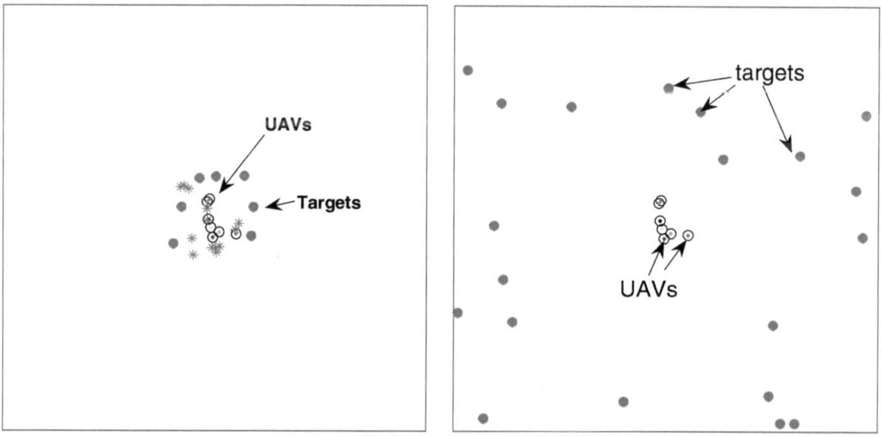

Fig. 6. Battle field with 20 targets for proximity factors $\rho = 0.625$ and $\rho = 0.11$, while the sensor radius $s_r = 10$

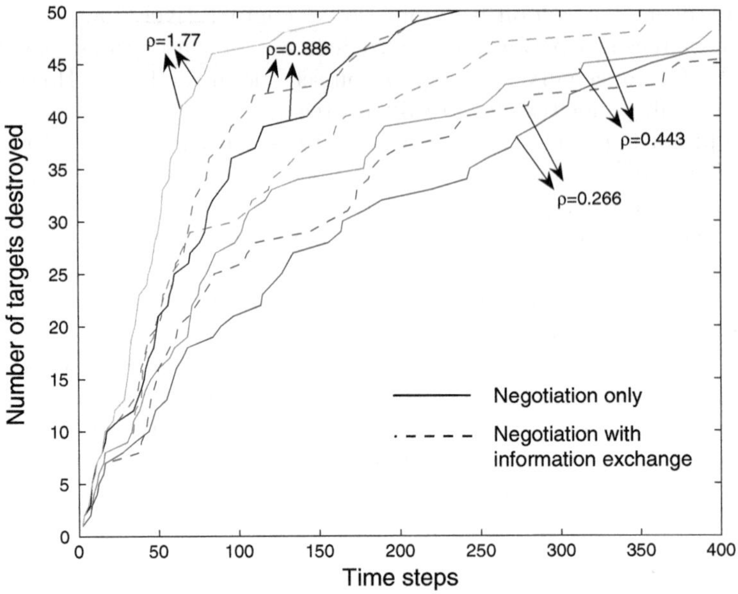

Fig. 7. Number of targets destroyed for different proximity factors

schemes. From the figure we can see that for lower proximity factors the number of targets destroyed are low as compared to the number of targets destroyed in the higher proximity factor case. When the proximity factor is small, the effect of target information sharing during decision-making by the agents that have targets in their sensor range is significant. For $\rho = 0.266$, we can see from the figure that the performance of negotiation with target information based task allocation is better than that using negotiation only. Here, the target information broadcast plays a crucial role in enhancing the performance. Similar kind of effect can be seen for $\rho = 0.443$. However, for $\sigma = 0.886$, the negotiation based task allocation is better than that with target information exchange. This is due to the fact that the additional information about distant targets makes the agent choose distant targets to attack rather than perform search in its own neighborhood. This causes UAVs to miss nearer targets outside its sensor range. For $\rho = 1.77$, the performance is the same for both the negotiation schemes. Since the proximity factor is high, all the agents can sense all the targets hence there is no improvement in performance with information exchange. It should be noted that the amount of information broadcast also plays a crucial role in the performance of the task allocation. Hence, there is always a tradeoff between how much of information should be broadcast and the performance.

5 Search using Game Theoretic Strategies

In the previous section we have seen search task to be a part of other tasks to be carried out by the UAVs. However, there are applications like search and surveillance missions where search is the only task that has to be carried out. By search we mean that the UAVs are deployed in an unknown region to collect information about the region.

Consider an unknown region over which a search mission has to be carried out. Based on the *a priori* knowledge of the search space, an uncertainty map is constructed. The uncertainty map is discretized into cells. Here, we discretize the map into a grid of hexagonal cells, as they offer the flexibility to move in any direction while expending the same amount of energy. The uncertainty map constitutes real numbers between 0 and 1 associated with each cell in the search space. These numbers represent the uncertainty with which the location of the target is known in that cell. An uncertainty value of 0 would imply that everything is known about the cell (that is, one can say with certainty whether a target is located in that cell or not). On the other hand, an uncertainty value of 1 would imply that nothing can be said about the location of the target in that cell.

One of the motivations for modeling a search problem in a game theoretical framework arises from the fact that this framework gives the flexibility of using two different solution concepts: one based on cooperation between players and the other based upon non-cooperation. Application of these notions to the economics had to take into account the fact that players are not inherently altruistic, thus making the cooperative framework somewhat untenable, unless the cooperation is enforced by a third party. On the other hand, in the non-cooperative framework it has been shown that in repeated games, cooperation automatically emerges as the best noncooperative solution and hence the notion of cooperation is inherent and enforceable in the non-cooperative framework. Although when we consider cooperation between automated agents that are devoid of any selfish motive and have only a common goal in mind, it is more logical to use a cooperative framework, in our work we show that the non-cooperative framework is almost equally effective and is no more computationally time consuming than the cooperative framework. There are other reasons too, related to the specific problem structure, which justifies the usage of the non-cooperative framework. For instance, when the sensor performance is unreliable or noisy, or due to ineffective communication the uncertainty map of each agent changes with time unknowingly to the other agents, leading to different uncertainty map for different agents. In such situations, the cooperative decision making mechanism breaks down. Here we show that when this is the case, the non-cooperative Nash strategies perform better than the cooperative strategies.

5.1 N-person Game Model

The strategies that we propose for N agents/search agents is based on a game theoretical model. We use q-step look ahead planning [29], where q determines the depth of the exploratory search to obtain optimal strategies.

The objective of the agents is to select their next action or path at time t in order to maximize their benefits (that is, maximize uncertainty reduction). This problem can be modelled as a N-person non-zero sum game with each agent as a player and the set of paths available to each agent as the set of strategies.

Another approach to decision-making in this situation, without the benefit of communication between agents, is to make some assumption on the behavior of other agents in the search space. So, a player/agent may consider the rest of the $N-1$ players to be one single player. Hence, we can model the N-person game as one player playing against the rest of $N-1$ players taken together as a single player (a coalition of $N-1$ players). Here, we describe both the models.

The payoff to each agent can be expressed in terms of search effectiveness functions. Every cell has an uncertainty value associated with it. Let $\mathcal{P}_i^q(C_{s_i})$, $i \in \{1, 2 \ldots, N\}$ be the set of all possible paths of length q, for an agent A_i, emanating from cell C_{s_i}. A path $P_i^j(C_{s_i}) \in \mathcal{P}_i^q(C_{s_i})$, $j = 1, 2, \ldots, |\mathcal{P}_i^q(C_{s_i})|$, is an ordered set of cells $P_i^j(C_{s_i})$, defined as,

$$P_i^j(C_{s_i}) = \left\{C_i^{j,1}, C_i^{j,2}, C_i^{j,3}, \ldots, C_i^{j,q} \mid C_i^{j,k} \in \mathcal{C},\ C_i^{j,1} = C_{s_i},\ C_i^{j,k+1} \in \mathcal{N}(C_i^{j,k})\right\} \tag{28}$$

where, $\mathcal{C}$ is the collection of all cells, C_{s_i} is the current position of A_i, and $\mathcal{N}(C_i^{j,k})$ is the set of all neighboring cells of $C_i^{j,k}$.

Let the uncertainty value of a cell C^k at time t, as perceived by A_i, be $U_i(C^k, t)$. Given a path $P_i^j(C_{s_i})$ of agent A_i, suppose A_i is at cell C^l at time t then the reduction of uncertainty associated with C^l, and the subsequent updated value of uncertainty, is evaluated as follows:

Case 1: Only A_i is in cell C^l at time t, then

$$v_i(t) = U_i(C^l, t)\beta_i \tag{29}$$
$$U_i(C^l, t+1) = U_i(C^l, t)(1 - \beta_i) \tag{30}$$

where, $v_i(t)$ is the benefit that agent A_i would obtain (that is, the amount of uncertainty reduction that it will achieve) when it visits cell C^l.

Case 2: When more than one agent visits cell C^l at time t, let $\mathcal{A}$ represent this set of agents. Then,

$$v_i(t) = \tilde{\beta}\hat{\beta}_i U_i(C^l, t) \tag{31}$$

$$U_i(C^l, t+1) = U_i(C^l, t) - \sum_{i=1}^{N} v_i(t) \tag{32}$$

where,

$$\tilde{\beta} = 1 - \prod_{j \in \mathcal{A}} (1 - \beta_j); \hat{\beta}_i = \frac{\beta_i}{\sum_{j \in \mathcal{A}} \beta_j} \tag{33}$$

So, given N routes $P_1, P_2, P_3, \ldots, P_N$ of the N agents, where P_i is any $P_i^j \in \mathcal{P}_i^q$ (we drop the C_{s_i} argument from the path notation $P_i^j(C_{s_i})$ as well as from $\mathcal{P}_i^q(C_{s_i})$, the set of all possible paths, for simplicity), the reduction in uncertainty achieved by A_i at each step t ($t = 1, 2, \ldots, q$) is given by $v_i(t)$ and is computed using Case 1 or Case 2 as the case may be. Note that this computation has to be done simultaneously for all the agents. The total benefit to A_i due to path P_i is

$$m^i(P_1, \ldots, P_N) = \sum_{t=1}^{q} v_i(t) \tag{34}$$

which represents the payoff obtained by A_i as the agents choose strategies $P_1, P_2, \ldots, P_N$. The functions $m^i : \prod_{i=1,\ldots,N} \mathcal{P}_i^q \rightarrow \mathbb{R}^+$, from the set of paths to the uncertainty reduction value, are called the search effectiveness functions.

5.2 Solution Concepts

The decision to choose a particular path that would provide the maximum information gain (or uncertainty reduction) can be based on various strategies. We consider the following strategies: Noncooperative Nash strategy, coalitional Nash strategy, security strategy, cooperative strategy, greedy strategy, and globally optimal strategy. The Nash, security, coalitional Nash, and greedy strategies do not require any kind of communication to arrive at an optimal decision, while cooperative and globally optimal strategies require communication to implement the decision making process.

(i) Noncooperative Nash Equilibrium Strategy: When the agents do not communicate with each other to decide on their future action at time t, and each agent assumes that the other $N - 1$ agents take actions that are beneficial to them, then we can use the concept of noncooperative Nash equilibrium.

(ii) Coalitional Nash Strategy: This is similar to the non-cooperative Nash equilibrium strategy, except that each agent assumes the other $N - 1$ agents to form a coalition and take actions jointly that are jointly beneficial to them.

(iii) Security Strategy: This strategy becomes relevant when, as before, the agents do not communicate with each other and each agent assumes the other $N - 1$ agents to be adversaries. In such a situation the best strategy for the agent is to secure its minimal benefit. Hence, it is logical for the agent to use security strategy that would guarantee a minimal payoff.

(iv) Cooperative Strategy: The agents communicate with each other and decide collectively (jointly) to take the best possible action. This is also the centralized decision making case.

(v) Greedy Strategy: Agents do not communicate among themselves and use greedy strategy. An agent does not consider the effect of the possible actions of the others agent and selects an action that yields the maximum benefit to itself. This strategy is used for comparison purposes only.

(vi) Globally Optimal Strategy: The game theoretical strategies are based on local information up to q steps. Hence, the solution is optimal for these q steps and not globally optimal. We can obtain a globally optimal solution by making q equal to the largest possible number of steps in an agent's search path. This requires huge computational time and also increases the computational complexity as the domain of the search effectiveness function increases. We will not use this strategy but, for the interested researcher, some heuristic algorithms to implement such strategies are discussed in [30].

Non-cooperative N-person Nash Equilibrium Strategy

We define a non-cooperative N-person game in normal form for N agents [31]. A N-person game consists of N search effectiveness functions $m^i, i = 1, \ldots, N$. The ordered N- tuple of real numbers $(m^1(P_1, \ldots, P_N), \ldots, m^N(P_1, \ldots, P_N))$ denotes the payoff to each agent respectively. The players do not cooperate with each other and arrive at their decisions independently. In such a situation the equilibrium solution can be stated as: An N-tuple of strategies $\{P_1^*, P_2^*, \ldots, P_N^*\}$ with $P_i^* \in \mathcal{P}_i^q$ is said to constitute a noncooperative (Nash) equilibrium solution for an N-person nonzero-sum game, if the following N inequalities are satisfied for all $P_i \in \mathcal{P}_i^q, i \in N$.

$$\begin{aligned} m^{1*} &\triangleq m^1(P_1^*, P_2^*, P_3^*, \ldots, P_N^*) \geq m^1(P_1, P_2^*, \ldots, P_{N-1}^*, P_N^*) \\ m^{2*} &\triangleq m^2(P_1^*, P_2^*, \ldots, P_N^*) \geq m^2(P_1^*, P_2, P_3^*, \ldots, P_{N-1}^*, P_N^*) \\ &\vdots \qquad \vdots \qquad \vdots \\ m^{N*} &\triangleq m^N(P_1^*, P_2^*, \ldots, P_N^*) \geq m^N(P_1^*, P_2^*, P_3^*, \ldots, P_{N-1}^*, P_N) \end{aligned} \tag{35}$$

The N-tuple $(m^{1*}, m^{2*}, \ldots, m^{N*})$ is known as a noncooperative (Nash) equilibrium outcome of the N-person game in normal form. The pure strategy Nash equilibrium may not exist always. In this case we need to compute mixed strategies which guarantee a solution to the noncooperative game.

Mixed Strategies : A mixed strategy for a player is a probability distribution on the space of its pure strategies. An allowable strategy for A_i is to choose P_i^1 with probability (w.p.) y_1^i, P_i^2, w.p. $y_2^i, \ldots$, $P_i^{|\mathcal{P}_i^q|}$ w.p. $y_{|\mathcal{P}_i^q|}^i$, so that,

$$\sum_{k=1}^{|\mathcal{P}_i^q|} y_k^i = 1, \quad \text{and } 0 \leq y_k^i \leq 1 \tag{36}$$

An N-tuple $\{y^{i*} \in Y^i, i \in N\}$ is said to constitute a mixed strategy non-cooperative (Nash) equilibrium solution for a N-person game in normal form, if the following N inequalities are satisfied for all $y^j \in Y^j, j \in N$:

$$
\begin{aligned}
J^{1*} &\triangleq \sum_{\mathcal{P}_1^q}\sum_{\mathcal{P}_2^q}\cdots\sum_{\mathcal{P}_N^q} y_{P_1}^{1*} y_{P_2}^{2*}\cdots y_{P_N}^{N*} m^1(P_1,P_2,\ldots,P_N) \\
&\geq \sum_{\mathcal{P}_1^q}\sum_{\mathcal{P}_2^q}\cdots\sum_{\mathcal{P}_N^q} y_{P_1}^{1} y_{P_2}^{2*}\cdots y_{P_N}^{N*} m^1(P_1,P_2,\ldots,P_N) \\
J^{2*} &\triangleq \sum_{\mathcal{P}_1^q}\sum_{\mathcal{P}_2^q}\cdots\sum_{\mathcal{P}_N^q} y_{P_1}^{1*} y_{P_2}^{2*}\cdots y_{P_N}^{N*} m^2(P_1,P_2,\ldots,P_N) \\
&\geq \sum_{\mathcal{P}_1^q}\sum_{\mathcal{P}_2^q}\cdots\sum_{\mathcal{P}_N^q} y_{P_1}^{1*} y_{P_2}^{2} y_{P_3}^{3*}\cdots y_{P_N}^{N*} m^2(P_1,P_2,\ldots,P_N) \\
&\vdots \qquad\qquad \vdots \qquad\qquad \vdots \\
J^{N*} &\triangleq \sum_{\mathcal{P}_1^q}\sum_{\mathcal{P}_2^q}\cdots\sum_{\mathcal{P}_N^q} y_{P_1}^{1*} y_{P_2}^{2*}\cdots y_{P_N}^{N*} m^N(P_1,P_2,\ldots,P_N) \\
&\geq \sum_{\mathcal{P}_1^q}\sum_{\mathcal{P}_2^q}\cdots\sum_{\mathcal{P}_N^q} y_{P_1}^{1*} y_{P_2}^{2*}\cdots y_{P_{N-1}}^{N-1*} y_{P_N}^{N} m^N(P_1,P_2,\ldots,P_N)
\end{aligned} \tag{37}
$$

The noncooperative Nash equilibrium outcome of a N-person game in mixed strategies is given by the N-tuple $\{J^{1*},\ldots,J^{N*}\}$. If there exists an inner mixed strategy solution then, such a solution $\{y^{i*} \in \breve{Y}^i; i \in N\}$ of an N-person game in normal form satisfies the following set of equations:

$$
\begin{aligned}
&\sum_{\mathcal{P}_2^q}\cdots\sum_{\mathcal{P}_N^q} y_{P_2}^{2*} \;\cdots\; y_{P_N}^{N*}\{m^1(P_1,\ldots,P_N) - m^1(P_1^l,P_2,\ldots,P_N)\} = 0, \\
&\qquad\qquad P_1 \in \mathcal{P}_1^q, \quad P_1 \neq P_1^l, \\
&\sum_{\mathcal{P}_1^q}\sum_{\mathcal{P}_3^q}\cdots\sum_{\mathcal{P}_N^q} y_{P_1}^{1*} y_{P_3}^{3*} \;\cdots\; y_{P_N}^{N*}\{m^2(P_1,\ldots,P_N) - m^2(P_1,P_2^l,\ldots,P_N)\} = 0, \\
&\qquad\qquad P_2 \in \mathcal{P}_2^q, \quad P_2 \neq P_2^l, \\
&\qquad\qquad \vdots \qquad\qquad \vdots \\
&\sum_{\mathcal{P}_1^q}\cdots\sum_{\mathcal{P}_{N-1}^q} y_{P_1}^{1*} \;\cdots\; y_{P_{N-1}}^{N-1*}\{m^N(P_1,\ldots,P_N) - m^N(P_1,\ldots,P_{N-1},P_N^l)\} = 0, \\
&\qquad\qquad P_N \in \mathcal{P}_N^q, \quad P_N \neq P_N^l
\end{aligned} \tag{38}
$$

where, P_i^l is any one of the search paths in $\mathcal{P}_i^q$, and $\breve{Y}^i$ is the interior of Y^i. If the inner mixed strategy solution does not exist then the above formulation may not yield a feasible solution. In that case, we may have to choose some other algorithm. The domain of the search effectiveness function increases with increase in q and also increase in number of players. Hence, solving the algebraic equations becomes computationally time consuming. In order to reduce computational time we use the concept of domination [31].

Dominating Strategies : There are certain strategies for an agent which yield less profit than other strategies. For instance, consider agent A_i choosing path P_i^k that has higher benefit than path P_i^l, for all possible combination of the paths of the rest of the $N-1$ agents. Then, we can eliminate path P_i^l without affecting the equilibrium solution. Since the objective of the searchers is to maximize their benefits, we are eliminating a strategy with lower benefit. In general, for A_i, considering the search effectiveness function m^i, we say that path P_i^k dominates path P_i^l, if

$$m^i(P_1, \ldots, P_i^k, \ldots, P_N) \geq m^i(P_1, \ldots, P_i^l, \ldots, P_N), \quad \forall\, P_j \in \mathcal{P}_j^q,\ j \neq i \tag{39}$$

and if, for at least one j, the strict inequality holds. Eliminating dominated strategies will reduce the computational time required to compute the mixed equilibrium strategy. The concept of dominating strategies in non-zero sum games as formulated above is similar to the dominating strategies as formulated for zero sum games [31]. The dominating strategies concept is applicable only for noncooperative games and not for cooperative and security strategies.

Coalitional Nash Strategies

In this model, we assume that agent A_i is playing against the coalition of the rest of the $N-1$ agents. The game is modelled as a bimatrix game which consists of two search effectiveness matrices, $M^{1,i} = \{m_{kl}^{1,i}\}$ and $M^{2,i} = \{m_{kl}^{2,i}\}$. The matrix $M^{1,i}$ represents the benefit obtained by agent A_i. Every element $m_{kl}^{1,i} = V_i(P_i^k, \hat{P})$, $\hat{P} = \{(P_1, P_2, \ldots, P_N) | P_j \neq P_i, j = 1, \ldots, N\}$, represents the benefit obtained by agent A_i choosing path $P_i^k \in \mathcal{P}_i^q$ while the coalition chooses a strategy l, $l = 1, 2, \ldots, |\prod_{j=1, j\neq i}^{N} \mathcal{P}_j^q|$. The matrix $M^{2,i}$ represents the benefit obtained by the coalition and every element $m_{kl}^{2,i} = \sum_{j=1, j\neq i}^{N} V_j(P_i^k, \hat{P})$. The agent A_i assumes the coalition to be a single player. The players (A_i and the coalition) arrive at their decisions independently. In such a situation the equilibrium solution can be stated as: A pair of strategies {row k^*, column l^*} is said to constitute a noncooperative (Nash) equilibrium solution to the bimatrix game, if the following pair of inequalities are satisfied, $\forall\, k = 1, 2, \ldots, |\mathcal{P}_i^q|$ and $\forall\, l = 1, 2, \ldots, |\prod_{j=1, j\neq i}^{N} \mathcal{P}_j^q|$

$$m_{k^*l^*}^{1,i} \geq m_{kl^*}^{1,i}, \quad m_{k^*l^*}^{2,i} \geq m_{k^*l}^{2,i} \tag{40}$$

The agent A_i considers strategy k^* as its equilibrium strategy. Each agent computes the two search effectiveness matrices and considers k^* as its equilibrium strategy. The dimension for the search effectiveness matrix is

$$|\mathcal{P}_i^q| \times |\prod_{\substack{j=1\\j\neq i}}^{N} \mathcal{P}_j^q| \tag{41}$$

The pure strategy Nash equilibrium may not always exist, in which case we have to use mixed strategy equilibrium. The main disadvantage of using the earlier model is that, if there is no inner mixed strategy Nash solution then we may not be able to find a feasible solution. However, in this model we can directly use the bilinear programming method to compute the mixed strategies equilibrium.

A pair $\{y^*, z^*\}$ constitutes a mixed-strategy Nash Equilibrium solution to a bimatrix game $(M^{1,i}, M^{2,i})$ if, and only if, their exists a pair (f^*, g^*) such that $\{y^*, z^*, f^*, g^*\}$ is a solution of the following bilinear programming problem:

$$\min_{y,z,f,g} [-y^{'} M^{1,i} z - y^{'} M^{2,i} z + f + g] \tag{42}$$

subject to

$$- M^{1,i} z \geq -f \cdot 1_{|\mathcal{P}_i^q|}, \quad - M^{2,i^{'}} z \geq -g \cdot 1_{|\prod_{\substack{j=1\\j\neq i}}^{N} \mathcal{P}_j^q|}$$

$$y \geq 0, \; z \geq 0, \; y^{'} \cdot 1_{|\mathcal{P}_i^q|} = 1, \; z \cdot 1_{|\prod_{\substack{j=1\\j\neq i}}^{N} \mathcal{P}_j^q|} = 1 \tag{43}$$

where $1_{|\mathcal{P}_1^q(C_{s_1})|}$ and $1_{|\prod_{\substack{j=1\\j\neq i}}^{N} \mathcal{P}_j^q|}$ are column vectors of dimensions $|\mathcal{P}_i^q|$ and $|\prod_{\substack{j=1\\j\neq i}}^{N} \mathcal{P}_j^q|$, with all elements equal to 1.

Security Strategy

In security strategy the individual agents try to secure their minimal profits assuming adversarial behavior of the other players. For this purpose, the coalitional form given above is the ideal framework to obtain security strategies. Then, agent A_i chooses a 'row k^*' whose smallest entry is no smaller than the smallest entry of any other row, which implies

$$\underline{V}(M^{1,i}) = \max_k \min_l m_{kl}^{1,i}, \quad k^* = \arg\max_k \{\min_l m_{kl}^{1,i}\} \tag{44}$$

where, $k = 1, 2, \ldots, |\mathcal{P}_i^q|$, and l represents a particular combination of strategies used by the $N - 1$ agents and

$$l = 1, 2, \ldots, |\prod_{\substack{j=1\\j\neq i}}^{N} \mathcal{P}_j^q| \tag{45}$$

Further, k^* is the security strategy for A_i, and $\underline{V}(M^i)$ is the guaranteed payoff to A_i. Every agent computes the security strategy individually and adopts the route given by k^*.

Cooperative Strategy

In this strategy the agents communicate during the decision process. The agents jointly choose a strategy such that the joint payoff of the game is maximized. Each agent computes the same search effectiveness function M and decides it's q step look ahead path using the function M. For N agents, let the search effectiveness function be $M = m^1(P_1, \ldots, P_N) + \ldots + m^N(P_1, \ldots, P_N)$ which represents the joint payoff due to all the agents' actions. A N-tuple of strategies $(P_1^*, \ldots, P_N^*)$ is said to be a cooperative strategy, if the following condition is satisfied:

$$m^1(P_1^*, \ldots, P_N^*) + \ldots + m^N(P_1^*, \ldots, P_N^*) = M(P_1^*, \ldots, P_N^*) \geq$$
$$M(P_1, \ldots, P_N) = m^1(P_1, \ldots, P_N) + \ldots + m^N(P_1, \ldots, P_N), \forall\ P_i \in \mathcal{P}_i^q \quad (46)$$

The cooperative strategy used in game theory involves communication between the players to coordinate their actions and arrive at a mutually acceptable decision. The drawback of cooperative strategy, when used in economics, where the players are selfish by nature, is that player may violate the mutually decided upon agreement to earn larger benefits at the cost of others. In our scenario, since the agents are automated, they can be assumed to be altruistic and hence they do not violate the decided upon agreement. Hence, using the search effectiveness functions, each agent can also compute the cooperative strategy without explicit communication between the agents. The equilibrium solution remains the same whether communication is present or not, provided that all the agents possess the same uncertainty map.

Greedy Strategy

In this case, the agents do not communicate among themselves and use a greedy strategy to determine their future actions. This is similar to the non-cooperative strategy used in [32]. The agent chooses a path P_i^k with a look ahead policy of q, using the following relation:

$$m^i(P_i^k) \geq m^i(P_i^j), \quad \forall\ \ j = 1, 2, \ldots, |\mathcal{P}_i^q| \quad (47)$$

where $m^i(P_i^k)$ is the benefit obtained by agents A_i using path P_i^k and it is evaluated using Eqn. (29).

Selection of Strategies

When there are multiple solutions, the selection of strategies by players becomes a crucial issue. The security and greedy strategies are straightforward to implement. If there exists multiple security or greedy strategies, any one of them will guarantee the same payoff level. In fact, for security strategies the actual payoff is bound to be higher for the players so long as they stick to their security strategies. In the case of multiple cooperative strategies, since all players communicate with each other during the decision process, they can decide to adopt a strategy which is beneficial to the overall team goal. One can also devise some protocol to automate this selection so that communication between agents can be dispensed with. But when multiple solutions occur for pure or mixed strategy Nash equilibrium, the agents have to select one of them. Since every agent can evaluate the search effectiveness function of all the other agents, they can jointly select a solution whose joint payoff is maximum. The selection of solution does not involve any communication with the other agents, but uses the available data through evaluation of search effectiveness functions. The solution method of choosing a strategy that would maximize the agents benefit is common for all the agents. When a mixed strategy equilibrium exists then agents can make a choice based on maximum likelihood or by random number generation. Here, we choose the maximum likelihood method.

5.3 Simulation Results

For the purpose of simulation, a region composed of hexagonal grids of size 30×30 is considered. We consider five agents with randomly located initial positions in the search space. We initially assume a perfect information case where each agent has the same uncertainty map throughout the search operation, although it is not a necessary condition. A typical uncertainty map is shown in Figure 8 along with the initial positions of the searchers. The percentage of uncertainty in a cell is proportional to the size of the grey area in the cell. The total uncertainty in the search space is defined as the sum of the uncertainties in all the cells.

The uncertainty map is updated at every search step in time. The simulation is carried out for look ahead step lengths of $q = 1$ and $q = 2$. The agents' uncertainty reduction factors are assumed to remain constant throughout the

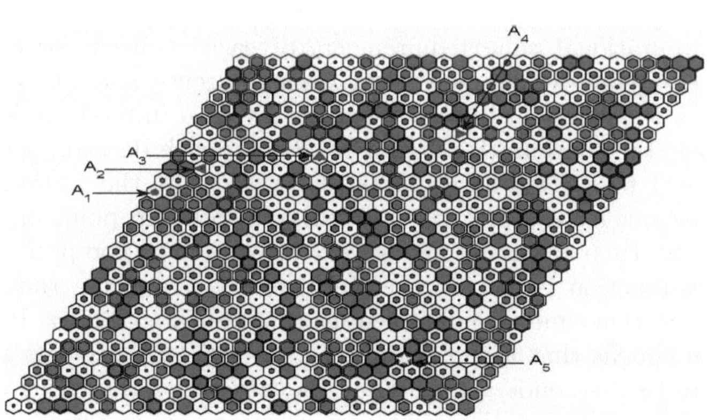

Fig. 8. A typical uncertainty map for 30×30 hexagonal grid

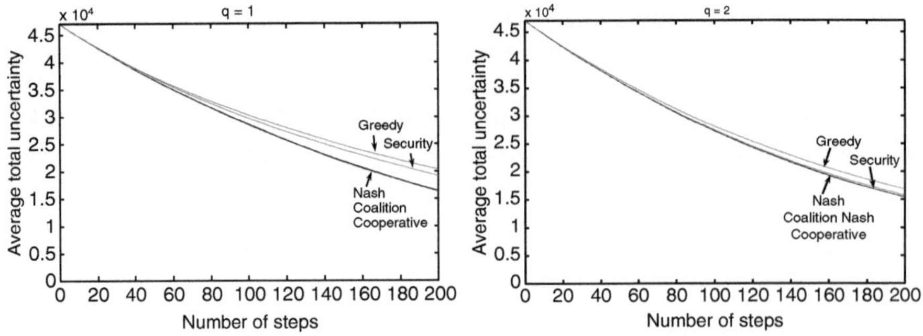

Fig. 9. Performance of various strategies for $q = 1$ and $q = 2$ averaged over 50 maps and with same initial searcher positions

search operation and have values $\beta_1 = 0.5$, $\beta_2 = 0.4$, $\beta_3 = 0.6$, $\beta_4 = 0.8$, and $\beta_5 = 0.7$. We will study the performance of various game theoretical strategies on total uncertainty reduction in a search space.

The simulation was carried out for 50 different uncertainty maps with the same initial placement of agents and same total uncertainty in each map. The positions of the searchers are as shown in Figure 8 and the total initial uncertainty in each map is assumed to be 4.75×10^4. The average total uncertainty is the average of the total uncertainty for the 50 maps at each step, computed up to a total of 200 search steps.

Figure 9 shows the comparative performance of various strategies with different look ahead policies of $q = 1$ and $q = 2$. We can see that the average total uncertainty reduces with each search step. The cooperative, noncooperative Nash, and coalitional Nash strategies perform equally well and they are better than the other strategies. From this figure we can see that for all the search strategies, look ahead policy of $q = 2$ performs better than $q = 1$, which is expected.

However, with the increase in look ahead policy length the computational time also increases significantly. Figure 10 gives the complete information on the computational time requirements of each strategy for $q = 1$ and $q = 2$. Since we consider 50 uncertainty maps, 5 agents, and 200 search steps, there are 5×10^4 number of decision epochs involved in the complete simulation. We plot the computational time needed by each decision epoch, where (i-1) $\times 10^3 + 1$ to $i \times 10^3$ decision epochs (marked on the vertical axis) are the decisions taken for searching the i-th map. So each point on the graph represents the time taken by the search algorithm to compute the search effectiveness function (wherever necessary) and arrive at the route decision. These computation times are obtained using a dedicated 3 GHz, P4 machine. All decision epochs that take computation time $\leq 10^{-3}$ seconds are plotted against time 10^{-3} seconds. The last plot in each set of graphs shows the distribution of computation times for various strategies in terms of the total number of decision epochs that need computation time less than the value on the horizontal axis. These plots reveal important information about the computational effort that each strategy demands.

Finally, we carried out another simulation to demonstrate the utility of the Nash strategies when the perceived uncertainty maps of the agents are different from the actual uncertainty map. For this it was assumed that the uncertainty reduction factors (β) of the agents fluctuate with time due to fluctuation in the performance of their sensor suites due to environmental or other reasons. Each agent knows its own current uncertainty reduction factor perfectly but assumes that the uncertainty reduction factors of the other agents to be the same as their initial value. This produces disparity in the uncertainty map between agents and from the actual uncertainty map which evolves according to the true β values as the search progresses. The variation in the value of β for the five agents are shown in Figure 11.

In this situation the total uncertainty reduction is as shown in Figure 12, which shows that both the Nash strategies, which do not make any assumption

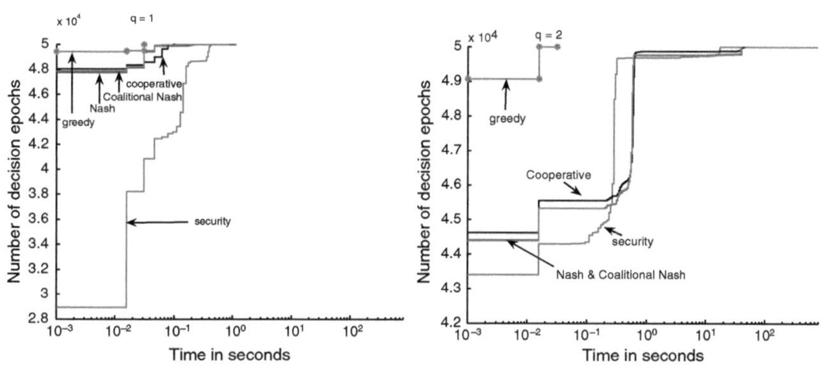

Fig. 10. Computational time of various strategies for $q = 2$ for random initial uncertainty maps

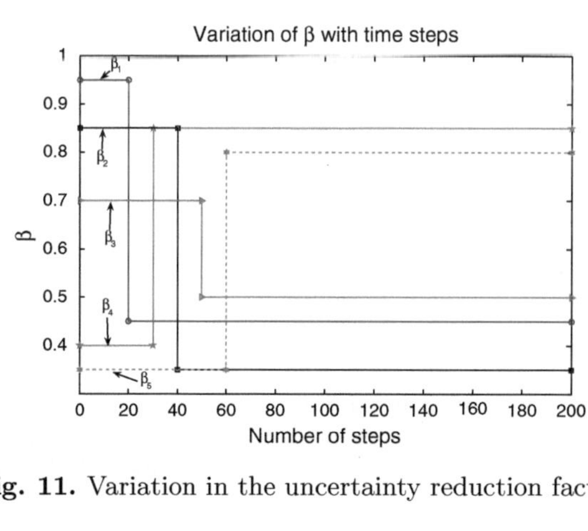

Fig. 11. Variation in the uncertainty reduction factors

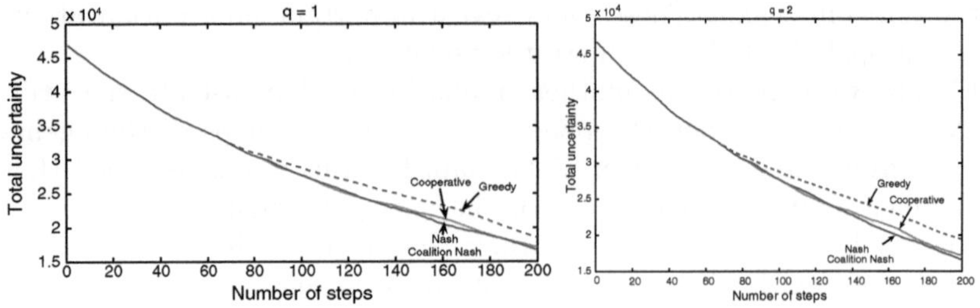

Fig. 12. Performance in the non-ideal case with varying β

about the other agents' actions, perform equally well and are also better than the cooperative strategy which assumes cooperative behavior from the other agents.

6 Conclusions

In this chapter, we addressed the problem of task allocation among autonomous UAVs operating in a swarm using concepts from team theory, negotiation, and game theory, and showed that effective and intelligent strategies can be devised from these well-known theories to solve complex decision-making problems in multi-agent systems. The role of communication between agents was explicitly accounted for in the problem formulation. This is one of the first use of these concepts to multi-UAV task allocation problems and we hope that this framework and results will be a catalyst to further research in this challenging area.

Acknowledgements

This work was partially supported by the IISc-DRDO Program on Advanced Research in Mathematical Engineering.

References

1. C. Schumacher, P. Chandler, S. J. Rasmussen: Task allocation for wide area search munitions via iterative netowrk flow, *AIAA Guidance, Navigation, and Control Conference and Exhibit*, August, Monterey, California, 2002, AIAA 2002–4586
2. J.W. Curtis and R. Murphey: Simultaneaous area search and task assignment for a team of cooperative agents, *AIAA Guidance, Navigation, and Control Conference and Exhibit*, August, Austin, Texas, 2003, AIAA 2003–5584

3. P.B. Sujit, A. Sinha, and D. Ghose: Multi-UAV task allocation using team theory, *Proc. of the IEEE Conference on Decision and Control*, Seville, Spain, December 2005, pp. 1497–1502
4. P.B. Sujit and D. Ghose, Multiple agent search in an unknown environment using game theoretical models, *Proc. of the American Control Conference*, Boston, pp. 5564–5569, 2004
5. P.B. Sujit and D. Ghose: Search by UAVs with flight time constraints using game theoretical models, *Proc. of the AIAA Guidance Navigation and Control Conference and Exhibit*, San Francisco, California, August 2005, AIAA-2005-6241
6. P.B. Sujit, A. Sinha and D. Ghose: Multiple UAV Task Allocation using Negotiation, *Proceedings of Fifth International joint Conference on Autonomous Agents and Multiagent Systems*, Japan, May. 2006 (to appear)
7. P.B. Sujit and D. Ghose: Multi-UAV agent based negotiation scheme, *Proc. of the American Control Conference*, Portland, Oregon, June 2005, pp. 2995–3000
8. P.B. Sujit and D. Ghose: A self assessment scheme for multiple-agent search, *Proc. of the American Control Conference*, Minneapolis, June 2006 (to appear)
9. K.E. Nygard, P.R. Chandler, M. Pachter: Dynamic network flow optimization models for air vehicle resource allocation, *Proc. of the the American Control Conference*, June 2001, Arlington, Texas, pp. 1853–1858
10. C. Schumacher, P. Chandler, M. Pachter, L.S. Pachter: UAV task assignment with timing constraints, *AIAA Guidance, Navigation, and Control Conference and Exhibit*, August, Austin, Texas, 2003, AIAA 2003–5664
11. C. Schumacher and P. Chandler: UAV task assignment with timing constraints via mixed-integer linear programming, *AIAA Unmanned Unlimited Technical Conference, Workshop and Exhibit*, Chicago, Illinois, Sept. 2004, AIAA-2004-6410
12. M. Alighanbari and J. How: Robust decentralized task assignment for cooperative UAVs, *AIAA Guidance, Navigation, and Control Conference and Exhibit*, Keystone, Colorado, Aug. 21–24, 2006
13. M. Darrah, W. Niland and B. Stolarik: UAV cooperative task assignments for a SEAD mission using genetic algorithms, *AIAA Guidance, Navigation, and Control Conference and Exhibit*, Keystone, Colorado, Aug. 2006, AIAA-2006-6456
14. C. Schumacher, P.R. Chandler, S. J. Rasmussen, and D. Walker: Task allocation for wide area search munitions with variable path length, *Proc. of the American Control Conference*, June, Denver, Colorado, 2003, pp. 3472–3477
15. D. Turra, L. Pollini, and M. Innocenti: Real-time unmanned vehicles task allocation with moving targets, *AIAA Guidance, Navigation, and Control Conference and Exhibit*, Providence, Rhode Island, August 2004, AIAA 2004–5253
16. Y. Jin, A. A. Minai, M. M. Polycarpou: Cooperative real-time search and task allocation in UAV teams, *IEEE Conference on Decision and Control*, Maui, Hawaii, December 2003, Vol. 1 , pp. 7–12
17. M.B. Dias and A. Stentz: A free market architecture for distributed control of a multirobot system, *6th International Conference on Intelligent Autonomous Systems*, Venice, Italy, July 2000, pp. 115–122
18. M.B. Dias, R.M. Zlot, N. Kalra, and A. Stentz: Market-based multirobot coordination: A survey and analysis, *Technical report CMU-RI-TR-05-13*, Robotics Institute, Carnegie Mellon University, April 2005

19. B. Gerkey, and M.J. Mataric: Sold!: Auction methods for multi-robot control, *IEEE Transactions on Robotics and Automation*, Vol. 18, No. 5, October 2002, pp. 758–768
20. B. Gerkey, and M.J. Mataric: A formal framework for the study of task allocation in multi-robot systems, *International Journal of Robotics Research*, Vol. 23, No.9, Sep 2004, pp. 939–954
21. M.J. Mataric, G.S. Sukhatme, and E.H. Stergaard: Multi-robot task allocation in uncertain environments, *Autonomous Robots*, Vol. 14, 2003, pp. 255–263
22. P. Gurfil: Evaluating UAV flock mission performance using Dudeks taxonomy, *Proc. of the American Control Conference*, Portland, Oregon, June 2005, pp. 4679–4684
23. M. Lagoudakis, P. Keskinocak, A. Kleywegt, and S. Koenig: Auctions with performance guarantees for multi-robot task allocation, *Proc. of the IEEE International Conference on Intelligent Robots and Systems*, Sendai, Japan, September 2004, pp. 1957–1962
24. S. Sariel and T. Balch: Real time auction based allocation of tasks for multi-robot exploration problem in dynamic environments, *AAAI workshop on Integrating Planning into Scheduling*, Pittsburgh, Pennsylvania, July 2005, Eds. Mark Boddy, Amedeo Cesta, and Stephen F. Smith, pp. 27–33
25. J. Marschak: Elements for a theory of teams, *Management Science*, Vol. 1, No. 2, Jan 1955, pp. 127–137
26. R. Radner: The linear team: An example of linear programming under uncertainty, *Proc. of 2nd Symposium in Linear Programming*, Washington D.C, 1955, pp. 381–396
27. S. Kraus: Automated negotiation and decision making in multiagent environments, *Multi-Agent Systems and Applications*, Springer LNAI 2086, (Eds.) M.Luck, V. Marik, O. Stepankova, and R. Trappl, 2001, pp. 150–172
28. A. Rubinstein: Perfect equilibrium in a bargaining model, *Econometrica*, Vol. 50, No. 1, 1982, pp. 97–109
29. K. Passino, M. Polycarpou, D. Jacques, M. Pachter, Y. Liu, Y. Yang, M. Flint, and M. Baum: Cooperative control for autonomous air vehicles, *Cooperative Control and Optimization*, (R. Murphey and P. M. Pardalos, eds.), vol. 66, Kluwer Academic Publishers, 2002, pp. 233–271
30. R.F. Dell, J.N. Eagel, G.H.A. Martins, and A.G. Santos: Using multiple searchers in constrained-path, moving-target search problems, *Naval Research Logistics*, Vol. 43, pp. 463–480, 1996
31. T. Basar and G.J. Olsder: *Dynamic Noncooperative Game Theory*, Academic press, CA 1995
32. S. Ganapathy and K.M. Passino: Agreement strategies for cooperative control of uninhabited autonomous vehicles, *Proc. of the American Control Conference*, Denver, Colorado, 2003, pp. 1026–1031

Author Biographies

P.B. Sujit has received his Bachelor's Degree in Electrical Engineering from the Bangalore University, MTech from Visveswaraya Technological University, and PhD from the Indian Institute of Science, Bangalore. At present, he is a Post Doctoral Fellow at Brigham Young University, Provo, Utah. His research interests include multi-agent systems, cooperative control, search theory, game theory, economic models, and task allocation.

A. Sinha has received her Bachelor's Degree in Electrical Engineering from Jadavpur University, Kolkata, India, and MTech from Indian Institute of Technology, Kanpur, India. At present she is a graduate student at the Department of Aerospace Engineering, Indian Institute of Science, Bangalore, India. Her research interests include cooperative control of autonomous agents, team theory, and game theory.

D. Ghose is a Professor in the Department of Aerospace Engineering at the Indian Institute of Science, Bangalore, India. He obtained a BSc(Engg) degree from the National Institute of Technology (formerly the Regional Engineering College), Rourkela, India, in 1982, and an ME and a PhD degree, from the Indian Institute of Science, Bangalore, in 1984 and 1990, respectively. His research interests are in guidance and control of aerospace vehicles, collective robotics, multiple agent decision-making, distributed decision-making systems, and scheduling problems in distributed computing systems. He is an author of the book *Scheduling Divisible Loads in Parallel and Distributed Systems* published by the IEEE Computer Society Press (presently John Wiley). He is in the editorial board of the *IEEE Transactions on Systems, Man, and Cybernetics, Part A: Systems and Humans*, and the *IEEE Transactions on Automation Science and Engineering*. He has held visiting positions at the University of California at Los Angeles and several other universities. He is an elected fellow of the Indian National Academy of Engineering.

UAV Path Planning Using Evolutionary Algorithms

Ioannis K. Nikolos, Eleftherios S. Zografos, and Athina N. Brintaki

Department of Production Engineering and Management,
Technical University of Crete, University Campus,
Kounoupidiana, GR-73100, Chania, Greece
jnikolo@dpem.tuc.gr

Abstract. Evolutionary Algorithms have been used as a viable candidate to solve path planning problems effectively and provide feasible solutions within a short time. In this work a Radial Basis Functions Artificial Neural Network (RBF-ANN) assisted Differential Evolution (DE) algorithm is used to design an off-line path planner for Unmanned Aerial Vehicles (UAVs) coordinated navigation in known static maritime environments. A number of UAVs are launched from different known initial locations and the issue is to produce 2-D trajectories, with a smooth velocity distribution along each trajectory, aiming at reaching a predetermined target location, while ensuring collision avoidance and satisfying specific route and coordination constraints and objectives. B-Spline curves are used, in order to model both the 2-D trajectories and the velocity distribution along each flight path.

1 Introduction

1.1 Basic Definitions

The term *unmanned aerial vehicle* or UAV, which replaced in the early 1990s the term *remotely piloted vehicle* (RPV), refers to a powered aerial vehicle that does not carry a human operator, uses aerodynamic forces to provide vehicle lift, can fly autonomously or be piloted remotely, can be expendable or recoverable, and can carry a lethal or non lethal payload [1]. UAVs are currently evolving from being remotely piloted vehicles to autonomous robots, although ultimate autonomy is still an open question.

The development of autonomous robots is one of the major goals in Robotics [2]. Such robots will be capable of converting high-level specification of tasks, defined by humans, to low-level action algorithms, which will be executed in order to accomplish the predefined tasks. We may define as *plan* this sequence of actions to be taken, although it may be much more complicated than that. Motion planning (or trajectory planning) is one category of such

I.K. Nikolos et al.: *UAV Path Planning Using Evolutionary Algorithms*, Studies in Computational Intelligence (SCI) **70**, 77–111 (2007)
www.springerlink.com

problems. Besides the great variety of planning problems and models found in Robotics, some basic terms are common throughout the entire subject.

The *state space* includes all possible situations that might arise during the planning procedure. In the case of an UAV each state could represent its position in physical space, along with its velocity. The state space could be either discrete or continuous; motion planning is planning in continuous state spaces. Although its definition is an important component of the planning problem formulation, in most cases is implicitly represented, due to its large size [3].

Planning problems also involve the *time* dimension. Time may be explicitly or implicitly modeled and may be either discrete or continuous, depending on the planning problem under consideration. However, for most planning problems, time is implicitly modeled by simply specifying a path through a continuous space [3].

Each state in the state space changes through a sequence of specific actions, included in the plan. The connection between actions and state changes should be specified through the use of proper functions or differential equations. Usually, these actions are selected in a way to "move" the object from an initial state to a target or goal state.

A planning algorithm may produce various different plans, which should be compared and valued using specific criteria. These criteria are generally connected to the following major concerns, which arise during a plan generation procedure: *feasibility* and *optimality*. The first concern asks for the production of a plan to safely "move" the object to its target state, without taking into account the quality of the produced plan. The second concern asks for the production of optimal, yet feasible, paths, with optimality defined in various ways according to the problem under consideration [3]. Even in simple problems searching for optimality is not a trivial task and in most cases results in excessive computation time, not always available in real-world applications. Therefore, in most cases we search for suboptimal or just feasible solutions.

Motion planning usually refers to motions of a robot (or a collection of robots) in the two-dimensional or three-dimensional physical space that contains stationary or moving obstacles. A motion plan determines the appropriate motions to move the robot from the initial to the target state, without colliding into obstacles. As the state space in motion planning is continuous, it is uncountably infinite. Therefore, the representation of the state space should be implicit. Furthermore, a transformation is often used between the real world where the robots are moving and the space in which the planning takes place. This state space is called the *configuration space* (C-space) and motion planning can be defined as a search for a continuous path in this high-dimensional configuration space that ensures collision avoidance with implicitly defined obstacles. However, the use of configuration space is not always adopted and the problem is formulated in the physical space; especially in cases with constantly varying environment (as in most of UAV applications) the use of configuration space results in excessive computation time, which is not available in real-time in-flight applications. *Path planning* is the generation of a space path between an initial location and the desired destination that has an optimal or near-optimal performance under specific constraints [4]. A detailed description of motion and path planning theory and classic methodologies can be found in [2] and in [3].

1.2 Cooperative Robotics

The term *collective behavior* denotes any behavior of agents in a system of more than a single agent. *Cooperative behavior* is a subclass of collective behavior which is characterized by cooperation [5]. Research in cooperative Robotics has gained increased interest since the late 1980's, as systems of multiple robots engaged in cooperative behavior show specific benefits compared to a single robot [5]:

- Tasks may be inherently too complex, or even impossible, for a single robot to accomplish, or the performance is enhanced if using multiple agents, since a single robot, despite its capabilities and characteristics, is spatially limited.
- Building or using a system of simpler robots may be easier, cheaper, more flexible and more fault-tolerant than using a single more complicated robot.

In [5] cooperative behavior is defined as follows: Given some tasks specified by a designer, a multiple robot system displays cooperative behavior if, due to some underlying mechanism, i.e. the "mechanism of cooperation", there is an increase in the total utility of the system.

Geometric problems arise when dealing with cooperative moving robots, as they are made to move and interact with each other inside the physical 2D or 3D space. Such geometric problems include multiple-robot path planning, moving to and maintaining formation, and pattern generation [5].

According to Fujimura [6], path planning can be either *centralized* or *distributed*. In the first case a universal path planner makes all decisions. In the second case each agent plans and adjusts its path. Furthermore, Arai and Ota [7] allow for hybrid systems that are combinations of *on-line*, *off-line*, *centralized*, or *decentralized* path planners. According to Latombe [2], centralized planning takes into account all robots, while decoupled planning corresponds to independent computation of each robot's path. Methods originally used for single robots can be also applied to centralized planning. For decoupled planning two approaches were proposed: a) prioritized planning, where one robot at a time is considered, according to a global priority, and b) path coordination, where the configuration space-time resource is appropriately scheduled to plan the paths.

Cooperation of UAVs has gained recently an increased interest due to the potential use of such systems for fire fighting applications, military missions, search and rescue scenarios or exploration of unknown environments (space-oriented applications). In order to establish a reliable and efficient framework for the cooperation of a number of UAVs several problems have to be encountered:

- UAV task assignment problem: a number of UAVs is required to perform a number of tasks, with predefined order, on a number of targets. The requirements for a feasible and efficient solution include taking into account: task precedence and coordination, timing constraints, and flyable trajectories [8]. The task re-assignment problem should be also considered, in order to take into account possible failure of a UAV to accomplish its task. The task assignment problem is a well-known optimization problem; it is *NP*-hard and, consequently, heuristic techniques are often used.
- UAV path planning problem: a path planning algorithm should provide feasible, flyable and near optimal trajectories that connect starting with target points. The requirement for feasible trajectories dictates collision avoidance between the cooperating UAVs as well as between the vehicles and the ground. The requirement of flyable trajectories usually dictates a lower bound on the turn radius and speed of the UAVs [8]. Additionally, an upper bound for the speed of each UAV may be required. The path optimality can be defined in various ways, according to the mission assigned. However, a typical requirement is to minimize the total length of the paths.
- Data exchange between cooperating UAVs and data fusion: exchange of information between cooperating UAVs is expected to enhance the effectiveness of the team. However, in real world applications communication imperfections and constraints are expected, which will cause coordination problems to the team [9]. Decentralized implementations of the decision and control algorithms may reduce the sensitivity to communication problems [10].
- Cooperative sensing of the targets: the problem is defined as how to co-operate the UAV sensors in terms of their locations to achieve optimal estimation of the state of each target [11] (a target localization problem).
- Cooperative sensing of the environment: the problem is defined as how to cooperate the UAV sensors in order to achieve better awareness of the environment (popup threats, changing weather conditions, moving obstacles etc.). In this category we may include the coordinated search of a geographic region [12].

1.3 Path Planning for Single and Multiple UAVs

Compared to the path-planning problem in other applications, path planning for UAVs has some of the following characteristics, according to the mission [13, 14, 15]:

- Stealth, in order to minimize the probability of detection by hostile radar, by flying along a route which keeps away from possible threats and/or has a lower altitude to avoid radar detection.
- Physical feasibility, which refers to the physical (or technology) limitations from the use of UAVs, such as limited range, minimum turning angle, minimum and maximum speed etc.
- Performance of mission, which imposes special requirements, including maximum turning angle, maximum climbing/diving angle, minimum and/ or maximum flying altitude and specific approaching angle to the target point.
- Cooperation between UAVs in order to maximize the possibility of mission accomplishment.
- Real-time implementation, which asks for computationally efficient algorithms.

The characteristics above imply special issues that have to be considered for an efficient modeling of the (single or multiple) UAV path planning problems.

Path modeling:

The simpler way to model an UAV path is by using straight-line segments that connect a number of way points, either in 2D or 3D space [12, 15]. This approach takes into account the fact that in typical UAV missions the shortest paths tend to resemble straight lines that connect way points with starting and target points and the vertices of obstacle polygons. Although way points can be efficiently used for navigating a flying vehicle, straight-line segments connecting the corresponding way points cannot efficiently represent the real path that will be followed by the vehicle. As a result, these simplified paths cannot be used for an accurate simulation of the movement of the UAV in an optimization procedure, unless a large number of way points is adopted. In that case the number of design variables in the optimization procedure explodes, along with the computation time. The problem becomes even more difficult in the case of cooperating flying vehicles.

In [16], paths from the initial vehicle location to the target location are derived from a graph search of a Voronoi diagram that is constructed from the known threat locations. The resulting paths consist of line segments. These paths are subsequently smoothed around each way point, in order to provide feasible trajectories within the dynamic constraints of the vehicle. A great advantage of the Voronoi diagram approach is that it reduces the path planning problem from an infinite dimensional search, to a finite-dimensional graph search. This important abstraction makes the path planning problem feasible in near-real time, even for a large number of way points [16].

Vandapel et al. [17] used a network of free space bubbles to model the path of small scale UAVs, in order to solve the path planning problem of autonomous unmanned aerial navigation below the forest canopy. Using a priori aerial data scans of forest environments, they compute a network of free space bubbles, which form safe paths within the forest environment. Their approach is tailored to the problem of small scale UAVs and can be decomposed into two steps: 1) the scene made of 3-D points is segmented into three classes (ground, vegetation and tree trunk-branches). 2) A path planning algorithm explores the segmented environment and computes connected obstacle-free areas, which will subsequently form a network of tunnels intersecting at some locations.

An alternative approach is to model the UAV dynamics using the Dubins car formulation [18]. The UAV is assumed to fly with constant altitude, constant flight speed and to have continuous time kinematics [19]. This approach cannot efficiently model real world scenarios, which may include 3D terrain avoidance or following of stealthy routes. However, this approach seems to be sufficient enough for task assignment purposes to cooperating UAVs flying at safe altitudes [19, 8, 20].

B-Spline curves have been used for trajectory representation in 2-D environments (simulated annealing based path line optimization, combined with fuzzy logic controller for path tracking) [21], and in 3-D environments (Evolutionary Algorithm based path line optimization for a UAV over rough terrain) [22, 23]. B-Spline curves need a few variables (the coordinates of their control points) in order to define complicated 2D or 3D curved paths, providing at least first order derivative continuity. Each control point has a very local effect on the curve's shape and small perturbations in its position produce changes in the curve only in the neighborhood of the repositioned control point.

Cooperation Scenarios:

Path planning algorithms were initially developed for the solution of the problem of a single UAV. The increasing interest for missions involving cooperating UAVs resulted in the development of algorithms that take into account the special characteristics and constraints of such missions. The related works present various scenarios, formulations and approaches connected to cooperating UAV path planning problems. Some of the most representative scenarios are presented below.

Beard et al. [16] considered the scenario where a group of UAVs is required to transition through a number of known target locations, with a number of threats in the region of interest. Some threats are known a priori, some others "pop up" or become known only when a UAV flies near them. It is desirable to have multiple UAVs arrive on the boundary of each target's radar detection region simultaneously. Collision avoidance is ensured by supposing that individual UAVs fly at different pre-assigned altitudes. In this work the problem is decomposed in several sub-problems: a) The assignment problem of a number of UAVs to a number of targets in a way that each target has multiple UAVs assigned to it, with a high preference to specific targets. b) The determination for each team of UAVs assigned to a target of an estimated time over target that ensures simultaneous intercept and is feasible for all UAVs in the team. c) The determination of a path (specified via waypoints) that can be completed within the specified time over target, taking into account minimum and maximum velocity constraints. d) The transformation of the initial path into a feasible UAV trajectory. e) The development of controllers for each UAV to track their computed trajectory.

A simpler scenario is presented in [15], where the problem under consideration is to generate routes for cooperating UAVs in real time, which take into account the exposure of UAVs to the threats and enable the vehicles to arrive at their goal location simultaneously. Some of the threats are known a priori, some of them "pop up" or become known only when a UAV approaches to it. For each UAV are imposed minimum and maximum velocity constraints. The cooperation related constraints are: a) the simultaneous arrival of all UAVs at goal locations, and b) the collision avoidance between UAVs.

In [20] the motion-planning problem for a limited resource of mobile sensor agents (MSAs) is investigated, in an environment with a number of targets larger than the available MSAs. The MSAs are assumed to move much faster than the targets. In order to keep the targets in surveillance the members of the MSA team have to fly back and forth to update the targets' status. This *NP*-hard problem is essentially a combination of the problems of sensor resource management and robot motion planning. The problem is formulated as an optimization problem whose objective is to minimize the average time duration between two consecutive observations of each target.

In [12] the objective is to provide a coordinated plan for searching a geographic region, represented by a grid of cells, using a team of searchers. Each cell is characterized by its elevation and a cost parameter that corresponds to the danger of visiting it. Each vehicle carries a sensor, characterized by scan radius, angle and direction. The mission objective is the target coverage, i.e. the percentage of the region that must be scanned during the mission. Scans can be performed only from safe cells (the cell and all 8 neighbors should have been previously scanned). Additionally the path that connects scanning points should traverse through already scanned cells (a soft constraint). For safety reasons scanning paths are not allowed to be very close to each other.

Solution methodologies:

Path planning problems are actually multi-objective multi-constraint optimization problems, in most cases very complex and computationally demanding [24]. The problem complexity increases when multiple UAVs should be used. Various approaches have been reported for UAVs coordinated route planning, such as Voronoi diagrams [16], mixed integer linear programming [25, 26] and dynamic programming [27] formulations.

In [25, 26] mixed-integer linear programming (MILP) is used to solve tightly-coupled task assignment problems with timing constraints. The advantage to this approach is that it yields the optimal solution for the given problem. The primary disadvantage is the high computational time required.

In [16] the motion-planning problem was decomposed into a waypoint path planner and a dynamic trajectory generator. The path-planning problem was solved via a Voronoi diagram and Eppstein's k-best paths algorithm. The trajectory generator problem was solved via a real-time nonlinear filter that explicitly accounts for the dynamic constraints of the vehicle and modifies the initial path. This decomposition of the motion-planning problem has the advantage of decomposing a non-polynomial optimization problem into two sub-problems that can be computed in near-real time, with the disadvantage of providing a suboptimal solution [16].

Computational intelligence methods, such as Neural Networks [28], Fuzzy Logic [29] and Evolutionary Algorithms (EAs) [15, 23] (or, in some cases, a combination of them) have been successfully used in the development of algorithms that produce trajectories for guiding mobile robots in known, unknown or partially known environments.

During the past few years, it has been shown by many researchers that EAs are a viable candidate to solve path planning problems effectively and provide feasible solutions within a short time without demanding excessive computer power. The reasons behind choosing EAs as an optimization tool for the path-planning problem are their high robustness compared to other existing directed search methods, their ease of implementation in problems with a relatively high number of constraints, and their high adaptability to the special characteristics of the problem under consideration [23].

Traditionally, EAs have been used for the solution of the path-finding problem in ground based or sea surface navigation [30]. Commonly, the generated trajectory composed of straight line segments, connecting successive way points, that guided a mobile robot or a vehicle along a 2-D path on the earth's or sea's surface. The design variables used represented the coordinates of the way points, where the vehicle changes its direction. Other approaches took into account the time dimension by using design variables that also described the vehicle steady speed as it traversed a part of its path. When the vehicle's operational environment was partially known or dynamic, a feasible and safe trajectory was planned off-line by the EA, and the algorithm was used on-line whenever unexpected obstacles were sensed [31, 32]. EAs have been also used for solving the path-finding problem in a 3-D environment for underwater vehicles, assuming that the path is a sequence of cells in a 3-D grid [33, 34].

In [23] an EA based framework was utilized to design an off-line/on-line path planner for UAVs autonomous navigation. The path planner calculates a curved path line, represented using B-Spline curves in a 3-D rough terrain environment; the coordinates of B-Spline control points serve as design variables. The off-line planner produces a single B-Spline curve that connects the starting and target points with a predefined initial direction. The on-line planner gradually produces a smooth 3-D trajectory aiming at reaching a predetermined target in an unknown environment; the produced trajectory consists of smaller B-Spline curves smoothly connected with each other. For both off-line and on-line planners, the problem is formulated as an optimization one; each objective function is formed as the weighted sum of different terms, which take into account the various objectives and constraints of the corresponding problem. Constraints are formulated using penalty functions.

Changwen Zheng et al. [15] proposed a route planner for UAVs, based on evolutionary computation, which can be used to plan routes for either single or multiple vehicles. The flight route consists of straight-line segments, connecting the way points from the starting to the goal points. A real coded chromosome representation is used; for each way point its physical coordinates are used as design variables, along with a state variable, which provides information on the feasibility of the corresponding way point and the feasibility of the route segment connecting the point to the next one. The cost function of flight route penalizes the length of the route, penalizes flight routes at high altitudes and routes that come dangerously close to known ground threat sites. The imposed constraints on route segments are relevant to: minimum route leg length, maximum route distance, minimum flying height, maximum turning angle, maximum climbing/diving angle, simultaneous arrival at target location and no collision between vehicles. The route planning problem is formulated as the problem of minimization of the cost function under the aforementioned constraints.

In [8] a multi-task assignment problem for cooperating UAVs is formulated as a combinatorial optimization problem. A Genetic Algorithm is utilized for assigning the multiple agents to perform multiple tasks on multiple targets. The algorithm allows efficiently solving this *NP*-hard problem and, additionally, allows taking into account requirements such as task precedence and coordination, timing constraints, and flyable trajectories. The performance metric for the optimization problem is defined as the cumulative distance traveled by the vehicles to perform all of the required tasks; the objective is to minimize this metric subject to the above requirements. Integer encoding is used for the chromosomes, which are composed of two rows; the first row presents the assignment of a vehicle to perform a task on the target appearing on the second row. The algorithm was compared to a stochastic random search and a deterministic branch and bound search methods and found to provide near optimal solutions considerably faster than the other methods.

1.4 Outline of the Current Work

The following scenario was considered in this work: Assuming a number of UAVs at different known initial locations, the issue is to produce 2-D trajectories, with a desirable velocity distribution along each trajectory, reaching a common target under specific coordination and route constraints. The constraints and objectives refer to: minimum path lengths, collision avoidance between the flying vehicles and the ground, predefined minimum and maximum UAV velocity magnitudes during their flights, predefined safety distance between UAVs, near simultaneous arrival to the target and target approach from different directions.

This work is an extension of a previous one [35], which used Differential Evolution (DE) in order to find optimal paths of coordinated UAVs, with the paths being modeled with straight line segments. The main drawback of that approach was the need of a large number of segments for complicated paths, resulting in a large number of design variables and, consequently, generations to converge. In this work the Differential Evolution (DE) algorithm is combined with a Radial Basis Functions Network (RBFN), which serves as a surrogate approximation, in order to reduce the number of exact evaluations of candidate solutions. The candidate paths are modeled in the physical space and evaluated with respect to the physical (working) space. B-Spline curves are used for path line modeling, and complicated paths can be produced with a small number of control variables.

The rest of the chapter is organized as follows: section 2 contains B-Spline and Evolutionary Algorithms fundamentals; the solid terrain formulation, used for experimental simulations, is also presented. An off-line path planner for a single UAV will be briefly discussed in section 3, in order to introduce the concept of UAV path planning using Evolutionary Algorithms. Section 4 deals with the concept of coordinated UAV path planning using Evolutionary Algorithms. The problem formulation is described, including assumptions, objectives, constraints, objective function definition and path modeling. Section 5 presents the optimization procedure using a combination of Differential Evolution and a Radial Basis Functions Artificial Neural Network, which is used as a surrogate model in order to enhance the converge rate of Differential Evolution algorithm. Simulations results are presented in section 6, followed by discussion and conclusions in section 7.

2 B-Spline and Evolutionary Algorithms Fundamentals

2.1 B-Spline Curves

Straight-line segments cannot represent a flying objects path line, as it is usually the case with mobile robots, sea and undersea vessels. B-Splines are adopted to define the UAV desired path, providing at least first order derivative continuity. B-Spline curves are well fitted in the evolutionary procedure; they need a few variables (the coordinates of their control points) in order to define complicated curved paths. Each control point has a very local effect on the curve's shape and small perturbations in its position produce changes in the curve only in the neighborhood of the repositioned control point.

B-Spline curves are parametric curves, with their construction based on blending functions [36, 37]. Their parametric construction provides the ability to produce non-monotonic curves. If the number of control points of the corresponding curve is $n+1$, with coordinates $(x_0, y_0, z_0), \ldots, (x_n, y_n, z_n)$, the coordinates of the B-Spline curve may be written as

$$x(u) = \sum_{i=0}^{n} x_i \cdot N_{i,p}(u), \tag{1}$$

$$y(u) = \sum_{i=0}^{n} y_i \cdot N_{i,p}(u), \tag{2}$$

$$z(u) = \sum_{i=0}^{n} z_i \cdot N_{i,p}(u), \tag{3}$$

where u is the free parameter of the curve, $N_{i,p}(u)$ are the blending functions of the curve and p is its degree, which is associated with curve's smoothness ($p+1$ being its order). Higher values of p correspond to smoother curves.

The blending functions are defined recursively in terms of a *knot* vector $U = \{u_0, \ldots, u_m\}$, which is a non-decreasing sequence of real numbers, with the most common form being the *uniform non-periodic* one, defined as

$$u_i = \begin{cases} 0 & if \quad i < p+1 \\ i-p & if \quad p+1 \le i \le n \\ n-p+1 & if \quad n < i. \end{cases} \tag{4}$$

The blending functions $N_{i,p}$ are computed, using the knot values defined above, as

$$N_{i,0}(u) = \begin{cases} 1 & if \quad u_i \le u < u_{i+1} \\ 0 & \quad\quad otherwise, \end{cases} \tag{5}$$

$$N_{i,p}(u) = \frac{u - u_i}{u_{i+p} - u_i} N_{i,p-1}(u) + \frac{u_{i+p+1} - u}{u_{i+p+1} - u_{i+1}} N_{i+1,p-1}(u). \tag{6}$$

If the denominator of either of the fractions is zero, that fraction is defined to have zero value. Parameter u varies between 0 and $(n-p+1)$ with a constant step, providing the discrete points of the B-Spline curve. The sum of the values of the blending functions for any value of u is always 1.

The use of B-Spline curves for the determination of a flight path provides the advantage of describing complicated non-monotonic 3-dimensional curves with controlled smoothness with a small number of design parameters, i.e. the coordinates of the control points. Another valuable characteristic of the adopted B-Spline curves is that the curve is tangential to the control polygon at the starting and ending points. This characteristic can be used in order to define the starting or ending direction of the curve, by inserting an extra fixed point after the starting one, or before the ending control point. Figure 1 shows a quadratic 2-dimensional B-Spline curve ($p = 2$) with its control points and the corresponding control polygon.

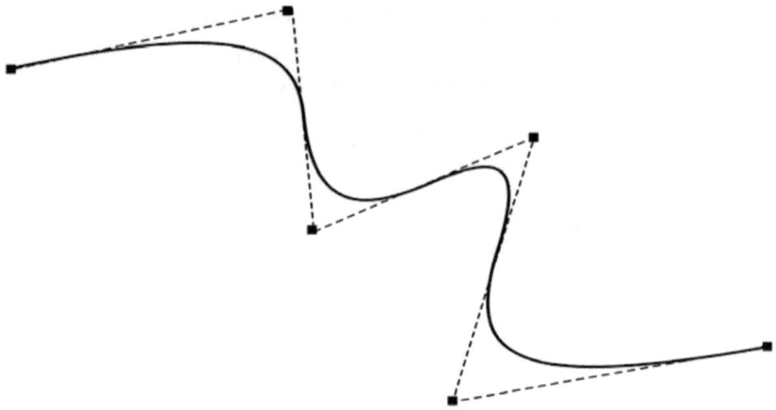

Fig. 1. A quadratic ($p = 2$) 2-dimensional B-Spline curve, produced using a uniform non-periodic knot vector, and its control polygon

2.2 Fundamentals of Evolutionary Algorithms (EAs)

EAs are a class of search methods with remarkable balance between exploitation of the best solutions and exploration of the search space. They combine elements of directed and stochastic search and, therefore, are more robust than directed search methods. Additionally, they may be easily tailored to the specific application of interest, taking into account the special characteristics of the problem under consideration [38, 39, 30].

The natural selection process is simulated in EAs, using a number (*population*) of individuals (candidate solutions to the problem) to evolve through certain procedures. Each individual is represented through *chromosome* - a string of numbers (bit strings, integers or floating point numbers), in a similar way with chromosomes in nature; it contains the design variables of the optimization problem. Each individual's quality is represented by a *fitness function* tailored to the problem under consideration.

Classic Genetic Algorithms (GAs) use binary coding for the representation of the *genotype*. However, floating point coding moves EAs closer to the problem space, allowing the operators to be more problem specific; this provides a better physical representation of the space constraints. Additionally, directed search techniques gain physical meaning and they are easily applicable.

In general, EA starts by generating, randomly, the initial chromosome population with their *genes* (the design variables in the case of floating point coding) taking values inside the desired constrained space of each design variable. The lower and higher constraints of each gene may be chosen in a way that specific undesirable solutions may be avoided. Although the shortening of the search space reduces the computation time, it may also lead to sub-optimal solutions, due to the lower variability between the potential solutions.

After the evaluation of each individual's fitness function, operators are applied to the population, simulating the according natural processes. Applied operators include various forms of *recombination*, *mutation* and *selection*, which are used in order to provide the next generation chromosomes. The first classic operator applied to the selected chromosomes is the *one-point crossover* scheme. Two randomly selected chromosomes are divided in the same (random) position, while the first part of the first one is connected to the second part of the second one and vice-versa. The crossover operator is used to provide information exchange between different potential solutions to the problem.

The second classic operator applied to the selected chromosomes is the uniform mutation scheme. This asexual operator alters a randomly selected gene of a chromosome. The new gene takes its random value from the constrained space, determined in the beginning of the process. The mutation operator is used in order to introduce some extra variability into the population.

The resulting intermediate population is evaluated and a fitness function is assigned to each member of the population. Using a selection procedure (different for each type of EA) the best individuals of the intermediate population (or the best individuals of the intermediate and the previous population) will form the next generation. The process of a new generation evaluation and creation is successively repeated, providing individuals with high values of fitness function.

2.3 The Solid Boundary Representation

In the simulation results that will be presented the terrain where UAVs fly is represented by a meshed 3-D surface, produced using mathematical functions of the form

$$
\begin{aligned}
z(x,y) = {} & \sin(y+a) + b \cdot \sin(x) + c \cdot \cos\left(d \cdot \sqrt{x^2+y^2}\right) \\
& + e \cdot \cos(y) + f \cdot \sin\left(f \cdot \sqrt{x^2+y^2}\right) + g \cdot \cos(y), \qquad (7)
\end{aligned}
$$

where a, b, c, d, e, f, g are constants experimentally defined, in order to produce either a surface with mountains and valleys (as shown in Fig. 2) or a maritime environment with islands close to each other (as shown in Fig. 6).

A graphical interface has been developed for the visualization of the terrain surface, along with the path lines [23]. The corresponding interface deals with different terrains produced either artificially or based on real geographical data, providing an easy verification of the feasibility and the quality of each solution. The path-planning algorithm considers the boundary surface as a group of quadratic mesh nodes with known coordinates.

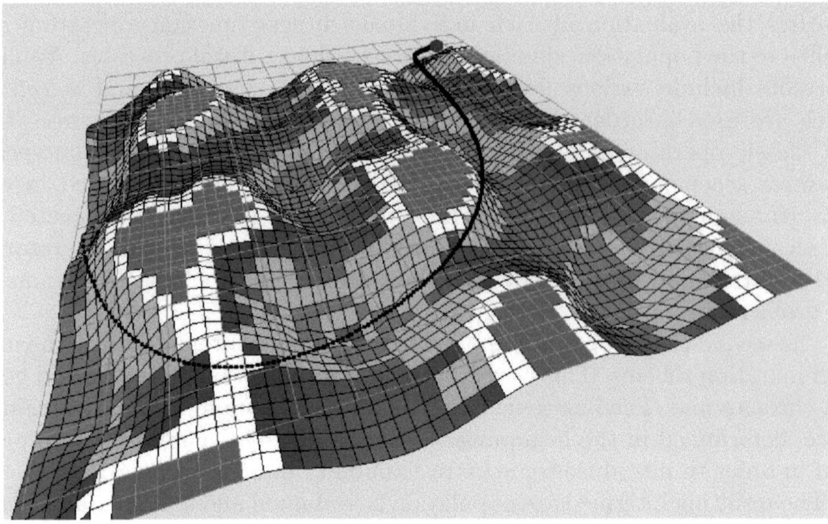

Fig. 2. A typical simulation result of the off-line path planner for a single UAV; the horizontal section of the terrain represents the imposed upper limit to the UAV flight. The starting position is marked with a circle

3 Off-line Path Planner for a Single UAV

The off-line path planner, discussed in detail in [23], will be briefly presented here, in order to introduce the concept of UAV path planning using Evolutionary Algorithms. The off-line planner generates collision free paths in environments with known characteristics and flight restrictions. The derived path line is a single continuous 3-D B-Spline curve, while the solid boundaries are interpreted as 3-D rough surfaces. The starting and ending control points of the B-Spline curve are fixed. A third point close to the starting one is also fixed, determining the initial flight direction. Between the fixed control points, free-to-move control points determine the shape of the curve, taking values in the constrained space. The number of the free-to-move control points is user-defined. Their physical coordinates are the genes of the EA artificial chromosome.

The optimization problem to be solved minimizes a set of four terms, connected to various objectives and constraints; they are associated with the feasibility and the length of the curve, a safety distance from the obstacles and the UAV's flight envelope restrictions. The objective function to be minimized is defined as

$$f = \sum_{i=1}^{4} w_i f_i. \tag{8}$$

Term f_1 penalizes the non-feasible curves that pass through the solid boundary. The penalty value is proportional to the number of discretized curve points located inside the solid boundary; consequently, non-feasible curves with fewer points inside the solid boundary show better fitness than curves with more points inside the solid boundary. Term f_2 is the length of the curve (non-dimensional with the distance between the starting and destination points) used to provide shorter paths.

Term f_3 is designed to provide flight paths with a safety distance from solid boundaries

$$f_3 = \sum_{i=1}^{nline} \sum_{j=1}^{nground} 1/\left(d_{i,j}/d_{safe}\right)^2 , \tag{9}$$

where *nline* is the number of discrete curve points, *nground* is the number of discrete mesh points of the solid boundary, $d_{i,j}$ is the distance between the corresponding nodes and curve points, while d_{safe} is a safety distance from the solid boundary. Term f_4 is designed to provide curves with a prescribed minimum curvature radius [23]. Weights w_i are experimentally determined, using as criterion the almost uniform effect of the last three terms in the objective function. Term $w_1 f_1$ has a dominant role in Eq. 8 providing feasible curves in few generations, since path feasibility is the main concern. The minimization of Eq. 8, through the EA procedure, results in a set of B-Spline control points, which actually represent the desired path.

Initially, the starting and ending path-line points are determined, along with the direction of flight. The limits of the physical space, where the vehicle is allowed to fly (upper and lower limits of their Cartesian coordinates), are also determined, along with the ground surface. The determined initial flight direction is used to compute the third fixed point close to the starting one; its position is along the flight direction and at a pre-fixed distance from the starting point.

The EA randomly produces a number of chromosomes to form the initial population. Each chromosome contains the physical coordinates of the free-to-move B-Spline control points. Using Eqs. 1 to 6, with a constant step of parameter u, a B-Spline curve is calculated for each chromosome of the population in the form of a sequence of discrete points. Subsequently, each B-Spline is evaluated, using the aforementioned criteria, and its objective function is calculated. Using the EA procedure, the population of candidate solutions evolves during the generations; at the last generation the population member with the smallest value of objective function is the solution to the problem and corresponds to the path line with the best characteristics according to the aforementioned criteria.

The simulation runs have been designed in order to search for path lines between "mountains". For this reason, an upper ceiling for flight height has been enforced, which is represented in the graphical environment by the horizontal section of the terrain. A typical simulation result is demonstrated in Fig. 2.

4 Coordinated UAV Path Planning

This section describes the development and implementation of an off-line path planner for Unmanned Aerial Vehicles (UAVs) coordinated navigation and collision avoidance in known static maritime environments. The problem formulation is described, including assumptions, objectives, constraints, objective function definition and path modeling.

4.1 Constraints and Objectives

The path planner was designed for navigation and collision avoidance of a small team of autonomous UAVs in maritime environments. Known and static environments are considered, characterized by the existence of a number of islands with short distances between them. The flight height is assumed to be almost constant, close to the sea-level, and the path-planning problem is formulated as a 2-D one. Having N UAVs launched from different known initial locations, the issue is to produce N 2-D trajectories, formed by B-Spline curves, with a desirable velocity distribution along each trajectory, aiming at reaching a predetermined target location, while ensuring collision avoidance either with the environmental obstacles or with the UAVs. Additionally the produced flight paths should satisfy specific route and coordination objectives and constraints. Each vehicle is assumed to be a point, while its actual size is taken into account by equivalent obstacle – ground growing.

The general constraint of the problem is the collision avoidance between UAVs and the ground. The route constraints are:

(a) Predefined initial and target coordinates for all UAVs
(b) Predefined initial and final velocity magnitudes for all UAVs, and
(c) Predefined minimum and maximum UAV velocity magnitudes during their flights.

Additionally, a single route objective is imposed: minimum path lengths, for maximizing the effective range of each vehicle. All three route constraints are explicitly taken into account by the optimization algorithm. The route objective is implicitly handled by the algorithm, through the definition of the objective function.

Besides route constraints and objective, coordination-relative constraints and objectives are imposed, which are implicitly handled by the algorithm, through the objective function definition. The coordination objectives used in this work are the following:

(a) Each UAV should arrive at the target, using a different path and a different approach vector, but the time of arrival for all UAVs should be as close as possible.

(b) Approaching the target from different directions. All angles between successive approaching directions should be as equal as possible, in order to assure an almost uniform distribution of UAVs around the target during their approach, for maximizing the probability of mission accomplishment.

The single coordination constraint is defined as keeping a minimum safety distance between UAVs, in order to ensure:

(a) collision avoidance between UAVs, and
(b) a spatial separation between the corresponding flight corridors, which, for some missions, increases the probability of survival.

4.2 Path Modeling Using B-Spline Curves

In this work each path is constructed using a B-Spline curve. Although the resulting curve in the physical space should be a 2-D one, 3-D B-Spline curves are utilized for the construction of each path. The two dimensions are used for the production of the x, y coordinates in the physical space of motion (horizontal plane), while the 3rd dimension corresponds to the velocity c along the path. For this reason, each B-Spline control point is defined by 3 numbers, corresponding to $x_{k,j}, y_{k,j}, c_{k,j} (k = 0, \ldots, n, j = 1, \ldots, N, N$ being the number of UAVs, while $n + 1$ is the number of control points in each B-Spline curve, the same for all curves). In this way a smooth variation of velocity c is defined along the path. The first ($k = 0$) and last ($k = n$) control points of the control polygon are the initial and target points of the jth UAV, which are predefined by the user. The corresponding velocities $c_{0,j}, c_{n,j}$ (launch and approaching velocities) are also predefined by the user.

The control polygon of each B-Spline curve is defined by successive straight line segments. For each segment, its length $seg\_length_{k,j}$, and its direction $seg\_angle_{k,j}$ are used as design variables ($k = 1, \ldots, n-1, j = 1, \ldots, N$). Design variables $seg\_angle_{k,j}$ are defined as the difference between the direction (in deg.) of the current segment and the previous one. For the first segment of each control polygon $seg\_angle_{1,j}$ is measured from x-axis. Additionally, the UAVs' velocities $c_{k,j}$ at each control point are used as design variables, except for the starting and target points (where they are predefined).

Using $seg\_length_{k,j}$ and $seg\_angle_{k,j}$ the coordinates of each B-Spline control point $x_{k,j}$ and $y_{k,j}$ can be easily calculated. The use of $seg\_length_{k,j}$ and $seg\_angle_{k,j}$ as design variables instead of $x_{k,j}$ and $y_{k,j}$ was adopted for two reasons. The first reason is the fact that abrupt turns of each flight path can be easily avoided by explicitly imposing short lower and upper bounds for the $seg\_angle_{k,j}$ design variables. The second reason is that by using the proposed design variables a better convergence rate was achieved compared to the case with the B-Spline control points' coordinates as design variables. The latter observation is a consequence of the shortening of the search space, using the proposed formulation. The lower and upper boundaries of each independent design variable are predefined by the user. Velocity boundaries depend on the flight envelope of each UAV. For the first segment of each control polygon $seg\_angle_{1,j}$ upper and lower boundaries can be selected such as to define an initial flight direction. Additionally, by selecting lower and upper boundaries for the rest of $seg\_angle_{k,j}$ variables close to 0 degrees (for example $-30°$ to $30°$), abrupt turns may be avoided.

4.3 Objective Function Formulation

The optimum flight path calculation for each UAV is formulated as a minimization problem. The objective (cost) function to be minimized is formulated as the weighted sum of five different terms

$$f = \sum_{i=1}^{5} w_i f_i, \tag{10}$$

where w_i are the weights and f_i are the corresponding terms described below.

Term f_1 corresponds to the single route objective of short flight paths and is defined as the sum of the non-dimensional lengths of all N flight paths (B-Spline curves)

$$f_1 = \sum_{j=1}^{N} l_j, \tag{11}$$

where l_j is the non-dimensional length of the jth path, given as

$$l_j = \frac{L_j}{\sqrt{(x_{target} - x_{0,j})^2 + (y_{target} - y_{0,j})^2}} - 1. \tag{12}$$

In Eq. 12 L_j is the length of the jth path, x_{target}, y_{target} are the coordinates of the target point and $x_{0,j}$, $y_{0,j}$ are the coordinates of the jth starting point. In Eq. 12, for the calculation of the non-dimensional length l_j, the distance between the starting and target points is subtracted, in order to obtain zero f_1 value for straight line paths.

Term f_2 is a penalty term, designed in order to materialize the general constraint of collision avoidance between UAVs and the ground. All N flight paths are checked whether or not pass through each one of the M ground obstacles. Discrete points are taken along each B-Spline path and they are checked whether or not they lie inside an obstacle. If this is true for a discrete point of the path line, a constant penalty is added to term f_2. Consequently, term f_2 is proportional to the number of discrete points that lie inside obstacles. Additionally, for each path line, a high penalty is added in case that even one discrete point of the corresponding path lies inside an obstacle.

Term f_3 was designed in order to take into account the second coordination objective, i.e. the target approach from different directions. For each flight

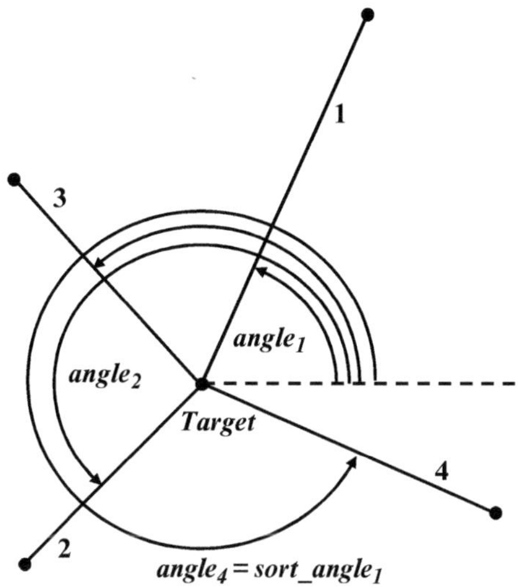

Fig. 3. Definition of azimuth angles, calculated for the last control polygon segment of each flight path

path the opposite to the flight direction azimuth angle of the last B-Spline control polygon segment is calculated as (Fig. 3)

$$angle_j = \begin{cases} \arctan(\Delta y/\Delta x) & if \quad \Delta y \geq 0 \; and \; \Delta x \geq 0 \\ 2\pi - \arctan(\Delta y/\Delta x) & if \quad \Delta y < 0 \; and \; \Delta x \geq 0 \\ \pi + \arctan(\Delta y/\Delta x) & if \quad \Delta x < 0 \end{cases} \tag{13}$$

$$\Delta y = y_{n-1,j} - y_{n,j}, \; \Delta x = x_{n-1,j} - x_{n,j}.$$

All calculated azimuth angles $angle_j$, $(j = 1, \ldots, N)$ are sorted in a descending order and stored as variables $sort\_angle_j$. An additional variable $sort\_angle_{N+1}$ is calculated as

$$sort\_angle_{N+1} = sort\_angle_1 - 2\pi. \tag{14}$$

Subsequently, the deference between two successive $sort\_angle_j$ is calculated as

$$\Delta sort\_angle_j = sort\_angle_j - sort\_angle_{j+1}, \; j = 1, \ldots, N, \tag{15}$$

where $\Delta sort\_angle_j$ is the angle between two successive flight paths, connected to the target point (Fig. 4). We define opt_angle as

$$opt\_angle = 2\pi/N. \tag{16}$$

Variable opt_angle denotes the optimum angle between successive B-Spline flight paths as UAVs are approaching the target, in order to have uniform distribution of UAVs around the target.

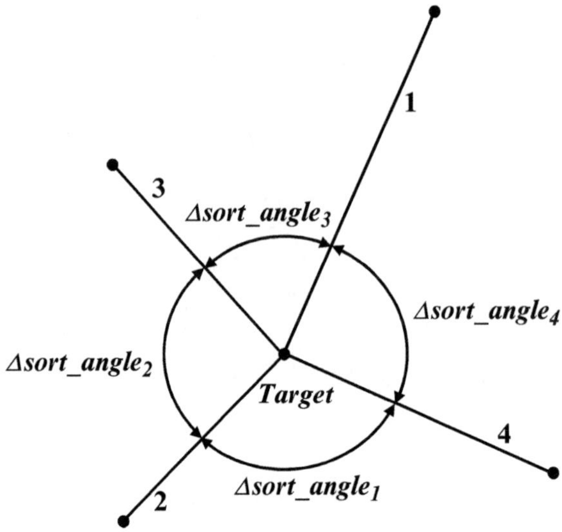

Fig. 4. Definition of $\Delta sort\_angle_j$

Term f_3 is then calculated as:

$$f_3 = \frac{\sum_{j=1}^{N} |opt\_angle - \Delta sort\_angle_j|}{ref\_angle}. \qquad (17)$$

In Eq. 17, ref_angle is a small reference angle which is used to provide a non-dimensional form of f_3 and takes a value equal to $\pi/20$.

Term f_4 is relevant to the single coordination constraint (keep a safety distance between UAVs), while term f_5 is relevant to the first coordination objective (arrival at target with minimum time intervals). For their calculation, a flight simulation is needed. Each candidate solution is defined by the corresponding design variables. Then the coordinates of all B-Spline control points are computed, while the coordinates and the velocities at the starting and target points are predefined by the user. Assuming a simultaneous launching of all UAVs at $t = 0$, a simulation of their flights is performed. According to B-Spline theory [36, 37], each curve is constructed in the physical space by giving specific values to the u parameter in the parametric space. Taking a constant increment of u, discrete points are computed along each curve, with the coordinates and velocity provided by the B-Spline function. Having the x, y coordinates and the UAV velocity in each discrete point, the time needed by the UAV to reach the next point can be easily computed. In this way, starting from the initial point at $t = 0$, a time of arrival can be assigned to each discrete point along each path. The time of arrival to the target for each UAV is stored in variable t_curr_j.

Taking a constant time step, linear interpolations between successive discrete points are performed, and the position of each UAV is calculated for a specific time step. Subsequently, the distances between all UAVs are calculated in each time step and in case that a distance is less than a predefined safety distance d_{safe}, a penalty is added to term f_4.

Term f_5 is calculated as

$$f_5 = \sum_{j=1}^{N} \left(t\_\max - t\_curr_j\right) / t\_\max \tag{18}$$

where t_max is the time of arrival of the last UAV. As the main objective is to obtain feasible paths, weights in Eq. 10 were experimentally determined in order term $w_2 f_2$ dominate the rest.

5 The Optimization Procedure

5.1 Differential Evolution Algorithm

In this work, Differential Evolution (DE) [40, 41] is used as the optimization tool. DE is an extremely simple to implement EA, which has demonstrated better convergence performance than other EAs. Differential Evolution algorithm represents a type of Evolutionary Strategy, especially formed in such a way, so that it can effectively deal with continuous optimization problems, often encountered in engineering design, being a recent development in the field of optimization algorithms. The classic DE algorithm evolves a fixed size population, which is randomly initialized. After initializing the population, an iterative process is started and at each iteration (generation), a new population is produced until a stopping condition is satisfied. At each generation, each element of the population can be replaced with a new generated one. The new element is a linear combination between a randomly selected element and a difference between two other randomly selected elements. Below a more analytical description of the algorithm's structure is presented.

Given an objective function

$$f_{obj}(X) : \mathrm{R}^{n_{param}} \to \mathrm{R}, \tag{19}$$

the optimization target is to minimize the value of this objective function by optimizing the values of its parameters (design variables)

$$X = \left(x_1, x_2, \ldots, x_{n_{param}}\right), \; x_j \in \mathrm{R}, \tag{20}$$

where X denotes the vector composed of n_{param} objective function parameters (design variables). These parameters take values between specific upper and lower bounds

$$x_j^{(L)} \le x_j \le x_j^{(U)}, \; j = 1, \ldots, n_{param}. \tag{21}$$

The DE algorithm implements real encoding for the values of the objective function's parameters. In order to obtain a starting point for the algorithm, an initialization of the population takes place. Often the only information available is the boundaries of the parameters. Therefore the initialization is established by randomly assigning values to the parameters within the given boundaries

$$x_{i,j}^{(0)} = r \cdot \left(x_j^{(U)} - x_j^{(L)}\right) + x_j^{(L)}, \; i = 1, \ldots, n_{pop}, \; j = 1, \ldots, n_{param}, \tag{22}$$

where r is a uniformly distributed random value within range [0, 1]. DE's mutation operator is based on a triplet of randomly selected different individuals. A new parameter vector is generated by adding the weighted difference vector between the two members of the triplet to the third one (the donor). In this way a perturbed individual is generated. The perturbed individual and the initial population member are then subjected to a crossover operation that generates the final candidate solution

$$x_{i,j}^{\prime(G+1)} = \begin{cases} x_{C_i,j}^{(G)} + F \cdot \left(x_{A_i,j}^{(G)} - x_{B_i,j}^{(G)}\right) \; if \; (r \le C_r \vee j = k) \; \forall j = 1, \ldots, n_{param} \\ x_{i,j}^{(G)} \qquad otherwise\,, \end{cases} \tag{23}$$

where $x_{C_i,j}^{(G)}$ is called the "donor", G is the current generation,

$$\begin{array}{l} i = 1, \ldots, n_{pop}, \; j = 1, \ldots, n_{param} \\ A_i \in [1, \ldots, n_{pop}]\,, \; B_i \in [1, \ldots, n_{pop}]\,, \; C_i \in [1, \ldots, n_{pop}] \\ A_i \neq B_i \neq C_i \neq i \\ C_r \in [0, 1]\,, \; F \in [0, 1+]\,, \; r \in [0, 1]\,, \end{array} \tag{24}$$

and k a random integer within [1, n_{param}], chosen once for all members of the population. The random number r is seeded for every gene of each chromosome. F and C_r are DE control parameters, which remain constant during the search process and affect the convergence behaviour and robustness of the algorithm. Their values also depend on the objective function, the characteristics of the problem and the population size.

The population for the next generation is selected between the current population and the final candidates. If each candidate vector is better fitted than the corresponding current one, the new vector replaces the vector with which it was compared. The DE selection scheme is described as follows (for a minimization problem)

$$X_i^{(G+1)} = \begin{cases} X_i^{\prime(G+1)} \; if \; f_{obj}\left(X_i^{\prime(G+1)}\right) \le f_{obj}\left(X_i^{(G)}\right) \\ X_i^{(G)} \qquad otherwise\,. \end{cases} \tag{25}$$

A new scheme [42] to determine the donor for mutation operation is used, for accelerating the convergence rate. In this scheme, donor is randomly selected (with uniform distribution) from the region within the "hyper-triangle", formed by the three members of the triplet. With this scheme the donor comprises the local information of all members of the triplet, providing a better starting-point for the mutation operation that result in a better distribution of the trial-vectors. As it is reported in [42], the modified donor scheme accelerated the DE convergence rate, without sacrificing the solution precision or robustness of the DE algorithm.

The random number generation (with uniform probability) is based on the algorithm presented in [43], which computes the remainder of divisions involving integers that are longer than 32 bits, using 32-bit (including the sign bit) words. The corresponding algorithm, using an initial seed, produces a new seed and a random number. In each different operation inside the DE algorithm that requires a random number generation, a different sequence of random numbers is produced, by using a different initial seed for each operation and a separate storage of the corresponding produced seeds. By using specific initial seeds for each operation, it is ensured that the different sequences differ by 100,000 numbers.

5.2 Radial Basis Function Network for DE Assistance

Despite their advantages, EAs ask for a considerable amount of evaluations. In order to reduce their computational cost several approaches have been proposed, such as the use of parallel processing, the use of special operators and the use of surrogate models and approximations. Surrogate models are auxiliary simulations that are less physically faithful, but also less computationally expensive than the expensive simulations that are regarded as "truth". Surrogate approximations are algebraic summaries obtained from previous runs of the expensive simulation [44, 45]. Such approximations are the low-order polynomials used in Response Surface Methodology [46, 47], the kriging estimates employed in the design and analysis of computer experiments [48], and the various types of Artificial Neural Networks [45]. Once the approximation has been constructed, it is typically inexpensive to use.

DE has been demonstrated to be one of the most promising novel EAs, in terms of efficiency, effectiveness and robustness. However, its convergence rate is still far from ideal, especially when it is applied in optimization problems with time consuming objective functions. In order to enhance the convergence rate of DE algorithm, an approximation model is used for the objective function, based on a Radial Basis Functions Artificial Neural Network [49]. In general a RBFN (Fig. 5), is a three layer, fully connected feed-forward network, which performs a nonlinear mapping from the input space to the hidden space ($\mathrm{R}^L \rightarrow \mathrm{R}^M$), followed by a linear mapping ($\mathrm{R}^M \rightarrow \mathrm{R}^1$) from the hidden to the output space (L is the number of input nodes, M is the number of hidden nodes, while the output layer has a single node).

The corresponding output $yy(xx)$, for an input vector xx=[xx_1, xx_2,...,xx_L] is given

$$yy\,(xx) = \sum_{i=1}^{M} w_i \cdot \varphi_i\,(xx). \tag{26}$$

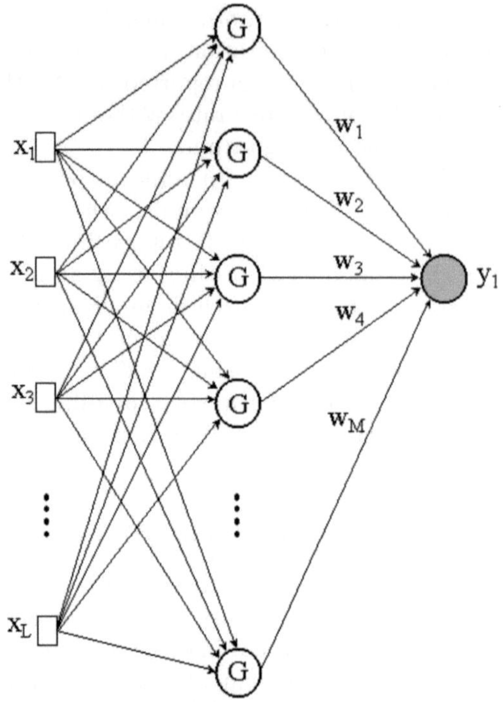

Fig. 5. A Radial Basis Function Artificial Neural Network

where $\varphi_i(xx)$ is the output of the ith hidden unit

$$\varphi_i(xx) = G(\|xx - cc_i\|), \; i = 1, \ldots, M. \tag{27}$$

The connections (weights) to the output unit (w_i, i=1,...,M) are the only adjustable parameters. The RBFN centers in the hidden units cc_i, i=1,...,M are selected in a way to maximize the generalization properties of the network. The nonlinear activation function G in our case is chosen to be the Gaussian radial basis function

$$G(u, \sigma) = \exp\left(-u^2/\sigma^2\right), \tag{28}$$

where σ is the standard deviation of the basis function.

The selection of RBFN centers plays an important role for the predictive capabilities and the generalization of the network. There are several strategies that can be adopted concerning the selection of the radial-basis functions centers in the hidden layer, while designing a RBFN. Haykin refers to the following [49]: a) Random selection of fixed centers, which is the simplest approach and the selection of centers from the training data set is a sensible choice, given that the latter is adequately representative for the problem at hand. b) Self-organized selection of centers, where appropriate locations for the centers are estimated with the use of a clustering algorithm whose assignment is to partition the training set in homogeneous subsets. c) Supervised selection of centers, which is the most generalized form of a RBFN since the location of the centers undergo a supervised learning process along with the rest of the network's free parameters.

The standard process is to select the input vectors in the training set as RBFN centers. In this case results $M=NR$, where NR is the number of training data. For large training sets (resulting in large M values) this choice is expected to increase storage requirements and CPU cost. Additionally, the $M=NR$ choice could lead to over-fitting and/or bad generalization of the network. The proposed solution [49, 45] is the selection of $M<NR$ and consequently the search for sub-optimal solutions, which will provide a better generalizing capability to the network.

As far as training is concerned, there are two different approaches, the direct and the iterative learning. In our case the first approach was adopted. The direct learning process is based on a matrix formulation of the governing equations of RBF network. The presentation of the network with the NR input patterns allows the formulation of a $(NR \times M)$ matrix H, which becomes square in the special case when $NR=M$. Each line in the interpolation matrix H corresponds to a learning example and each column to a RBFN center. The output unit values result in the form of the matrix product:

$$H\,(NR \times M)\,w\,(M \times 1) = yy\,(NR \times 1)\,, \tag{29}$$

where yy is the desired output vector as provided by the training dada set, and w is the synaptic weights vector, which consists of M unknowns to be computed.

A possible way for inverting H is through the Gram-Schmidt technique. H is first decomposed as

$$H = Q\,R, \tag{30}$$

with Q and R being $(NR \times M)$ and $(M \times M)$ matrices respectively, where R is upper triangular and

$$Q^{\mathrm{T}}Q = diag\,(1, 1, \ldots, 1)\,. \tag{31}$$

After the computation of Q and R matrices, the weights vector can be computed using back-substitution in

$$R(M \times M)w(M \times 1) = Q^{T}(M \times NR)yy(NR \times 1) \tag{32}$$

There are several reasons why one should choose RBFN as the approximation model; Haykin [49] offers comparative remarks for RBFNs and Multilayer Perceptrons (MLPs). However, the main reason for choosing RBFNs is their compatibility with the adopted local approximation strategy, as it is described in the subsequent section. The use of relatively small numbers of training patterns i.e. small networks, helps creating local range RBFNs. That in turn allows the inversion of matrix H to use almost negligible CPU time and the approximation error is kept very small. We should keep in mind that the computing cost associated with the use of neural networks is the cost of training the networks, whereas the use of a trained network to evaluate a new individual adds negligible computation cost [45].

5.3 Using RBFN for Accelerating DE Algorithm

In each DE generation, during the evaluation procedure, each trial vector must be evaluated and then compared with the corresponding current vector, in order to select the better-fitted between them to pass to the next generation. The concept is to replace the costly exact evaluations of trial vectors with fast inexact approximations, and at the same time maintain the robustness of the DE algorithm. During the evaluation phase, each trial vector is pre-evaluated, using the approximate model. If it is pre-evaluated as lower-fitted (higher objective function in minimization problems) than the corresponding vector of the current population, then no further exact evaluation is needed and the current vector is transferred to the next generation, while the trial vector is abandoned. In case the trial vector is pre-evaluated as better fitted than the corresponding current vector, then an exact re-evaluation takes place after the pre-evaluation, along with a new comparison between the two vectors. If the trial vector is still better-fitted than the current vector, then the trial vector passes to the next generation. Otherwise the current vector is the one that will pass to the next generation. Additionally, a small percentage of the candidate solutions, are selected with uniform probability to be exactly evaluated, without taking into account their performance provided by the approximation model. In the first two generations, all vectors are exactly evaluated. According to the afore mentioned procedure, only exactly evaluated trial vectors have the opportunity to pass to the new generation, so the current population always comprises exactly evaluated individuals. In this way, one part of the comparison (the current vector) is always an exact-evaluated vector, and this enhances the robustness of the procedure.

The result of each evaluation (exact or inexact), along with the corresponding chromosome, are stored in a database. In order to have a local approximation model, only the best-fitted individuals of database entries are used in each generation to re-train the RBFN. In this way the approximation model evolves with the population and uses only the useful information for approximating the objective function. The surrogate model predictions replace exact and costly evaluations only for the less-promising individuals, while the more-promising ones are always exactly evaluated.

6 Simulation Results

The same artificial environment was used for all the test cases considered, with different starting and target points. The (experimentally optimized) settings of the Differential Evolution algorithm were as follows: *population size* = 50,

$F = 0.99$, $C_r = 0.85$. The algorithm was defined to terminate after 700 generations, although feasible solutions can be reached in less than 30 generations. The large number of generations was used in order to compare the convergence behavior between the original DE algorithm and the RBFN assisted one. For the 4 test cases presented here, 3 free-to-move control points were used for each B-Spline path, resulting in a total number of control points equal to 5 for each B-Spline curve (along with the fixed starting and target points). For 3 different paths (corresponding to 3 UAVs) and 3 free-to-move control points for each path, a total number of 27 design variables are needed ($seg\_length_{k,j}$, $seg\_angle_{k,j}$ and $c_{k,j}$, for each path$j$ and each control point k).

Figures 6 to 9 present simulation results for the four different test cases, using the RBFN assisted DE. For all test cases safety distance d_{safe} was set equal to 12.5% of the length of each side of the rectangular terrain. For all test cases, term f_4 of the cost function converged to zero, indicating no violation of the safety distance constraint. Concerning the time intervals between the first and the last arrival to the target, for all the test cases considered this time interval was kept less than about 3% of the flight duration (0.71% for the 1st case, 3.08% for the 2nd case, 1.33% for the 3rd case and 1.41% for the 4th case). As it can be observed, term f_3 of the fitness function managed to produce uniform distribution of UAVs around the target for all cases considered. Even for the fourth test case a uniform distribution of UAV paths around the target was achieved, although the target point was positioned very close to an obstacle (island coast).

As it has been already stated, the main reason for introducing the RBFN surrogate model was to speed-up the optimization procedure. However, as

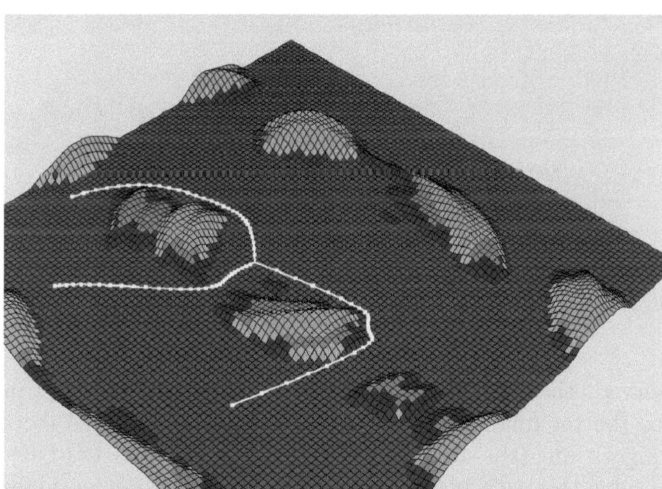

Fig. 6. The first test case for the coordinated UAV path planning

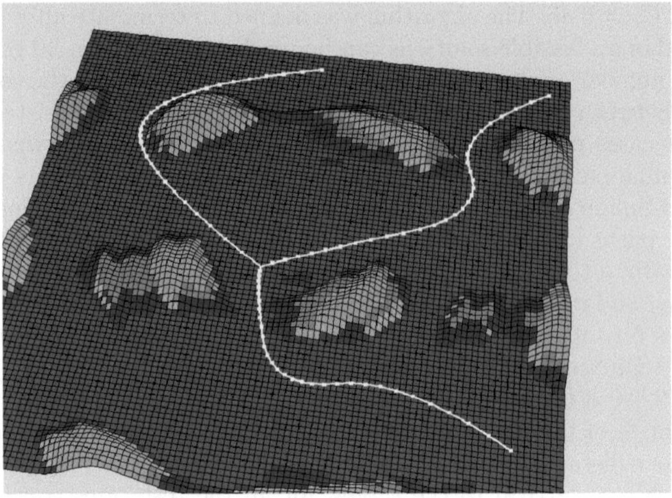

Fig. 7. The second test case for the coordinated UAV path planning

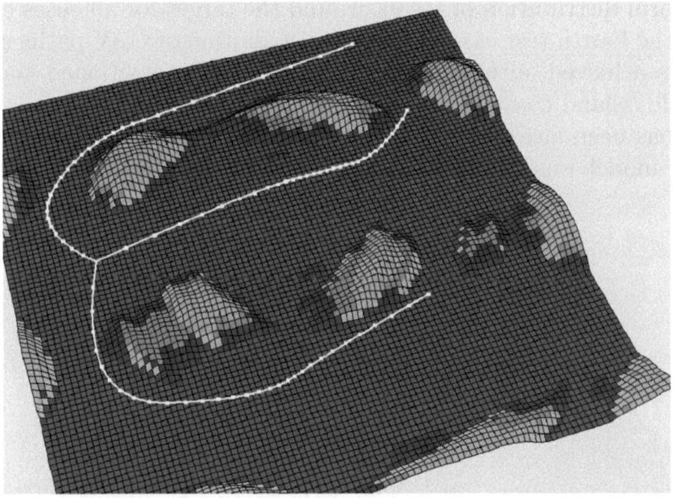

Fig. 8. The third test case for the coordinated UAV path planning

it was observed, the introduction of RBFN assistance resulted in a deeper convergence (better final value of fitness function), compared to the original DE. Both algorithms (the original DE and the RBFN assisted DE) were used in order to solve the path planning problem for the aforementioned four test cases, using the same parameters. In order to compare the effect of RBFN

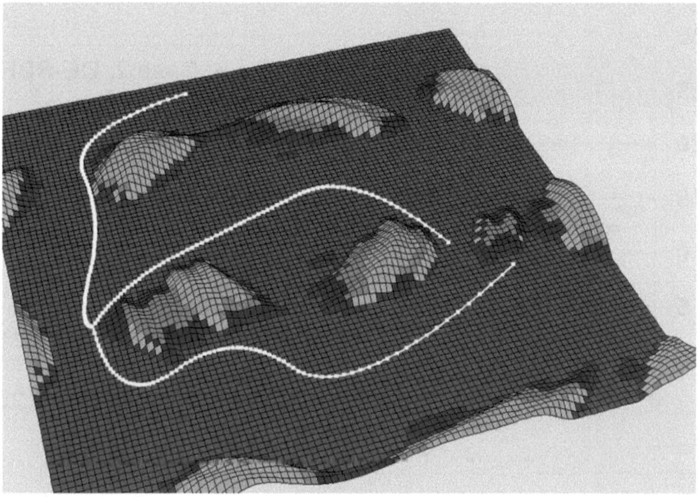

Fig. 9. The fourth test case for the coordinated UAV path planning

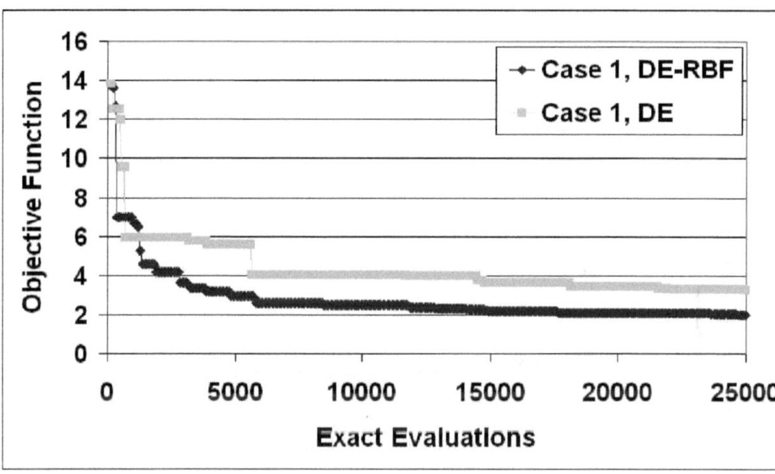

Fig. 10. Convergence histories for the first test case, with and without the use of the RBFN assistance

surrogate model, the convergence histories for the four test cases of the DE algorithm (with and without the RBFN assistance) are presented in Fig. 10 to 13. As it can be observed, the adoption of the approximation model resulted in a considerable reduction in the number of exact evaluations for a specific fitness value. This reduction, for all cases considered, reached or exceeded 80% of the number of exact evaluations with respect to the case without

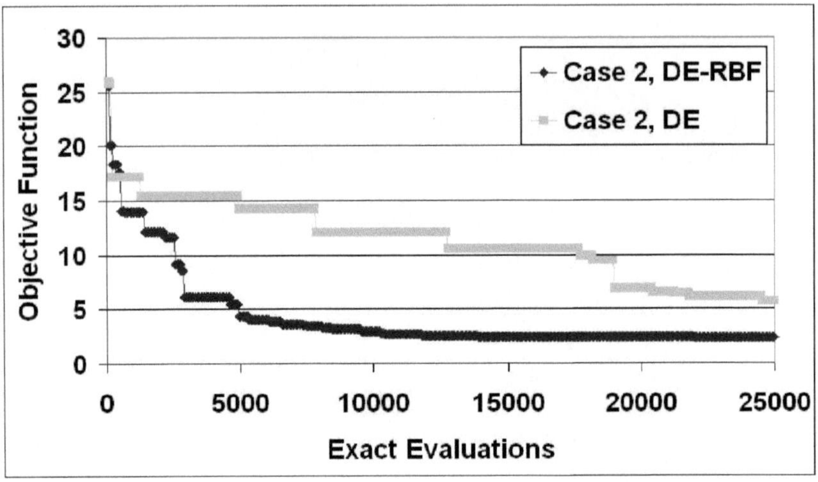

Fig. 11. Convergence histories for the second test case, with and without the use of the RBFN assistance

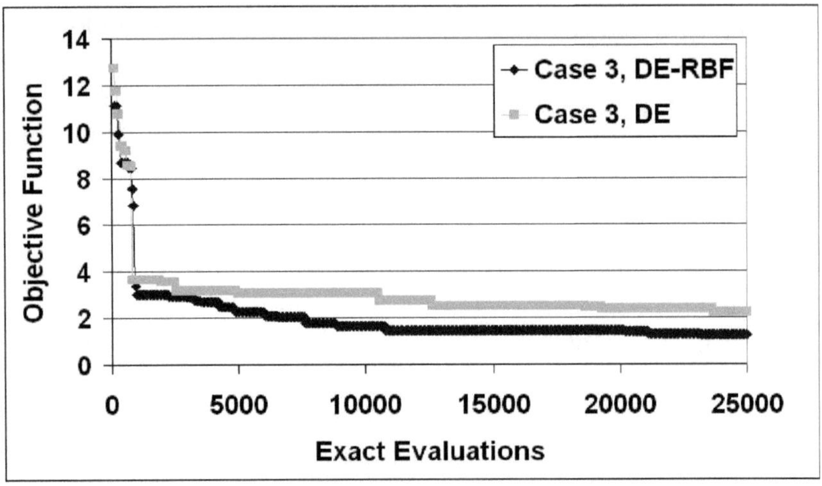

Fig. 12. Convergence histories for the third test case, with and without the use of the RBFN assistance

RBFN assistance. As the introduction of the RBFN has a minor effect on the computation time, this 80% reduction in the number of exact evaluations results in a speedup factor approximately equal to 5 to the whole computation procedure, which is very important for real world applications of such kind.

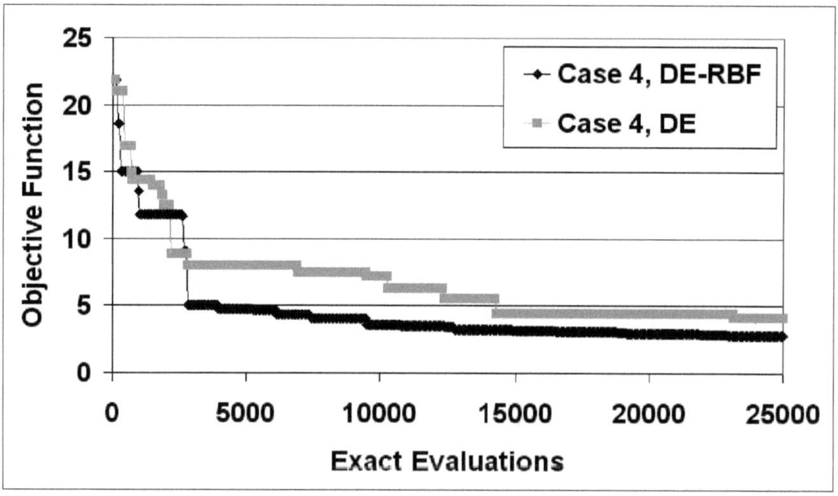

Fig. 13. Convergence histories for the fourth test case, with and without the use of the RBFN assistance

7 Conclusions

This work is an extension of a previous one, which used Differential Evolution in order to find optimal paths of coordinated UAVs, with the paths being modeled with straight line segments. Although very satisfactory results were achieved, the main drawback of the previous approach was the need of a large number of segments for complicated paths, resulting in a large number of design variables. However, as the number of design variables increases, the dimensionality of the optimization problem also increases; consequently, much more generations are needed for a converged solution, which is not always affordable for real world applications.

In this work an off-line path planner for UAVs coordinated navigation and collision avoidance in known static maritime environments was presented. The problem was formulated as a single-objective optimization one, with the objective function being the weighted sum of different terms, which correspond to various objectives and constraints of the problem. B-Spline curves were adopted in order to model the 2-D flight paths, as they provide the ability to produce complicated paths with a small number of control variables. In this way the number of design variables, and the dimensionality of the optimization problem, can be kept small. The velocity distribution along each flight path was also modeled using the B-Spline formulation. A Radial Basis Function Artificial Neural Network was introduced in the Differential Evolution algorithm (the optimizer) to serve as a surrogate model and decrease the number of costly exact evaluations of the objective function. The RBF Network managed to considerably reduce the DE computation time and to provide deeper convergence to the optimization procedure.

The path planner was tested in a simulated environment, and the simulation results demonstrated the ability of the algorithm to produce near optimal paths without violating the imposed constraints. The adoption of the B-Spline formulation provided the ability to keep the number of design variables as small as possible, and at the same time produce reasonable and smooth paths, without abrupt turns.

Future work will be focused on the development of an on-line path planner for coordination of a team of UAVs. The methodology that will be used for the planner will be a combination of this work and the work presented in [23], where an on-line path planner for a single UAV was presented, which is able to gradually produce B-spline paths in an unknown 3D environment.

7.1 Trends and challenges

The military market for UAVs has demonstrated a strong positive trend during the past decade, with the corresponding commercial market showing a similar behavior, although not so strong [1]. This trend is expected to continue, as the technology provides new solutions to the problems of autonomous navigation of UAVs, and new ideas are emerging about the roles and tasks that can be assigned to UAVs. The trend is supported by the large number of research teams that are working in the field. In particular, the field of UAV cooperation gained increased interest during the past years due to the advantages of using a team of UAVs instead of a single one to accomplish a complicated mission. However, because of the youth of the field, the research has been "scenario" oriented, and a rigorous formalism is still missing. Work is still needed in the direction of clarifying: a) the assumptions about the systems of cooperating UAVs, b) the terminology used to describe the various problems under consideration, c) the different categories of working scenarios, d) the objectives and constraints for each problem, e) the "best practices" that can be adopted for specific problems or sub-problems.

The research in the field of cooperating UAVs is highly interdisciplinary, and knowledge from different science and technology fields is needed, even from the beginning of the formulation of the problem under consideration. It would be highly beneficially for the researchers, especially for those working in more theoretical fields, to collaborate with possible users of the systems or methodologies under development. Some of the problems, relevant to UAV cooperation, that gain high interest are: a) the assignment of multiple tasks to a team of UAVs, b) the path planning problem of cooperating UAVs in the presence of various cooperation and mission constraints, c) information exchange and data fusion between the cooperating vehicles, d) cooperative sensing of targets, e) cooperative sensing of the environment, f) centralized versus distributed coordination methodologies, especially for cases with communication problems between the vehicles, g) on-line mission rescheduling and task reassignment for fault-tolerant systems.

References

1. Newcome, L.R.: Unmanned Aviation, a Brief History of Unmanned Aerial Vehicles. AIAA (2004)
2. Latombe, J.-C.: Robot Motion Planning. Kluwer Academic Publishers (1991)
3. LaValle, S.M.: Planning Algorithms. Cambridge University Press (2006)
4. Gilmore, J.F.: Autonomous vehicle planning analysis methodology. Proceedings of the Association of Unmanned Vehicles Systems Conference. Washington, DC (1991) 503–509
5. Uny Cao, Y., Fukunaga, A.S., Kahng, A.B.: Cooperative Mobile Robotics: Antecedents and Directions. Autonomous Robots 4 (1997) 7-27
6. Fujimura, K.: Motion Planning in Dynamic Environments. Springer-Verlag, New York, NY, (1991)
7. Arai, T. and Ota, J. 1992. Motion planning of multiple robots. Proceedings of the IEEE/RSJ IROS (1992) 1761–1768
8. Shima, T., Rasmussen, S.J., Sparks, A.G.: UAV Cooperative Multiple Task Assignments using Genetic Algorithms. Proceedings of the 2005 American Control Conference, June 8-10, Portland, OR, USA (2005)
9. Shima, T., Rasmussen, S.J., Sparks, A.G.: UAV Team Decision and Control using Efficient Collaborative Estimation. Proceedings of the 2005 American Control Conference, June 8-10, Portland, OR, USA (2005)
10. Mitchell, J.W. and Sparks, A.G.: Communication Issues in the Cooperative Control of Unmanned Aerial Vehicles. Proceedings of the Forty-First Annual Allerton Conference on Communication, Control, & Computing (2003)
11. Schumacher, C.: Ground Moving Target Engagement by Cooperative UAVs. Proceedings of the 2005 American Control Conference, June 8-10, Portland, OR, USA (2005)
12. Moitra, A., Mattheyses, R.M., Hoebel, L.J., Szczerba, R.J., Yamrom, B.: Multivehicle reconnaissance route and sensor planning. IEEE Transactions on Aerospace and Electronic Systems, 37 (2003) 799–812
13. Bortoff, S.: Path planning for UAVs. Proceedings of the Amer. Control Conf., Chicago, IL, (2000) 364–368
14. Szczerba, R.J., Galkowski, P., Glickstein, I.S., and Ternullo, N.: Robust algorithm for real-time route planning. IEEE Transactions on Aerospace Electronic Systems 36 (2000) 869–878
15. Zheng, C., Li, L., Xu, F., Sun, F., Ding, M.: Evolutionary Route Planner for Unmanned Air Vehicles. IEEE Transactions on Robotics 21 (2005) 609–620
16. Beard, R.W., McLain, T.W., Goodrich, M.A., Anderson, E.P.: Coordinated target assignment and intercept for unmanned air vehicles. IEEE Transactions on Robotics and Automation, 18 (2002) 911–922
17. Vandapel, N., Kuffner, J., Amidi, O.: Planning 3-D Path Networks in Unstructured Environments. Proceedings of the IEEE International Conference on Robotics and Automation, ICRA (2005)
18. Dubins, L.: On curves of minimal length with a constraint on average curvature, and with prescribed initial and terminal position. American Journal of Math. 79 (1957) 497–516.
19. Shima, T., Schumacher, C.: Assignment of cooperating UAVs to simultaneous tasks using Genetic Algorithms. AIAA Guidance, Navigation, and Control Conference and Exhibit, San Francisco (2005)

20. Tang, Z., and Ozguner, U.: Motion Planning for Multi-Target Surveillance with Mobile Sensor Agents. IEEE Transactions on Robotics 21 (2005) 898-908
21. Martinez-Alfaro H., and Gomez-Garcia, S.: Mobile robot path planning and tracking using simulated annealing and fuzzy logic control. Expert Systems with Applications 15 (1988) 421–429
22. Nikolos, I.K., Tsourveloudis, N., and Valavanis, K.P.: Evolutionary Algorithm Based 3-D Path Planner for UAV Navigation. CD-ROM Proceedings of the 9th Mediterranean Conference on Control and Automation, Dubrovnik, Croatia (2001)
23. Nikolos, I.K., Valavanis, K.P., Tsourveloudis, N.C., Kostaras, A.: Evolutionary Algorithm based offline / online path planner for UAV navigation. IEEE Transactions on Systems, Man, and Cybernetics – Part B: Cybernetics 33 (2003) 898–912
24. Mettler, B., Schouwenaars, T., How, J., Paunicka, J., and Feron E.: Autonomous UAV guidance build-up: Flight-test Demonstration and evaluation plan. Proceedings of the AIAA Guidance, Navigation, and Control Conference, AIAA-2003-5744 (2003)
25. Richards, A., Bellingham, J., Tillerson, M., and How., J.: Coordination and control of UAVs. Proceedings of the AIAA Guidance, Navigation and Control Conference, Monterey, CA, (2002)
26. Schouwenaars, T., How, J., and Feron, E.: Decentralized Cooperative Trajectory Planning of multiple aircraft with hard safety guarantees. Proceedings of AIAA Guidance, Navigation, and Control Conference and Exhibit, AIAA-2004-5141 (2004)
27. Flint, M., Polycarpou, M., and Fernandez-Gaucherand, E.: Cooperative Control for Multiple Autonomous UAV's Searching for Targets. Proceedings of the 41st IEEE Conference on Decision and Control (2002)
28. Gomez Ortega, J., and Camacho, E.F.: Mobile Robot navigation in a partially structured static environment, using neural predictive control. Control Eng. Practice 4 (1996) 1669–1679
29. Kwon, Y.D., and Lee, J.S.: On-line evolutionary optimization of fuzzy control system based on decentralized population. Intelligent Automation and Soft Computing 6 (2000) 135–146
30. Michalewicz, Z.: Genetic Algorithms + Data Structures = Evolution Programs. Springer Publications (1999)
31. Smierzchalski, R.: Evolutionary trajectory planning of ships in navigation traffic areas. Journal of Marine Science and Technology 4 (1999) 1–6
32. Smierzchalski, R., and Michalewicz Z.: Modeling of ship trajectory in collision situations by an evolutionary algorithm. IEEE Transactions on Evolutionary Computation 4 (2000) 227–241
33. Sugihara, K., and Smith, J.: Genetic Algorithms for Adaptive Motion Planning of an Autonomous Mobile Robot. Proceedings of the 1997 IEEE International Symposium on Computational Intelligence in Robotics and Automation, Monterey, California (1997) 138–143
34. Sugihara, K., and Yuh, J.: GA-based motion planning for underwater robotic vehicles. UUST-10, Durham, NH (1997)
35. Nikolos, I.K., Brintaki, A.: Coordinated UAV Path Planning Using Differential Evolution. Proceedings of the 13th Mediterranean Conference on Control and Automation, IEEE, Limassol, Cyprus (2005)

36. Piegl, L., Tiller, W.: The NURBS Book. Springer (1997)
37. Farin, G.: Curves and Surfaces for Computer Aided Geometric Design, A Practical Guide. Academic Press (1988)
38. Goldberg, D.E.: Genetic Algorithms in Search, Optimization and Machine Learning. Addison-Wesley (1989)
39. Holland, J.H.: Adaptation in Natural and Artificial Systems. The MIT Press (1992)
40. Storn, R., and Price, K.: DE - a Simple and Efficient Adaptive Scheme for Global Optimization over Continuous Space. ICSI, Technical Report TR-95-012 (1995)
41. Price, K.V., Storn, R.M., Lampinen, J.A.: Differential Evolution, a Practical Approach to Global Optimization. Springer-Verlag, Berlin Heidelberg (2005)
42. Hui-Yuan F., Lampinen J., Dulikravich G.S.: Improvements to Mutation Donor Formulation of Differential Evolution. Proceedings of EUROGEN 2003 conference on Evolutionary Methods for Design, Optimization and Control, Applications to Industrial and Societal Problems, CIMNE, Barcelona (2003)
43. Marse, K. and Roberts, S.D.: Implementing a portable FORTRAN uniform (0,1) generator. Simulation (1983) 41–135
44. Torczon, V., Trosset, M.W.: Using Approximations to Accelerate Engineering Design Optimization. NASA/CR-1998-208460, ICASE Report No. 98-33 (1998)
45. Giannakoglou, K.C.: Design of optimal aerodynamic shapes using stochastic optimization methods and computational intelligence. Progress in Aerospace Sciences 38 (2002) 43–76
46. Myers, R.H., Montgomery, D.C.: Responce Surface Methodology: Progress and Product in Optimization Using Designed Experiments. Wiley – Interscience, New York (1995)
47. Shyy, W., Papila, N., Vaidynathan, R., Tucker, K.: Global Design Optimization for Aerodynamics and Rocket Propulsion Components. Prog. Aerospace Sci. 37 (2001) 59–118
48. Ratle, A.: Optimal Sampling Strategies for Learning a Fitness Model. Proceedings of the 1999 Congress on Evolutionary Computation (CEC99), Washington DC, USA (1999)
49. Haykin, S.: Neural Networks, a Comprehensive Foundation. Second Edition, Prentice Hall (1999)

Evolution-based Dynamic Path Planning for Autonomous Vehicles

Anawat Pongpunwattana and Rolf Rysdyk

Autonomous Flight Systems Laboratory
University of Washington, Seattle, WA 98195

Planning is an essential element of autonomous systems. This work presents a dynamic path planning algorithm for an unmanned autonomous vehicle to execute a set of assigned tasks in a changing environment. This problem comprises path planning and task sequencing. The approach adopted here is to solve these subproblems simultaneously using an evolutionary planning algorithm and a stochastic model of the environment. During the mission, the planner replans and adapts the path in response to changes in the environment. Simulation results demonstrate that the path planning algorithm can compute feasible effective solutions to path planning problems. These include planning with timing constraints and dynamic planning with moving targets and obstacles. The vehicle is able to autonomously travel from the initial location to the goal location while avoiding obstacles and performing the assigned tasks.

1 Introduction

The path planning algorithm presented here was developed as a part of the Evolution-based Cooperative Planning System (ECoPS) [19]. The ECoPS is a distributed system for real-time task and path planning for a team of autonomous vehicles. The planning algorithms are based on the combination of a market-based planning architecture and optimization techniques called Evolutionary Computation (EC). The planning system was successfully demonstrated for the Defense Advanced Research Projects Agency under the Mixed Initiative Control of Automa-teams program. In related work, it is scheduled for flight testing on Seascan Unmanned Aerial Vehicles (UAVs) manufactured by The Insitu Group in combination with autonomous Unmanned Surfaces Vehicles (USVs), see Figure 1.

The overall goal of this work is to increase autonomy of unmanned vechicles. A vehicle is called autonomous if it has the ability to plan its own actions using the acquired information about its environment to accomplish its tasks.

A. Pongpunwattana and R. Rysdyk: *Evolution-based Dynamic Path Planning for Autonomous Vehicles*, Studies in Computational Intelligence (SCI) **70**, 113–145 (2007)
www.springerlink.com

Fig. 1. The Insitu Group Seascan UAV (Top) and the Northwind Marine Seafox USV (Bottom)

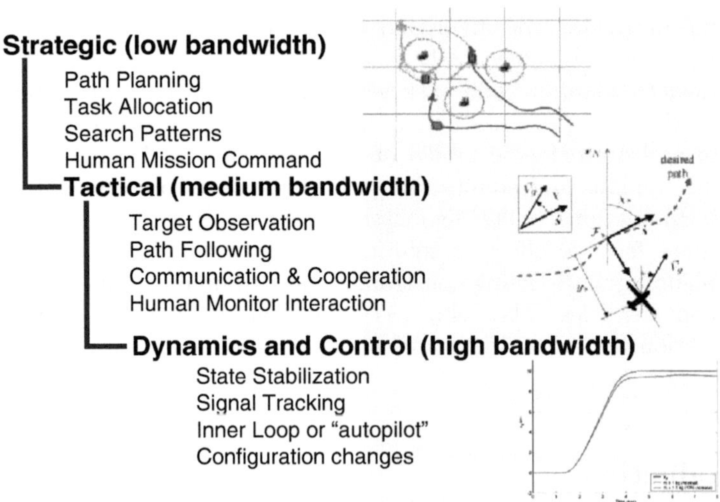

Fig. 2. Hierarchy of vehicle autonomy

Figure 2 shows three layers in the design of autonomy for air vehicles. The stability and control, as well as the guidance aspects, of air vehicles are well established, but strategic decision making is not. A major challenge in improving autonomy of unmanned vehicles is strategic planning. The focus of this work is on path planning. This problem is not just planning a path between two target locations, but an optimal path to visit all the targets. This is essentially a mission planning problem for a single vehicle. The problem is composed of path planning and task sequencing between various target locations.

There are several approaches to path planning. A popular one is graph-based search. In this approach, as described by Murphy [16], the environment in which the vehicle operates is discretized and represented by a graph composed of a number of nodes linked together with arcs. Each node corresponds to a location in the environment and an arc links two adjacent nodes. There is a cost associated with each arc. A path in this graph consists of a series of connected arcs. One advantage of this approach is the compactness of the topological map of the environment represented by the graph. The path planning problem is to find the optimal path from one node to another in the graph. The planning algorithm minimizes the total cost which is the sum of the cost of each arc in the path. A graph search algorithm such as Dijkstra's algorithm or the A* algorithm [17] can be used to find the optimal path. Thrun [25], Mata and Mitchell [15], Mandow [14], and Bander [2] have all worked on this graph-based planning concept and applied it to many scenarios.

Another approach to path planning is *probabilistic roadmap planning* (PRM). It is an efficient method to compute collision-free paths for vehicles with many degrees of freedom [12]. This method consists of two phases, a building phase and a query phase. The building phase is the construction of a graph called roadmap. The nodes in the roadmap are collision-free configurations and the edges linking the nodes are collision-free paths which enable a robot to move from one configuration to another. The query phase is finding a path between an initial and goal configurations by connecting these configuration nodes to the road map and searching the roadmap for a sequence of edges linking the two nodes. This method was originally developed for holonomic robots in a static environment. Overmars [18] applied this technique to simple holonomic robots and non-holonomic robots having constrained kinematics and high degrees of freedom. LaValle and Kuffner [13] proposed a randomized path planning technique similar to PRM for robots having high degree-of-freedom with both kinematic and dynamic constraints. Song et al. [24] proposed a new method of building and querying probabilistic roadmaps to improve the performance of the planning process.

The common drawback of the above approaches is the requirement to repeatedly update the graph representing the environment if it is changing rapidly. Evolution-based approach is suitable in this situation. It does not require a graph-based representation of the environment. Furthermore, evolutionary algorithms are continual adaptation techniques favorable for dynamic path planning. They can run continuously during the execution of the plans and handle changes in the operating environment and the vehicle capabilities. Evolution-based techniques respond to the changing environment quickly because they do not have to recompute the entire plans. The current plan is adapted in response to the changes. Works in this approach have been investigated by several researchers [9], [26], [6], [10], and [20].

Dynamic path planning is typically solved using a hierarchical approach such as Brumitt [3] and Chien et al. [7]. In such approach, the planning system is divided into two levels, task sequencing and local path planning. The path planner provides optimal local paths for traveling between each pair of target locations and between the current vehicle location and all target locations. These optimal paths must also satisfy all the given constraints such as avoiding obstacles in the environment. The cost information of this set of local paths is then used during the evaluation of candidate overall paths. This process is performed by the mission planner which determines the optimal sequence of target locations for the vehicle to visit during the mission. The planning at this level is a combinatorial optimization problem. This approach is shown to be effective in a static environment. However, it may not be suitable in dynamic environments where the targets and obstacles may move during the mission. The changes in the environment force the planning system to frequently update the local paths prior to the optimization process for task sequencing. The process of updating the local paths is computationally extensive because it often needs to update most if not all of the local paths joining all target locations.

In this paper, we present an evolution-based dynamic path planning algorithm to compute paths for a vehicle to reach a set of targets while avoiding obstacles in the environment. The planning algorithm searches for the optimal path that maximizes an objective function and satisfies all the constraints imposed on the problem. The planner must be able to compute an effective path off-line prior to the start of the mission and dynamically adapt the path in response to changes in the environment during the mission.

The remaining sections are organized as follows. Section 2 describes the concept of dynamic path planning and a stochastic model of the system used to formulate the planning problem. In Section 3, we present a technique to approximate the probability of intersection of a vehicle and a site in the environment. This technique addresses the collision avoidance problem, either with static or moving objects such as other vehicles. The planning algorithm is presented in Section 4. Example problems of static and dynamic path planning are also given. Section 5 and 6 show the ability of the planning algorithm to handle timing constraints and a changing environment. Finally Section 7 provides a summary and conclusion of the work presented here.

2 Dynamic Path Planning

Dynamic path planning is a continual process which generates a new path by adapting previously computed paths for future motion. This concept is illustrated in Figure 3. We define a *spawn point* as a time where the planner updates the output trajectory. During the course of the mission, the planner continually computes an updated path which starts at the next closest spawn point. In the figure, the closest spawn point is located at time t_{s_p} which is the *execution time horizon* of the vehicle controller. This spawn point separates the committed section and the adapting section of the planned trajectory previously computed at time $t_{s_{p-1}}$. The committed section, which will not be altered, is a portion of the path for the time period $t_k < t < t_{s_p}$. The adapting section of the trajectory remains free to be further refined by the planner.

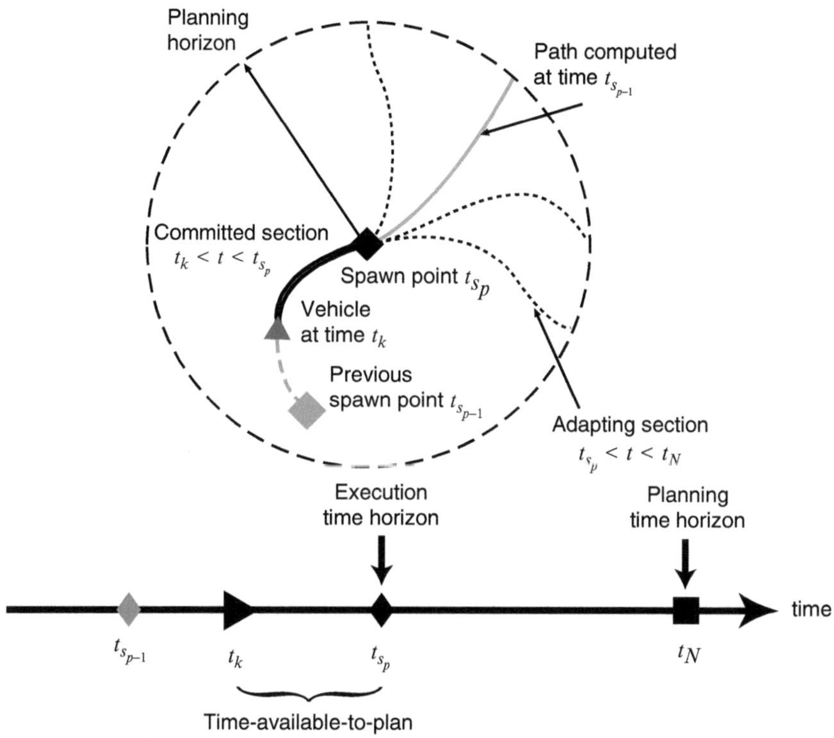

Fig. 3. Illustration of the concept of dynamic path planning. A vehicle (at time t_k) is moving along a trajectory previously computed at time $t_{s_{p-1}}$ which is shown as a gray line. The next closest spawn point at time t_{s_p} is shown as a black diamond

The time t_N at the end point of this trajectory segment is the *planning time horizon*. It can either be fixed or a variable whose value is specified by the planner. An update trajectory is sent to the vehicle's controller for execution every time the vehicle reaches a spawn point. The planner will start computing a new trajectory starting at the next spawn point. The time difference between two adjacent spawn points ($\Delta T_s = (t_{s_p} - t_{s_{p-1}})$) specifies the maximum time available to plan for the planner to update its path. In this manner, the planner can incorporate any new information about the environment that becomes available during the mission. The trade-off between time available to plan and adaptability is important. One would want the non-adapting committed sections of the planned path to be short in a highly dynamic environment. However, that shortens the time available to plan. The shorter time available to plan requires faster planning algorithms. Hence, dynamic planning algorithms must be fast enough to compute new plans within the specified time limit.

The concept of this dynamic planning process is very similar to that of the *Model-based Predictive Control* [4]. Figure 4 illustrates the block diagram of a

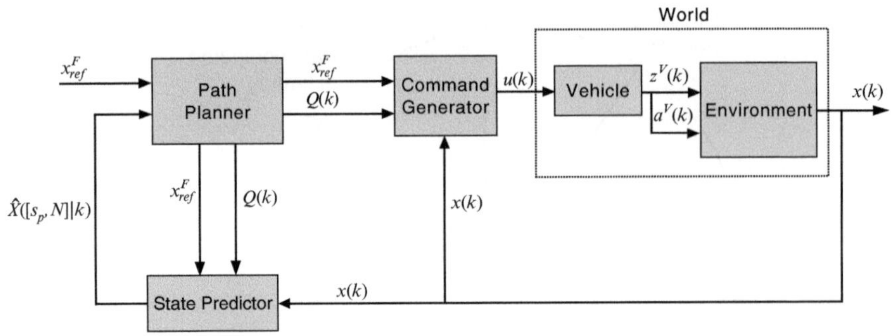

Fig. 4. Block diagram of a dynamic path planning system

dynamic path planning system. Like model-based predictive control systems, the current state information is periodically fed back to the state predictor. Given a set of assigned tasks, a candidate path and the current environment information, the state predictor provides the planner predicted future states of the system using a dynamic model. Using this prediction, the planner computes a path that optimizes an objective function which is a function of the future system states. The path together with the information of the assigned tasks is then sent to the command generator to generate commanded positions and payload actions for the vehicle. This sequence is repeated in the next time period. The quality of computed paths depends on the quality of the model. The difficulty of the problem hinges on how the objective function and the constraints on the input variables are defined. The rest of this section describes the model and the objective function used in this work. To simplify our notation in the following equations describing the model, any sampled signal $s(t_k)$ at time t_k, where k can assume any non-negative integer value, is written as $s(k)$.

The variables shown in Figure 4 have the following meanings:

$u(k)$ = commanded inputs to the vehicle at time t_k
$x(k)$ = states of the world at time t_k
$x^V(k)$ = states of the vehicle at time t_k
$x^E(k)$ = states of the environment at time t_k
$x^F(k)$ = task states at time t_k
x^F_{ref} = desired values of the task states
$z^V(k)$ = location of the vehicle at time t_k
$a^V(k)$ = action of the vehicle at time t_k
$Q(k)$ = trajectory for time $t_k \leq t \leq t_N$
$\hat{X}([s_p, N]|k)$ = predicted states of the world at time $t_{s_p}, t_{s_p}+1, \ldots, t_N$ given information observed at time t_k

In real-world applications, environment information is often known with a limited level of certainty. It is practical to account for this uncertainty in the planning algorithm by using a stochastic model. The model presented here is for 2-dimension problems, but it can be extended to 3-dimension problems. The system considered here consists of a vehicle and its environment. We call this system the *world*. The vehicle is assigned to perform N_T tasks.

For a planning time horizon t_N, the world model used to predict future states during the time $t_k < t \leq t_N$ can be written in a discrete form as

$$x(q+1) = f(x(q), u(q)), \quad q = k, k+1, \ldots, N-1 \tag{1}$$

where f is the state transition function, x is the state vector of the system which includes states of the vehicle x^V, states of the environment x^E, and task states x^F. That is $x = [x^V, x^E, x^F]^T$. We assume that the information of all states at time t_k is available to the planner except x^E which is known with a limited level of certainty. In the following equations, the expected value of a random variable y will be simply written as $\tilde{y}$.

The state of each task i, x_i^F, indicates whether the task is completed; $x_i^F = 1$ when the task is initially assigned, and $x_i^F = 0$ if it is completed. The states of the vehicle consists of its position z^V, velocity $\dot{z}^V$, and health state ξ^V. That is $x^V = [z^V, \dot{z}^V, \xi^V]^T$. The health state indicates if the vehicle is intact or destroyed; $\xi^V = 1$ if the vehicle is intact, and $\xi^V = 0$ if it is destroyed.

We define *sites* as any objects in the environment. The states of the environment are the states of all the sites. *Obstacles* are special types of sites with ability to change the states of vehicles if they become in contact. A *target* is defined as a site in the environment associated with a task. Thus, a site can be a target and an obstacle simultaneously, or it can be neither. The number of obstacles N_O plus the number of targets N_G is not necessary equal to the number of all the sites N_S. The states of the environment are composed of the position z^E, velocities $\dot{z}^E$ and health states ξ^E of all the sites. That is $x^E = [z^E, \dot{z}^E, \xi^E]^T$. The health state of site j indicates whether the site exists or not; $\xi_j^E = 1$ if the site exists, and $\xi_j^E = 0$ if it does not exist.

The input to the vehicle u includes a commanded position $\bar{z}^V$ and a commanded velocity $\bar{\dot{z}}^V$ of the vehicle, and a task assignment vector $\bar{d}^V$. That is $u = [\bar{z}^V, \bar{\dot{z}}^V, \bar{d}^V]^T$. Given a trajectory $Q(s_{p-1})$ previously computed at time $t_{s_{p-1}}$, the commanded position and velocity while $t_q \in (t_k, t_N]$ are given by

$$\begin{aligned} \bar{z}^V(q) &= h_x(Q(s_{p-1}), q) \\ \bar{\dot{z}}^V(q) &= h_v(Q(s_{p-1}), q) \end{aligned} \tag{2}$$

where Q represents the set of parameters needed to define the planned trajectory. The mapping functions h_x and h_v, and Q, depend on how the trajectory is encoded. The task assignment vector $\bar{d}^V$ is a $N_T \times 1$ vector whose element i is equal to one if $\left|x_{ref,i}^F - x_i^F\right| > 0$, otherwise it is zero.

Using the following equations, we describe the stochastic model used to predict the expected values of all system states $\tilde{x} = \left[\tilde{z}^V, \tilde{\dot{z}}^V, \tilde{\xi}^V, \tilde{z}^E, \tilde{\dot{z}}^E, \tilde{\xi}^E, \tilde{x}^F\right]^T$.

The feedback information of the states $x(k)$ known at time t_k and the commanded inputs $u(q) = [\bar{z}^V(q+1),\ \bar{\dot{z}}^V(q+1),\ \bar{d}^V(q+1)]^T$ for all $q \in \{k, k+1, \ldots, N-1\}$ is given. We assume that the vehicle's guidance system can follow its commanded trajectory. Thus, the predicted position and velocity of the vehicle is equivalent to the commanded inputs:

$$\begin{aligned} \tilde{z}^V(q+1) &\equiv \bar{z}^V(q+1) \\ \tilde{\dot{z}}^V(q+1) &\equiv \bar{\dot{z}}^V(q+1) \end{aligned} \tag{3}$$

for all $q \in \{k, k+1, \ldots, N-1\}$. Assuming the location of each obstacle is independent of the location of all other obstacles, the dynamic propagation of the expected health state $\tilde{\xi}^V$ of the vehicle is given by

$$\tilde{\xi}^V(q+1) = \tilde{\xi}^V(q) \prod_{j=1}^{N_o} \left(1 - \tilde{B}_j^v(q+1)\tilde{\xi}_j^O(q)\eta_j^O\right) \tag{4}$$

Here $\tilde{B}_j^v(q+1)$ is the probability that the vehicle collides or intersects with an obstacle j during the time $t_q < t \leq t_{q+1}$. The variable $\tilde{\xi}_j^O$ is the expected value of the health state of obstacle j, and η_j^O is the effectiveness of the obstacle j in destroying a vehicle if they make contact. The value of η_j^O is in the range $[0,\ 1]$. $\tilde{B}_j^v(q+1)$ is a function of $\tilde{z}^V, \tilde{\dot{z}}^V, \tilde{z}^E$ and the environment uncertainty parameter vector σ^x. The details of how to compute $\tilde{B}_j^v$ are given in Section 3.

The position of site $j \in \{1, 2, \ldots, N_S\}$ at time t_k is a random variable which can be written as

$$z_j^E(k) = \tilde{z}_j^E(k) + \varepsilon_j^x \tag{5}$$

where $\tilde{z}_j^E$ is the expected value of the position. ε_j^x is assumed to be a random variable with zero mean with a probability density function

$$\rho_j^x\left(x, \sigma_j^x\right) = \begin{cases} 1/(\pi\left(\sigma_j^x\right)^2), & \|x\| \leq \sigma_j^x \\ 0, & \|x\| > \sigma_j^x \end{cases} \tag{6}$$

Here σ_j^x is a given parameter specifying the uncertainty radius of site j. The area within the circle with center location $\tilde{z}_j^E$ and radius σ_j^x contains all possible locations of the site.

The velocity of site $j \in \{1, 2, \ldots, N_S\}$ known at time t_k is also a random variable expressed by

$$\dot{z}_j^E(k) = \tilde{\dot{z}}_j^E(k) + \varepsilon_j^v \tag{7}$$

where $\tilde{\dot{z}}_j^E$ is the expected value of the velocity and ε_j^v is assumed to be a random variable with zero mean and a probability density function

$$\rho_j^v\left(v, \sigma_j^v\right) = \begin{cases} 1/(\pi\left(\sigma_j^v\right)^2), & \|v\| \leq \sigma_j^v \\ 0, & \|v\| > \sigma_j^v \end{cases} \tag{8}$$

σ_j^v is a given parameter specifying the bound of the uncertainty in the velocity.

In the model presented here, each site is assumed to maintain constant velocity at all times. Therefore, the expected velocity of site j while $t_q \in (t_k, t_N]$ is given by

$$\dot{\tilde{z}}_j^E(q) = \dot{\tilde{z}}_j^E(k) \tag{9}$$

As a result, the dynamic propagation of the expected position of site j is given by

$$\tilde{z}_j^E(q+1) = \tilde{z}_j^E(q) + \dot{\tilde{z}}_j^E(k)\Delta t \tag{10}$$

for all $q \in \{k, k+1, \ldots, N-1\}$ and $\Delta t = t_{q+1} - t_q$. The dynamic equation of the uncertainty radius σ_j^x is given by

$$\sigma_j^x(q+1) = \sigma_j^x(q) + \sigma_j^v(k)\Delta t \tag{11}$$

The dynamic propagation of the expected value of the state of task i, $i \in \{1, 2, \ldots, N_T\}$, is described by

$$\tilde{x}_i^F(q+1) = \tilde{x}_i^F(q)\left(1 - \tilde{B}_v^i(q+1)\tilde{\xi}^V(q)\eta^V\right) \tag{12}$$

where $\tilde{B}_v^i$ is the probability that the path of vehicle v intersects the target location z_i^G associated with task i during the time $t_q < t \leq t_{q+1}$. The details of how to compute $\tilde{B}_v^i$ are given in Section 3. η^V is the effectiveness of the vehicle in performing the task. Using the Equations 3 to 12, we can compute the expected values of all states at time $t_q \in (t_k, t_N]$.

To formulate the planner's objective function, we define a variable $\tilde{R}_i(q)$ as the task score the vehicle will have at time t_q as a result of executing task i. This task score is used as a measure of success of the mission. The expected task score of task i for $q \in \{1, 2, \ldots, N\}$ is given by

$$\tilde{R}_i(q+1) = \tilde{R}_i(q) + \alpha_i^F(q)\left(\tilde{x}_i^F(q) - \tilde{x}_i^F(q+1)\right); \quad \tilde{R}_i(0) = 0 \tag{13}$$

Substituting Equation 12 into Equation 13, we obtain

$$\tilde{R}_i(q+1) = \tilde{R}_i(q) + \alpha_i^F(q)(\tilde{x}_i^F(q)\left(\tilde{B}_v^i(q+1)\tilde{\xi}^V(q)\eta^V\right) \tag{14}$$

where $\alpha_i^F(q)$ is the score weighting factor for task i. It can either be a constant or a time dependent function. This function is used to define a time window requirement for the vehicle to execute each task.

At any time $t_k < t_{s_p}$, the goal of the planner is to find a path that maximizes the predicted total score obtained by completing each task while minimizing the predicted operation cost during the time $t_{s_p} < t \leq t_N$. The objective function can be written as

$$\tilde{J} = \sum_{i=1}^{N_T}\left(\tilde{R}_i(N) - \tilde{R}_i(s_p)\right) - \tilde{C}(Q(s_p)) \tag{15}$$

where $\tilde{C}$ is the function used to compute the expected operation cost of the vehicle traveling along its planned path $Q(s_p)$ during time $t_{s_p} < t \leq t_N$. Substituting Equation 14 into Equation 15, the objective function becomes

$$\tilde{J} = \sum_{i=1}^{N_T} \sum_{q=s_p}^{N-1} \alpha_i^F(q)\tilde{x}_i^F(q)\left(\tilde{B}_v^i(q+1)\tilde{\xi}^V(q)\eta^V\right) - \tilde{C}(Q(s_p)) \tag{16}$$

The cost function $\tilde{C}$ is defined as

$$\tilde{C}(Q(s_p)) = \alpha^V\left(\tilde{\xi}^V(s_p) - \tilde{\xi}^V(N)\right) + \alpha^Q(1 - F(N)) \tag{17}$$

where α^V is the vehicle cost weighting factor, and α^Q is the path cost weighting factor. $F(N)$ is the ratio of the amount of fuel remaining in the vehicle's fuel tank at time t_N to the full capacity of the tank. The weighting factors α_i^F, α^V, and α^Q are parameters whose values are set based on the operator's assessment of the task accomplishment, cost of vehicle loss, and fuel consumption.

From the description of the stochastic world model provided above, we can see that the states of the world, which include task states, vehicle states, and states of the environment, are coupled. For example, the probability of success in executing a task at a time during the mission depends on the probability of survival of the vehicles at that time. This, in turn, depends on the states of the obstacles intersecting with the vehicles along the path. Hence, to evaluate the utility function $\tilde{J}$ given in Equation 16, it is necessary to run a simulation to predict the expected values of the world states at each of the discretized time steps.

3 Probability of Intersection

The position of each site i is assumed to be known with limited certainty. It is described by the expected value $\tilde{z}_i^E$ and a probability density function ρ_i^x. An intersection of a site and a vehicle is predicted probabilistically. This section presents the approach proposed by Rathbun and Capozzi [20] to approximate the probability of intersection of a vehicle and a site with limited knowledge about the location of the site.

In general, a collision or intersection of two objects has an effect on both objects. In our model, the effects of an intersection depend on the *effective ranges* of the site and the vehicle. We define the effective range of a site as the radius of the circular area where the site can affect the states of any vehicle within that area. The effective range of a vehicle is defined as the radius of the circular area where the vehicle can affect the states of any site being within that area. For any pair of site i with an effective range R_i^O and a path Q_v of a vehicle v with an effective range R_v^V, we define the *site-to-vehicle intersection region* as the region in space Z_i^v of all points within a distance R_i^O from the path Q_v,

$$Z_i^v = \{z : \|z - s\| < R_i^O, \forall s \in S\} \tag{18}$$

where S is the set of all points along the path Q_v. If the center location of the site, z_i^E, is within the site-to-vehicle intersection region, the states of the vehicle can be changed by the site. The *vehicle-to-site intersection region* is defined as the region in space Z_v^i of all points within a distance R_v^V from the path Q_v,

$$Z_v^i = \{z : \|z - s\| < R_v^V, \forall s \in S\} \tag{19}$$

If the center location of the site is within the vehicle-to-site intersection region, the states of the site can be changed by the vehicle.

Let denote ρ_i the probability density function of the position of site i in 2-dimension space. The probability that site i hits vehicle v is the integral of the density function over the site to vehicle intersection region:

$$\tilde{B}_i^v = \int \int_{Z_i^v} \rho_i(z) dx dy \tag{20}$$

The probability that vehicle v hits site i is the integral of the density function over the vehicle-to-site intersection region:

$$\tilde{B}_v^i = \int \int_{Z_v^i} \rho_i(z) dx dy \tag{21}$$

We do not know the solutions to the Equation 20 and 21 for a piecewise set of path segments generated by our path planner. Instead, a computationally efficient technique is used to approximate the probability of intersection. We call this method, Field Integral Approximation [20]. In this method, we define two *probability density fields* β_i^v and β_v^i related to the probability density function ρ_i of the site. The probability density field β_i^v is defined such that the probability that site i hits vehicle v can be approximated by integrating the probability density field along the path Q_v,

$$\tilde{B}_i^v = \int_{Q_v} \beta_i^v ds \tag{22}$$

Likewise, the probability density field β_v^i is defined the same way. The probability that vehicle v hits site i can then be written in the form

$$\tilde{B}_v^i = \int_{Q_v} \beta_v^i ds \tag{23}$$

The remaining question is how to determine the fields β_i^v and β_v^i that provide a good approximation of the exact probabilities. Work by Rathbun and Cappozi [20] suggests an approximation method for a site having constant probability density at all points located at the same distance from the site's expected center location. In this method, the probability density field at a distance r from the center location of the site is defined by a function which provides the correct probability of intersection given a path of a single revolution about the expected center location of the site at the distance r. The two probability density fields are given by

$$\beta_i^v(r) = \frac{p_i(r + R_i^O) - p_i(r - R_i^O)}{2\pi r} \tag{24}$$

and

$$\beta_v^i(r) = \frac{p_i\left(r + R_v^V\right) - p_i\left(r - R_v^V\right)}{2\pi r} \tag{25}$$

where $p_i(r)$ is the probability that the center location of site i is within a distance r of the expected location $\tilde{z}_i^E$,

$$p_i(r) = \int_0^r 2\pi y \rho_i^x(y) dy \tag{26}$$

where ρ_i^x is the probability density function of the position of the site. For $r < 0$, we define

$$p_i(r) = -p_i(|r|), \ r < 0 \tag{27}$$

For a site with a probability density function,

$$\rho_i^x(x, \sigma_i^x) = \begin{cases} 1/\left(\pi \left(\sigma_i^x\right)^2\right), & \|x - \tilde{z}_i^E\| \leq \sigma_i^x \\ 0, & \|x - \tilde{z}_i^E\| > \sigma_i^x \end{cases} \tag{28}$$

the probability distribution function $p(r)$ is given by

$$p_i(r) = \begin{cases} r^2/\left(\sigma_i^x\right)^2, & r \leq \sigma_i^x \\ 1, & r > \sigma_i^x \end{cases} \tag{29}$$

For simple types of paths, it may be possible to determine a closed form solutions to the Equation 22 and 23. For complex types of paths, the probability of site i encountering vehicle v during time $t_q < t \leq t_{q+1}$ can be approximated as a summation of the probability density field β_i^v along the path which can be written as

$$\tilde{B}_i^v(q+1) = \sum_{k=0}^{M-1} \beta_i^v(\|\tilde{z}_v^V(T_k) - \tilde{z}_i^E(T_k)\|)\dot{\tilde{z}}_v^V(T_k)\Delta T \tag{30}$$

Here $\tilde{z}_v^V(T_k)$ is the expected position of the vehicle moving along on the path Q_v at time t_{T_k} with $t_{T_0} = t_q$ and $t_{T_M} = t_{q+1}$. $\tilde{z}_i^E(T_k)$ is the expected center location of the site at time t_{T_k}. $\dot{\tilde{z}}_v^V(k)$ is the velocity of the vehicle at time t_{T_k}. ΔT is the time step of the numerical integration, and $\Delta T = T_{k+1} - T_k$. M is the number of steps used to discretize the time period. The probability of vehicle v hitting site i during time $t_q < t < t_{q+1}$ can be approximated by

$$\tilde{B}_v^i(q+1) = \sum_{k=0}^{M-1} \beta_v^i(\|\tilde{z}_v^V(T_k) - \tilde{z}_i^E(T_k)\|)\dot{\tilde{z}}_v^V(T_k)\Delta T \tag{31}$$

4 Planning Algorithm

In this section, we describe an evolutionary algorithm for solving the dynamic path planning problem presented in Section 2. This problem basically is iteratively finding a path represented by the variable $Q(s_p)$ that maximizes the objective function given in Equation 16.

The algorithm presented here can be used to compute both the static and dynamic path planning problems. Static or off-line path planning is a process used to compute an optimal path connecting the initial position of the vehicle to the goal location prior to the start of a mission. Dynamic path planning is an iterative process. This planning problem is dynamic because it changes as the vehicle moves along the last updated path. In a dynamic environment, the locations of objects in the environment may also change. We will first describe the algorithm for static path planning. Then we extend the static path planning algorithm to be used for dynamic path planning.

4.1 Algorithm for Static Planning

The path planning algorithm presented here is based on the algorithm described in Rathbun et al. [21]. An evolutionary algorithm is used to solve the underlying optimization problem. The general concept of the planning algorithm is illustrated in Figure 5. The algorithm runs in a loop which has three phases. It starts by randomly generating a population of encoded plans. Then it evaluates the fitness value of each plan. The next step is to select the best plan to be the candidate solution for this generation. Using a selection scheme, a portion of the plans in the population are selected to be the parents of the next generation based on their fitness values. The last step is to produce offspring from the parents selected in the previous phase. An offspring is generated by cloning a parent and applying a mutation mechanism to it or by crossover of two parents. This loop is run continuously to update the plan as the optimization process proceeds.

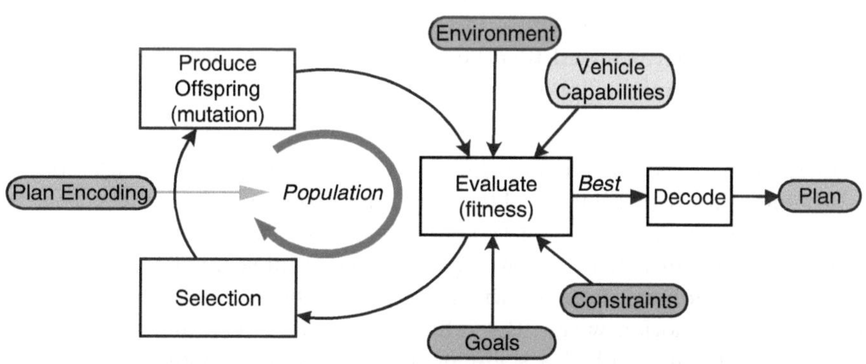

Fig. 5. Overview of evolutionary planning algorithms

The planning algorithm used in the Evolution-based Cooperative Planning System (ECoPS) is based on both Genetic Algorithms (GA) [11] and Evolutionary Programming (EP) [8]. Work by Capozzi [5] suggests that the algorithm combining features of both paradigms can improve the performance of the optimization process for path planning problems. The design of evolutionary path planning algorithms involves the following issues: path encoding, fitness evaluation, mutation mechanisms, and selection scheme. The following subsections describes these issues of path planning problems for a single autonomous vehicle. Extensions to multiple cooperating UAVs are given in [1].

Path Encoding

One of the key issues of applying evolutionary computation to a given problem is finding a good representation of each individual in the population which adequately represents a candidate solution to the problem. In path planning problems, the search space Ω which consists of the parameters of the chosen path representation must be defined. These parameters must be sufficient to specify a candidate path in spatial and/or temporal space. Examples of path representations are:

- *Sequence of waypoints:* A sequence of waypoints including the initial and goal locations are used to represent a path. A path represented by these waypoints is the linked straight-line segments connecting the initial location, all the waypoints in the sequence, and the goal location.
- *Sequence of velocity vectors:* A path is presented by a time sequence of velocity vectors whose elements are speed, heading and climb angle. Using a fixed time interval, this representation allows the planning algorithm to constrain the candidate solution to lie within the acceleration capabilities of the vehicle.
- *Sequence of maneuvers:* A path is represented by a time sequence of primitive maneuvers which are achievable by the vehicle. Examples of maneuvers are speed-up, slow-down, turn-left, turn-right, and so on. The time duration of each maneuver can be varied.
- *Sequence of motion primitives:* In this representation, a path is a chain of motion primitives linked together end to end. Examples of motion primitives are a straight line segment, constant radius curve, constant climb angle line segment.

In ECoPS, a path is encoded as a sequence of simple *segments* chained end-to-end, shown in Figure 6. The locations of the vehicle and the goal are shown as a blue triangle and a green circle respectively. In this approach, for 2-dimension problems, there are two elemental types of segments, straight lines and constant radius curves, shown in Figure 7. The segment parameters are limited to keep motion within the vehicle capabilities. Continuity is enforced between the joining end of a segment and the starting point, heading, and speed of another joining segment. We also enforce every path to end at the

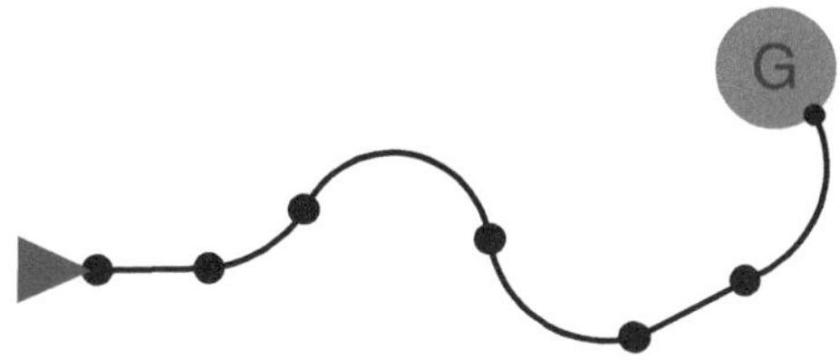

Fig. 6. An encoded path which is composed of a chain of connected segments

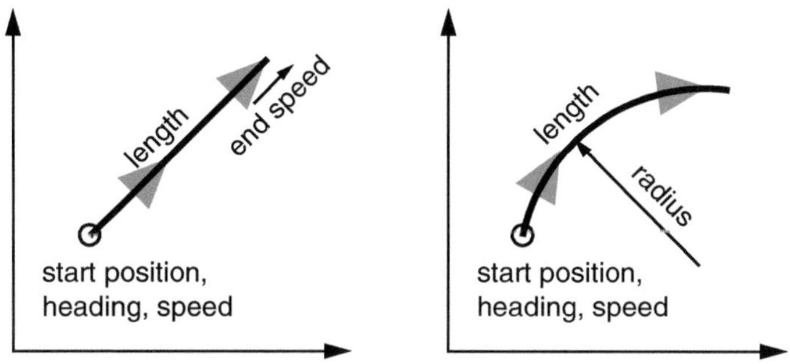

Fig. 7. Elemental path segment types

goal location by adding a number of segments at the end of the last segment to extend the path to the goal location. The *go-to-goal* segments are added to a new path after it is created.

Fitness Evaluation

Fitness of a candidate path is the value which represents the performance measure of the path based on the objectives given by the problem. The fitness of individuals in the population must be determined during the evaluation process. Individuals with higher fitness have more chance to survive during the selection process. The fitness function is a parameterized function which is used to evaluate fitness of each candidate path. Typically, the fitness function is a weighted linear combination of parameterized terms which represent the mission specifications and constraints. The quality of the resulting path depends heavily on the fitness function. Examples of basic components in the fitness function are:

- *RangeGoal:* Distance from the terminal point of a trial path to the goal location.
- *RangeTargets:* Distance of the closest approach between a trial path and the assigned targets.
- *ObstacleIntersection:* A measure of the degree to which a trial path intersects with any known obstacles in the field of operation.
- *FuelConsumption:* The amount of fuel required for the vehicle to travel along a trial path.
- *PathLength:* The cumulative length of a trial path.

In ECoPS, the fitness function is a complex function designed to capture the dynamics and uncertainties in the system. The fitness value of a candidate path is defined as the inverse of the *loss function* given by

$$L = \alpha^L \left(1 - \frac{\tilde{J}}{\alpha^F_{sum}}\right) \tag{32}$$

where $\tilde{J}$ is the objective function given in Equation 16. α^L is a scaling factor, and α^F_{sum} is defined by

$$\alpha^F_{sum} = \sum_{i=1}^{N_T} \alpha^F_{i,max} \tag{33}$$

and

$$\alpha^F_{i,max} = \max\{\alpha^F_i(k), k = 0, 1, \ldots, N\} \tag{34}$$

where $\alpha^F_i(k)$ is time-dependent score weighting function of task i at time step k. If there is no assigned task or all the tasks are completed, the loss function is given by

$$L = \alpha^L \left(1 - \frac{\tilde{J}}{\alpha^V}\right) \tag{35}$$

where α^V is the vehicle cost weighting factor.

Mutation Mechanisms

Mutation mechanisms are essential for the evolution process to improve the fitness of the candidate path generated in each generation and to eventually converge to an optimal solution. A mutation has the effect of randomly moving a candidate solution from one point in the search space to another. Therefore, an effective set of mutation mechanisms must be able to move a candidate solution from one point to any point in the search space by applying a series of these mutation mechanisms. Work by Rudolph [23] shows that this property of mutation mechanisms is necessary for an evolutionary algorithm to converge.

The selection of the types of mutation mechanisms depends on the population representation. For example, in the representation using waypoints, mutation can be done by randomly perturbing the physical location of various points in the waypoint sequence. Therefore, the mutation of the ith trial solution can be expressed as

$$x^{i+\mu}_k = x^i_k + G(0, \sigma^i), \quad k = \{1, 2, \ldots, c\} \tag{36}$$

where x^i_k is the kth waypoint in the original sequence, $x^{i+\mu}_k$ is the kth waypoint in the mutated sequence, μ is the number of parents in the population, and $G(0, \sigma^i)$ presents a Gaussian random variable with zero mean and standard deviation of σ^i. This perturbation is applied only to c waypoints chosen at random where $c \in \{0, 1, \ldots, l_i\}$ and l_i is the length of the waypoint sequence.

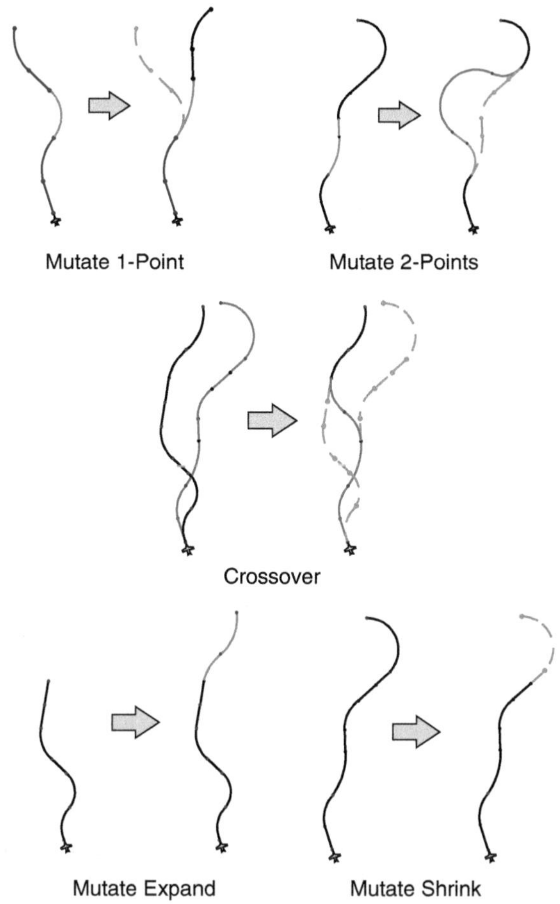

Fig. 8. Path mutation mechanisms

In ECoPS, offsprings are created either by crossing over two randomly selected individuals in the population, or by copying an individual and mutating it using one of the mutation mechanisms chosen at random. We have five mutation mechanisms which are illustrated in Figure 8. The mutation mechanisms change the parameters of some of the segments in a selected path but ensure that the mutated path is within the vehicle's capabilities. The list below explains each of the mutation mechanisms.

- *Mutate 1-Point:* randomly changes the parameters of one or more segments, and then re-locates all the following unchanged segments to enforce the end-point constraints. The first segment to be mutated and the number of segments to be mutated are selected at random.
- *Mutate 2-Points:* randomly changes the parameters of one or more segments, computes the new resultant end point for those segments (similar

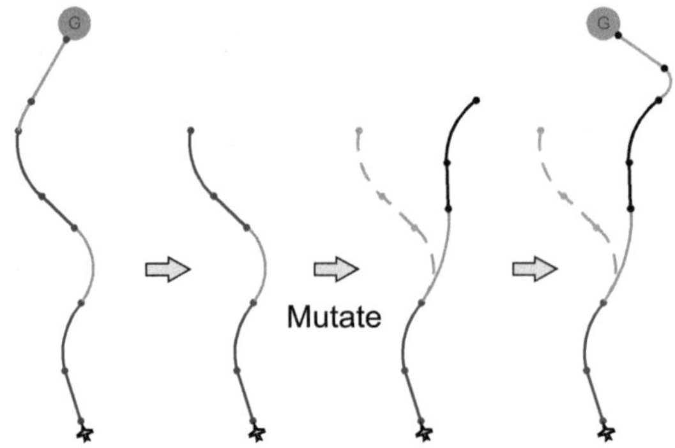

Fig. 9. Go-to-goal segments are added at the end of each mutated path to force the path to end at the goal location

to Mutate 1-Point), and then connects back to the start of another segment of the path further along. The beginning segment and the segment joined to are both chosen at random.
- *Crossover:* takes the starting segments of one path and the ending segments of another and joins the two sets of segments together.
- *Mutate Expand:* adds one or more randomly created segments onto the end of the path. All the original parts of the path are left untouched.
- *Mutate Shrink:* removes one or more segments from the end of the path. The number of segments to be removed are selected at random.

Some of these mutation mechanisms require a method to construct of joining segments which connect the ending point of a segment of a path being mutated to the starting point of another segment while maintaining the continuity of the whole path. The methods used to compute joining segments are presented in [19]. For a particular joining, all three methods are used, but the resulting path with the shortest total length is selected. Once a new path is created through the mutation process, a set of segments called *go-to-goal* segments are added to the mutated path. This modification makes all paths in the population end at the goal location. This procedure is illustrated in Figure 9.

Selection Scheme

Given a population of the size $\mu + \lambda$, the selection scheme is a mechanism for selecting μ individuals to be parents of the nth generation. These μ parents will be used to create λ offspring in the next mutation process. Many types of selection schemes can be used in the evolution process. In this research, a q-fold binary tournament selection scheme is chosen. Figure 10 illustrates the procedures of the selection scheme which can be described as follows.

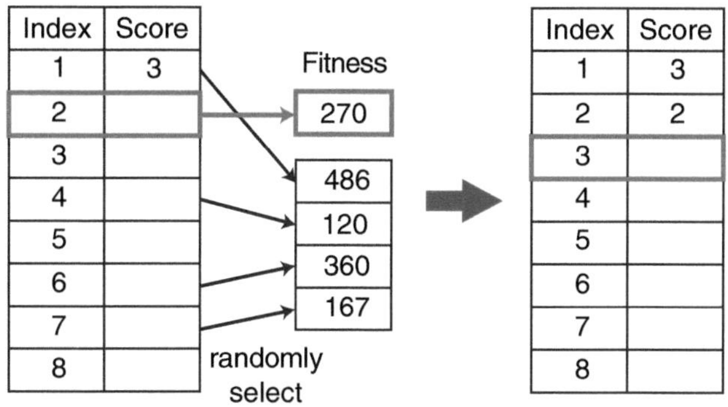

Fig. 10. Illustration of the tournament selection scheme. The plan with index number 2 is compared to other four randomly selected plans

For each individual $i \in \{1, 2, \ldots, (\mu + \lambda)\}$:

1. Draw $q \geq 2$ individuals randomly from the population (excluding individual i) with uniform probability $\frac{1}{\mu+\lambda-1}$. Denote these competitors by the indices $\{i_1, i_2, \ldots, i_q\}$.
2. Compare individual i's fitness against each of the competitors, $i_j, j \in 1, 2, \ldots, q$. Whenever the fitness of individual i is not worse than that of competitor i_j, individual i scores a point.

The score that each individual receives during the tournament is an integer in the range $[0, q]$. After the scores of all individuals are determined, the top μ individuals with the best scores are selected as the parents for the next generation.

Path Planning Example

To demonstrate the performance of the planning algorithm, we present the results of a static path planning problem. In this problem, the mission objective is to observe all the assigned targets and return safely to the goal location. The path planner runs the evolutionary algorithm with a population size of 30. The tournament selection algorithm selects 15 of them to be parents for the next evolution cycle. The score weighting function of each task is shown in Figure 11. Unless otherwise specified, the profile of this function is the same for this example and all other planning examples presented in other sections.

The performance of the the planning algorithm was validated using the Open Experimental Platform (OEP) simulation program developed by the Boeing Company. The OEP is designed for evaluation of planning and coordinated control methodologies in a Monte Carlo simulation. System parameters can be specified as constants or random variables with specific probability distributions. The OEP can simulate large numbers of vehicles and targets.

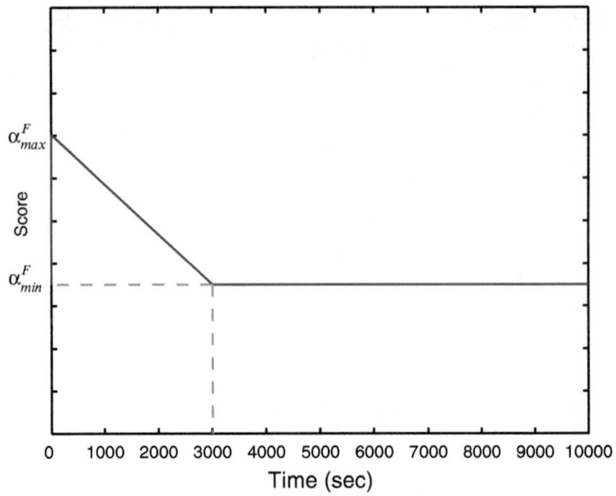

Fig. 11. Normal task score profile

Table 1. Planner parameters

Parameter	Description	Value
α^F_{min}	minimum task score weighting factor	3500
α^F_{max}	maximum task score weighting factor	7000
α^V	vehicle cost weighting factor	2500
α^Q	fuel cost weighting factor	300
α^L	loss function scaling factor	250

Table 2. Simulation parameters

Parameter	Description	Value	Unit
η^O	obstacle payload effectiveness	0.5	–
R^O	obstacle payload range	30	km
σ^O	obstacle uncertainty radius	20	km
σ^G	target uncertainty radius	20	km
η^V	vehicle payload effectiveness	0.7	–
R^V	vehicle payload range	20	km
V_a	vehicle speed	150	m/s
$Fuel_{init}$	vehicle initial fuel level	6.0	liters
$Fuel_{max}$	vehicle max fuel level	7.0	liters

It simulates effects caused by actions or events occurring during the simulation. Simulation entities are customizable and may be used to represent a variety of air and ground vehicles, targets, and buildings. Unless otherwise specified, the values of parameters listed in Table 1 and 2 are used in all of the simulations presented here.

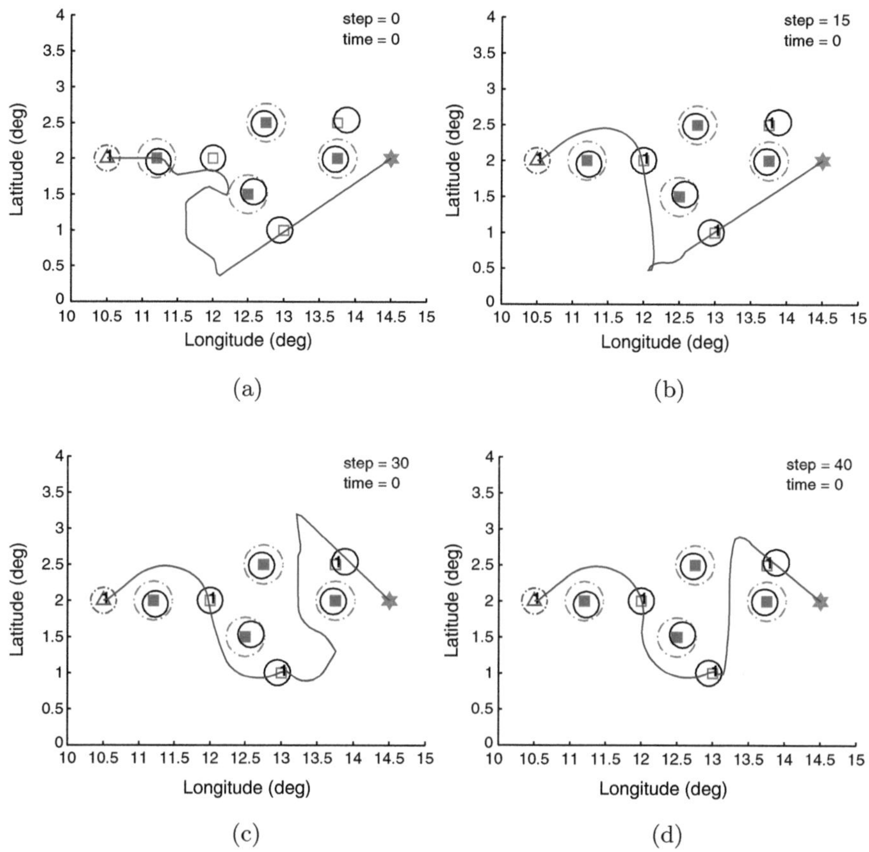

Fig. 12. Snapshots of off-line path planning with multiple targets

In this example problem, there are three targets and three obstacles. Figure 12 shows snapshots of the planning results for different generations in the evolution process. The vehicle is represented by a triangle with vehicle number on it. The dashed circle around the vehicle represents the range of the payload on-board the vehicle. This payload can be a sensor or offensive payload. The square markers represent actual locations of sites. Each of these square markers will have a vehicle number on it if the site is a target and assigned to that vehicle. A solid circle located near each square marker represents an area which covers all of the possible locations of the site represented by the square marker. Each filled square marker with a dashed circle around it represents a site with defensive capabilities which can destroy or change the health states of vehicles if they are within the area marked by the dashed circle. The goal location where the vehicle is required to be at the end of the mission is represented by a hexagram in the plots. This representation of the scenario in the plots is also used in all other planning examples. The results show the ability of the planner to generate an effective path to visit all the assigned targets and avoiding collision with the

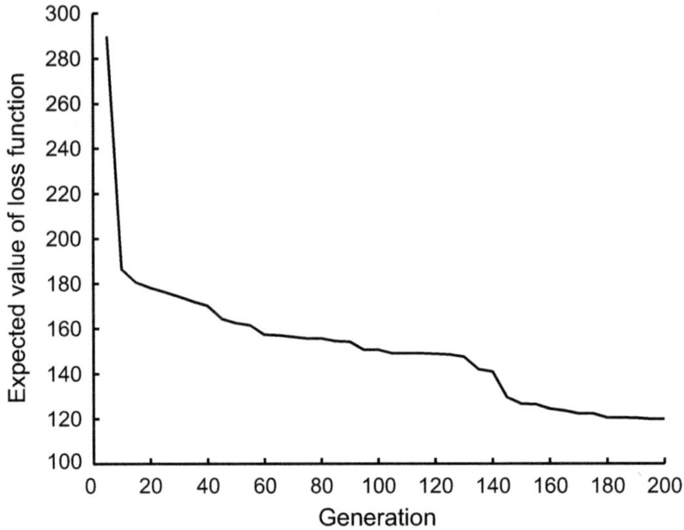

Fig. 13. Evolution of the loss function of the candidate path in each generation

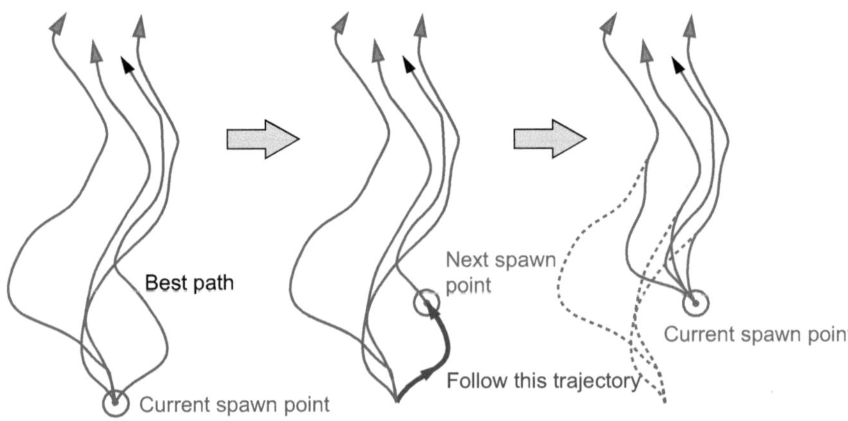

Fig. 14. Concept of dynamic path planning algorithm which retains the knowledge gained from the previous planning cycle

obstacles. Figure 13 shows that the expected value of the loss function decreases dramatically in the early generations. The path planner then fine tunes the resultant path in later generations.

4.2 Algorithm for Dynamic Planning

Dynamic path planning is a continuous process. A diagram describing the concept of the dynamic path planning is shown in Figure 14. The planning problem in each cycle is a similar problem to that in the previous cycle. This approach attempts to preserve some information of the past solutions and uses it as the basis to compute new solutions even though the new problem is slightly different from the previous problem. This takes an advantage of evolutionary algorithms that several candidate solutions are available at any time during the optimization process.

The planner of the vehicle continually updates its path while the vehicle is moving in the field of operation. The planning process starts with the static path planning process to generate an initial population $\mathbf{P}_0$ and find the first best candidate path $Q(0) \in \mathbf{P}_0$ depicted as the black path in Figure 14. The location of the first spawn point is at the desired vehicle position $\bar{z}^V(s_1)$ at time t_{s_1} specified by the path $Q(0)$. The following steps in the dynamic path planning algorithm are described below

1. Generate a new population $\mathbf{P}_{i+1}$ from the current population $\mathbf{P}_i$ by updating all the paths in the current population to begin at the location of the next spawn point. The paths are modified by removing a small number of segments from the start of the paths and adding other segments to join the paths to the spawn point.
2. Run the static planning algorithm continuously to update the population and to find the best candidate path.
3. Send the updated candidate path to the vehicle navigator once the vehicle reaches the current spawn point.
4. Update the estimates of the locations of sites in the environment.
5. Return to step 1

To demonstrate dynamic path planning, we revisit the scenario presented in the last planning example. Starting from the off-line planning result shown in frame (d) of Figure 12, the results of dynamic planning are shown in Figure 15. Frames in the figure show snapshots of the simulation at various simulation time steps. In this simulation, the vehicle is assumed to have an on-board radar which can improve the estimates of nearby sites' locations. The radar can detect a site within the range of 60 kilometers with 40 meters standard deviation. During the simulation, the planning algorithm updates the candidate path every 10 seconds of the simulation time. The size of the execution time horizon, which is the time difference between two consecutive spawn points, is 100 seconds. Since the scenario does not change during the simulation, the dynamically updated path is little different to the off-line planned path. The vehicle follows the path to successfully observe the first two targets, but misses the last target in its first attempt. Frame (c) of Figure 15 shows that the planner is able to quickly update the path to guide the vehicle back to the target and eventually observe the target as shown in Frame (d).

5 Planning with Timing Constraints

To incorporate time-of-execution specifications into the path planning problem, the task score weighting factor α_i^F used in the objective function is defined as a time-dependent function. This time-dependent score weighting function

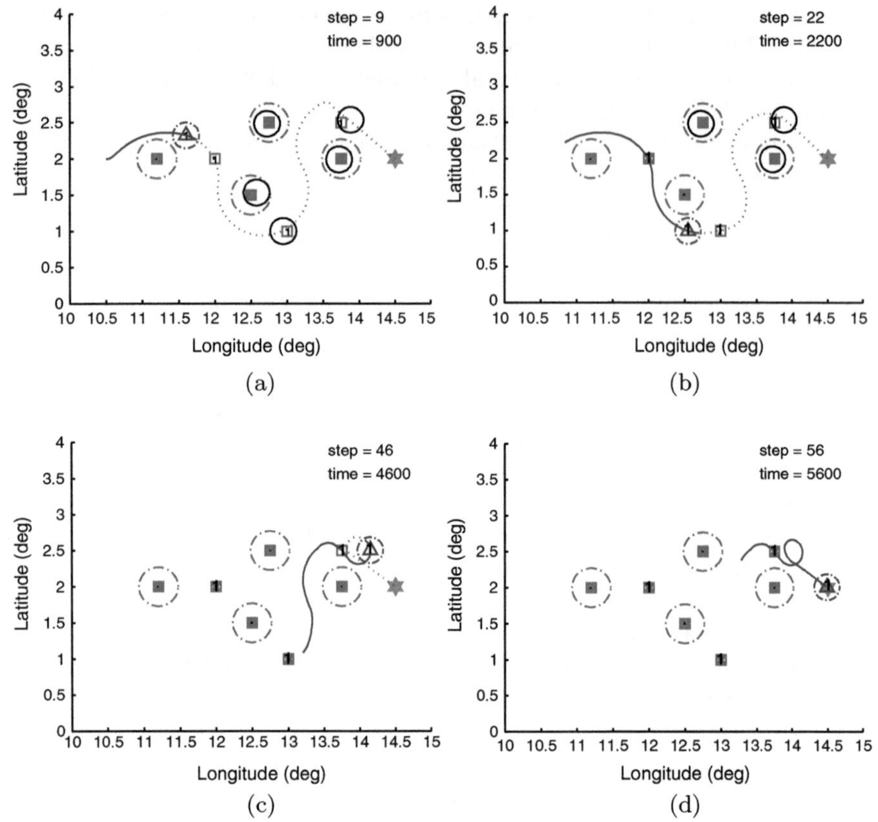

Fig. 15. Snapshots of dynamic path planning at different time steps. Each observed target is marked with a cross symbol

$\alpha_i^F(q)$ can be used to define a time window for the vehicles to execute each task. This function gives a high positive value during the time period in which we want the vehicle to perform the task. The function gives a small positive value or zero value during the time period in which executing the task does not meet the mission objectives.

In this section, we present an example showing the ability of the planner to generate paths which satisfy the imposed timing constraints. The mission objective is to observe a target site which is protected by a nearby defensive site. There are two vehicles each of which has its own path planner. The task of Vehicle 1, which has an offensive payload, is to destroy the defensive site before the beginning of the execution time window of the target site. Vehicle 2, which is equipped with a sensor payload, has to observe the target after the beginning of the the execution time window. That is at 2000 seconds after the mission starts. The duration of the execution time window is 500 seconds. Observing the target site after the expiration time of the execution time window yields a smaller task score. The profiles of score weighting functions of both tasks are given in Figure 16. Figure 17 and 18 show the results of an off-line path planning problem with timing constraints.

In this simulation, each vehicle is equipped with a planner which has identical knowledge of the environment and planning parameters. The static off-line planning result in Figure 17 shows that the planner of Vehicle 1 decides to go directly to the defensive site while Vehicle 2 takes a longer path to wait for the expiration of the the execution time period for reaching the target site.

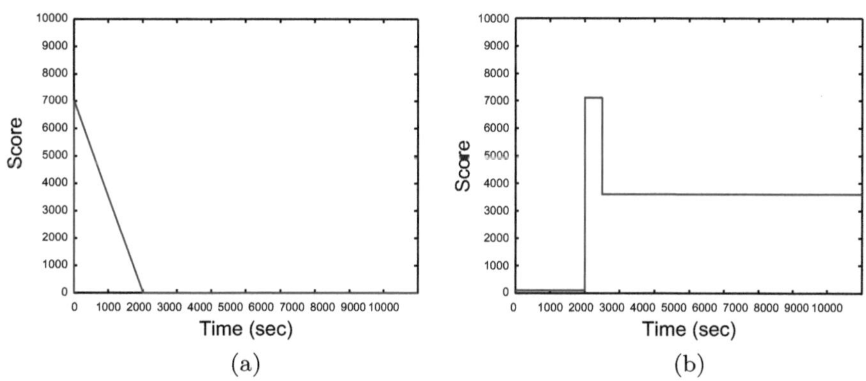

Fig. 16. Frame (a) shows the profile of the task score weighting function of the defensive site. Frame (b) shows the profile of the task score weighting function of the target

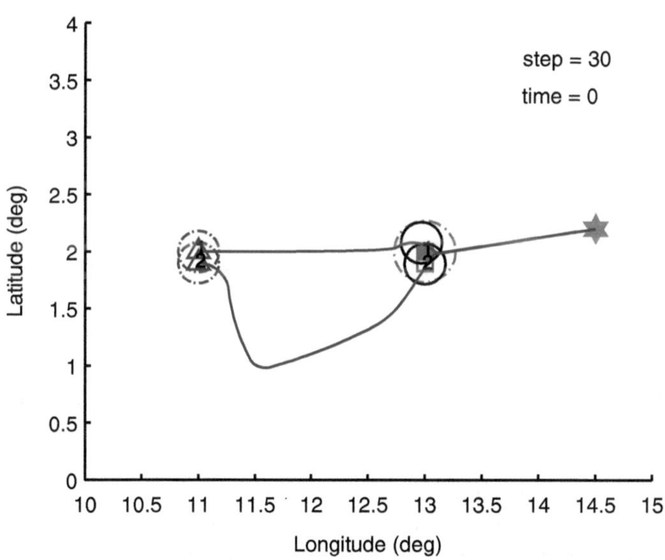

Fig. 17. Off-line path planning with execution time window

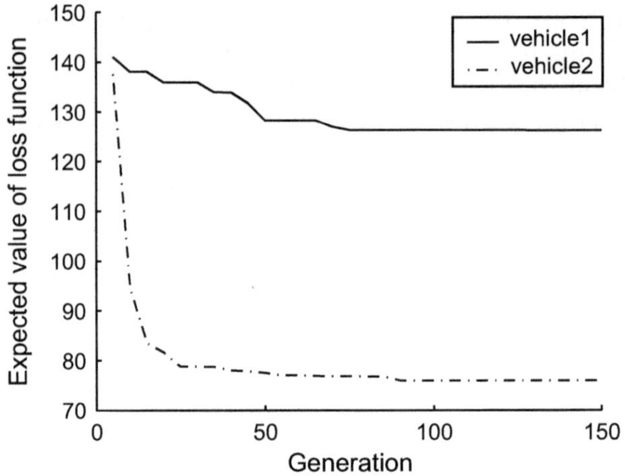

Fig. 18. Evolution of off-line path planning loss function with execution time window

To verify that the path planners are capable of generating plans with timing constraints, we ran a simulation starting with the off-line planning results. The dynamic planning simulation results are shown in Figure 19. Frame (b) of the figure shows that Vehicle 1 reaches the defensive site well before the simulation time 2000 seconds and successfully destroys the obstacle, although the vehicle is also destroyed. Frame (c) shows that Vehicle 2 reaches the target site at time 2200 seconds and successfully observes the target. If it is important for Vehicle 1 to survive, this can be insured by adjustment of the task score weighting function. However, the example illustrates the use of a vehicle in a *sacrificial* role.

6 Planning in Changing Environment

In a changing environment, obstacles and targets may move unexpectedly during the operation. Dynamic planning is essential in this situation. The planner must be capable of replanning during the mission and predicting future states of the sites in the environment. In ECoPS, the site locations and their uncertainties are predicted using Equation 10 and 11.

One advantage in using the approximation to the probability of intersection described in Equation 30 or 31 is the ease with which it can be extended to include moving sites. It is the form of the solution which is a summation over a defined function that allows for the simple inclusion of time into the equations. This approach accommodates the integration of uncertainties and dynamics of the environment into the model and the objective function.

This section provides two examples of planning in dynamic uncertain environments. The first example is a scenario with one moving target which is

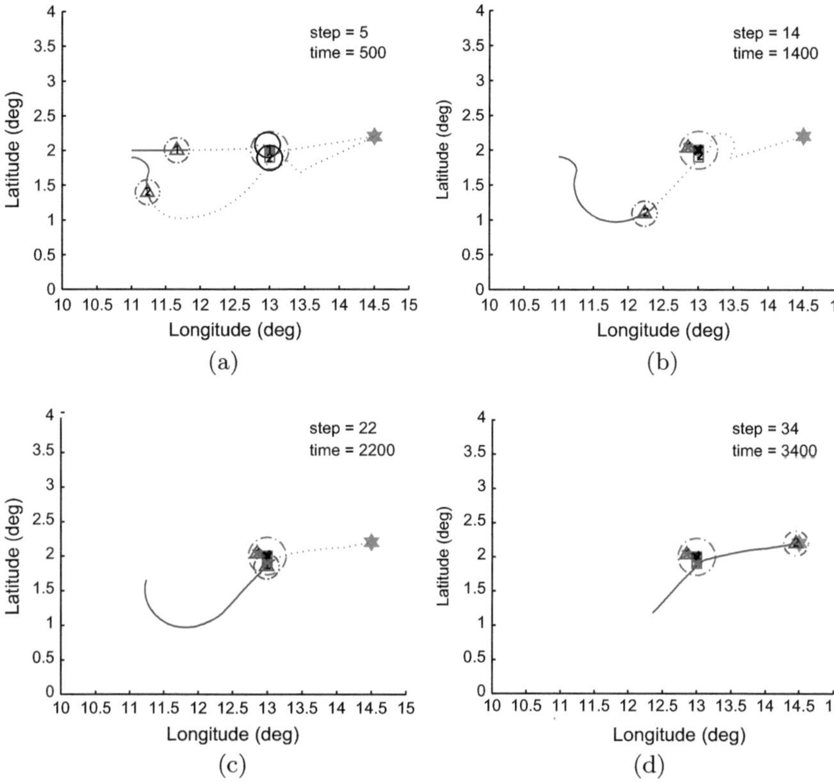

Fig. 19. Dynamic path planning with execution time window

initially located at position (14.0, 2.5) and later heads west at the speed of 300 kilometers/hour after the vehicle has been moving for 100 seconds. In this example, the radius of the uncertainty circle of each obstacle and target is 10 kilometers. During the off-line planning period, the planner does not have the knowledge that the target will move in the future. The off-line planning result is shown in Figure 20. During the mission, the planner will need to dynamically adapt its path to intersect with the predicted location of the target. Frame (a) and (b) of Figure 21 show that the planner is adapting the path to intersect the target at a predicted location. In this simulation, the planner has knowledge of the velocity of the target site. The planner decides to wait until the target moves past the area covered by the top-right defensive site, and the vehicle successfully observes the target as shown in frame (c). The expected value of the loss function during the simulation is shown in Figure 22. The spike in the plot is due to the unexpected movement of the target which causes the planner to temporarily lose track of the target. The value of the loss function drops near zero when the vehicle intersects and successfully observes the target.

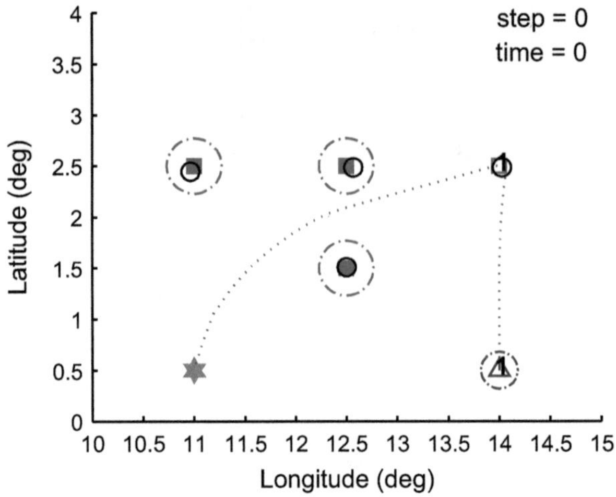

Fig. 20. Off-line path planning result. The planner have no knowledge that the target will move in the future

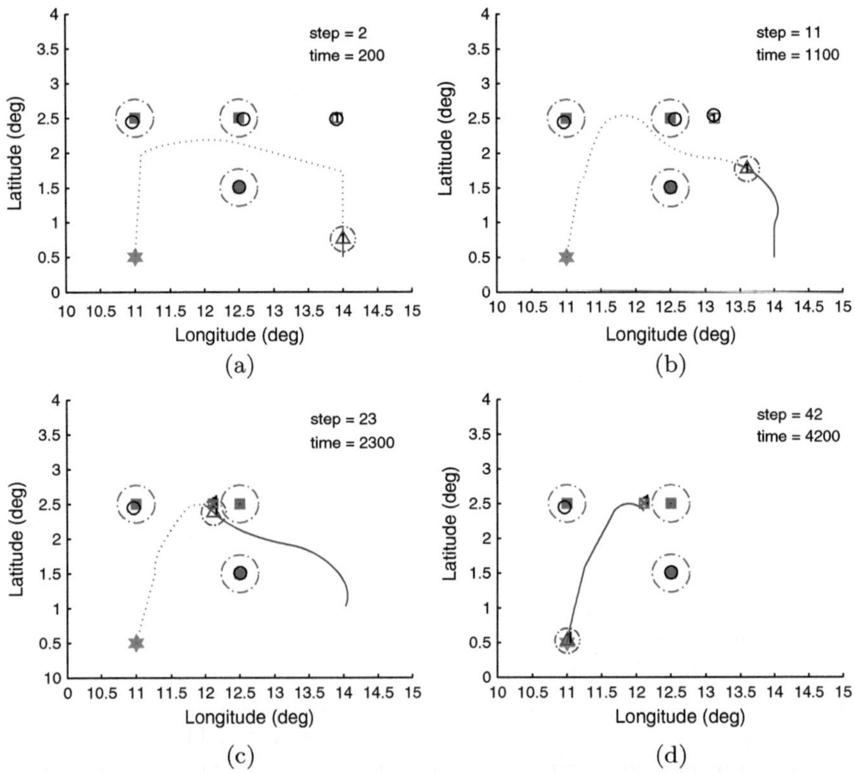

Fig. 21. Snapshots of dynamic path planning with a moving target

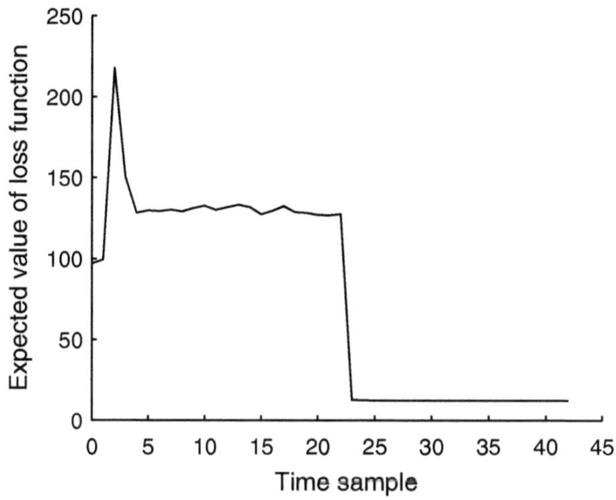

Fig. 22. Evolution of dynamic path planning with a moving target

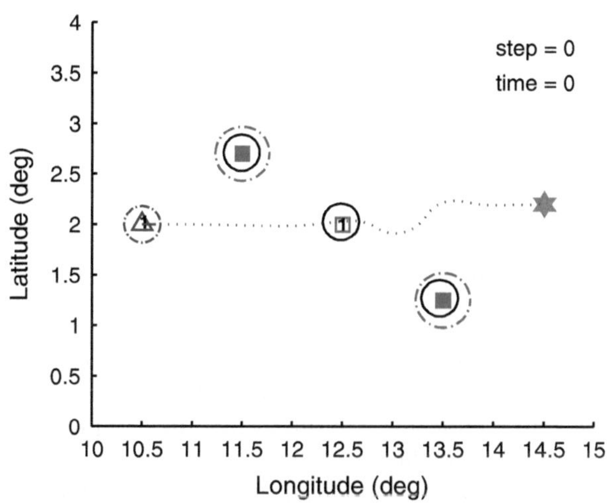

Fig. 23. Off-line path planning result. The planner does not have the knowledge that the obstacles will move in the future

The second example is a dynamic planning problem with moving obstacles. During the off-line planning, the planner does not have the knowledge that the obstacles will move in the future. The off-line planning result shown in Figure 23 illustrates that the planner is able to find a near-optimal path to go almost directly to the target and the goal location. Almost immediately after the vehicle starts moving from its initial location, the obstacles begin moving north and south to block the vehicle from getting in and out of the area where the target is located. Figure 24 shows snapshots of the dynamically generated path during the simulation. These snapshots show that the vehicle is able

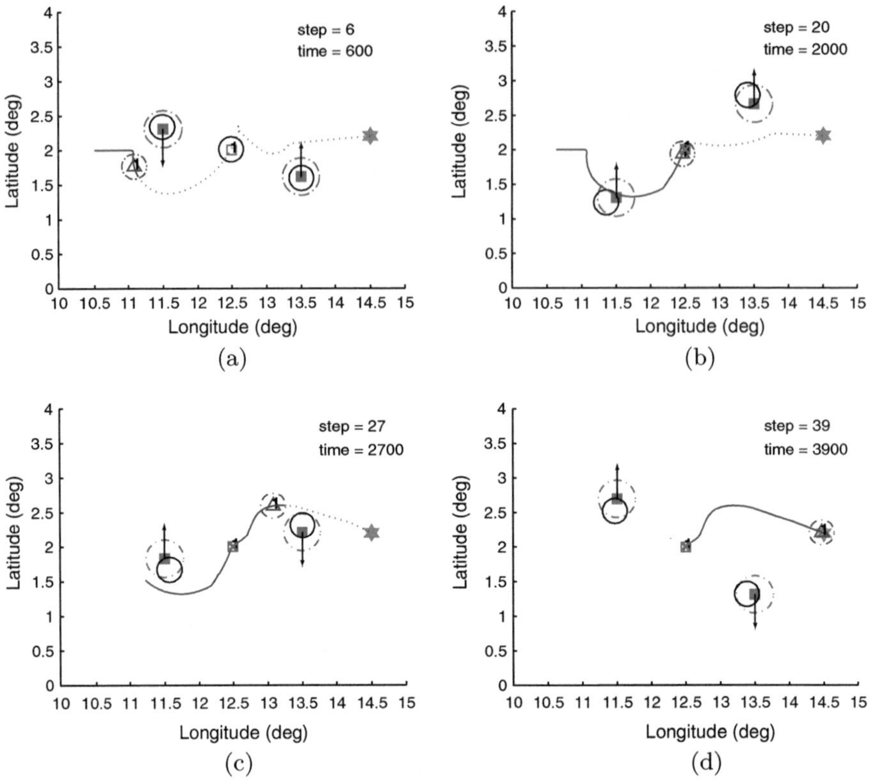

Fig. 24. Snapshots of dynamic path planning with moving obstacles

to avoid collision with the obstacles and successfully observe the target and finally reach the goal location. The expected value of the loss function at each time step is given in Figure 25. The first spike in the plot occurs when the obstacles start moving. The planner dynamically replans the path with a lower loss value according to the new updated information about the environment.

7 Conclusion

The goal of this work is to develop a dynamic path planning algorithm for autonomous vehicles operating in changing environments. The algorithm must be capable of replanning during the operation. We present the concept of dynamic path planning and a framework to solve the planning problem based on a model-based predictive control scheme. We describe a model used to predict the expected values of future states of the system. The model takes into account the uncertain information of the environment and the dynamics of the system.

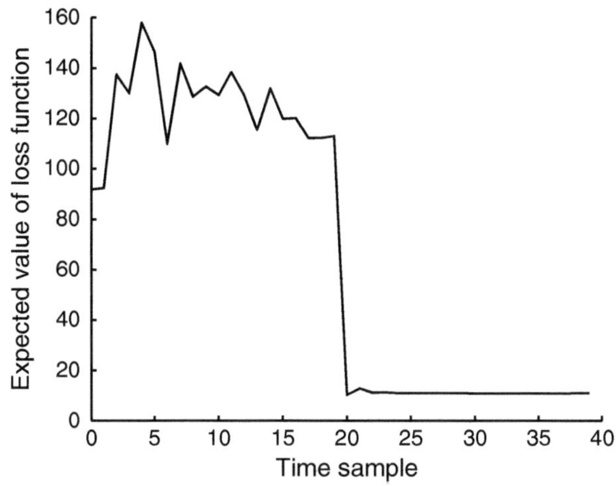

Fig. 25. Evolution of dynamic path planning with moving obstacles

We present an evolutionary algorithm suitable for dynamic path planning problems. The algorithm was developed as part of the Evolution-based Cooperative Planning System for teams of autonomous vehicles. The algorithm is used to find an optimal path that maximizes an objective function. This function is formulated using the stochastic world model to capture the dynamics and uncertainties in the system. The algorithm has been applied to path planning for UAVs in real wind fields and predicted icing conditions [22]. Extensions to apply this algorithm to the coupled task of the path planning problem for multiple cooperating vehicles are reported in [19].

Simulation results demonstrate that the path planning algorithm can provide computationally feasible effective solutions to all of the path planning problems which include planning with timing constraints and dynamic planning with moving targets and obstacles. Even though there are some uncertainties in the knowledge of the environment, the algorithm can generate feasible paths which are within the capabilities of the vehicle to complete all tasks and to avoid collision with the obstacles. During the mission, the planner is able to quickly adapt the path in response to changes in the environment.

8 Acknowledgments

The research presented in this paper is partially funded by the Washington Technology Center. The simulation software is provided by the Boeing Company. Professor Emeritus Juris Vagners at the University of Washington provided direction and advice for this research.

References

1. R. Rysdyk A. Pongpunwattana. Real-time planning for multiple autonomous vehicles in dynamic uncertain environments. *Journal of Aerospace Computing, Information, and Communication,* 1(12):580–604, 2004.
2. J. L. Bander and C. C. White. A heuristic search algorithm for path determination with learning. *IEEE Transactions of Systems, Man, and Cybernetics – Part A: Systems and Humans*, 28:131–134, January 1998.
3. B. L. Brumitt and A. Stentz. Dynamic mission planning for multiple mobile robots. In *Proceedings of the IEEE International Conference on Robotics and Automation*, Minneapolis, MN, April 1996.
4. E. F. Camacho and C. Bordons. *Model Predictive Control.* Springer, London, UK, 1999.
5. B. J. Capozzi. *Evolution-Based Path Planning and Management for Autonomous Vehicles.* PhD thesis, University of Washington, 2001.
6. B. J. Capozzi and J. Vagners. Evolving (semi)-autonomous vehicles. In *Proceedings of the AIAA Guidance, Navigation, and Control Conference*, Montreal, Canada, August 2001.
7. S. Chien et al. Using iterative repair to improve the responsiveness of planning and scheduling for autonomous spacecraft. In *IJCAI99 Workshop on Scheduling and Planning meet Real-time Monitoring in a Dynamic and Uncertain World*, Stockholm, Sweden, August 1999.
8. D. B. Fogel. *Evolutionary Computation: Toward a New Philosophy of Machine Intelligence.* IEEE Press, Piscataway, NJ, second edition, 2000.
9. D. B. Fogel and L. J. Fogel. Optimal routing of multiple autonomous underwater vehicles through evolutionary programming. In *Proceedings of the 1990 Symposium on Autonomous Underwater Vehicle Technology*, pages 44–47, Washington, DC, 1990.
10. C. Hocaoğlu and A. C. Sanderson. Planning multiple paths with evolutionary speciation. *IEEE Transctions on Evolutionary Computation*, 5(3):169–191, June 2001.
11. J. Holland. *Adaptation in Natural and Artificial Systems.* PhD thesis, University of Michigan, Ann Arbor, MI, 1975.
12. L. E. Kavraki et al. Probabilistic roadmaps for path planning in high-dimensional configuration spaces. *IEEE Transactions on Robotics and Automation*, 12(4):566–580, 1996.
13. S. M. LaValle and J. J. Kuffner. Randomized kinodynamic planning. In *Proceedings of IEEE International Conference on Robotics and Automation*, 1999.
14. A. Mandow et al. Multi-objective path planning for autonomous sensor-based navigation. In *Proceedings of the IFAC Workshop on Intelligent Components for Vehicles*, pages 377–382, 1998.
15. C. S. Mata and J. S. Mitchell. A new algorithm for computing shortest paths in weighted planar subdivisions. In *Symposium on Computational Geometry*, pages 264–273, 1997.
16. R. R. Murphy. *Introduction to AI Robotics.* MIT Press, 2000.
17. N. J. Nilsson. *Principles of Artificial Intelligence.* Tioga Publisher Company, Palo Alto, CA, 1980.
18. M. H. Overmars and P. Svestka. A probabilistic learning approach to motion planning. In Goldberg, Halperin, Latombe, and Wilson, editors, *Algorithmic*

Foundations of Robotics, The 1994 Workshop on the Algorithmic Foundations of Robotics, 1995.

19. A. Pongpunwattana. *Real-Time Planning for Teams of Autonomous Vehicles in Dynamic Uncertain Environments.* PhD thesis, University of Washington, 2004.
20. D. Rathbun and B. J. Capozzi. Evolutionary approaches to path planning through uncertain environments. In *AIAA 1st Unmanned Aerospace Vehicles, Systems, Technologies, and Operations Conference and Workshop*, Portsmouth, VA, May 2002.
21. D. Rathbun et al. An evolution based path planning algorithm for autonomous motion of a uav through uncertain environments. In *Proceedings of the AIAA 21st Digital Avionics Systems Conference*, Irvine, CA, October 2002.
22. J. C. Rubio. *Long Range Evolution-based Path Planning for UAVs through Realistic Weather Environments.* PhD thesis, University of Washington, 2004.
23. G. Rudolph. Convergence of evolutionary algorithms in general search spaces. In *Proceedings of the Third IEEE Conference on Evolutionary Computation*, pages 50–54, Piscataway, NJ, 1996.
24. G. Song et al. Customizing PRM roadmaps at query time. In *Proceedings of IEEE International Conference on Robotics and Automation*, 2001.
25. S. Thrun et al. Map learning and high-speed navigation in rhino. In *Artificial Intelligence and Mobile Robots.* MIT Press, Cambridge, MA, 1998.
26. J. Xiao et al. Adaptive evolutionary planner/navigator for mobile robots. *IEEE Transactions on Evolutionary Computation*, 1:18–28, April 1997.

Algorithms for Routing Problems Involving UAVs

Sivakumar Rathinam[1] and Raja Sengupta[2]

[1] University of California, Berkeley rsiva@berkeley.edu
[2] University of California, Berkeley raja@path.berkeley.edu

Abstract. Routing problems naturally arise in several civil and military applications involving Unmanned Aerial Vehicles (UAVs) with fuel and motion constraints. A typical routing problem requires a team of UAVs to visit a collection of targets with an objective of minimizing the total distance travelled. In this chapter, we consider a class of routing problems and review the classical results and the recent developments available to address the same.

1 Introduction

This chapter is dedicated to reviewing classical approaches and disseminating recent approaches on the resource allocation problems that arise in the use of Unmanned Aerial Vehicles (UAVs). Small autonomous UAVs are seen as ideal platforms for many applications such as monitoring a set of targets, mapping a given area, aerial surveillance, fire monitoring etc. A collection of small autonomous UAVs with the necessary sensors can potentially replace a manned vehicle in dangerous environments and warfare. A common mission that can be carried out by a group of UAVs is a surveillance operation where a set of destinations needs to be monitored. If the number of destinations to be visited are higher than the number of UAVs available, then the following resource allocation questions naturally arises:

1. How to partition the set of destinations into subsets such that each UAV gets a subset of destinations to monitor?
2. Given a subset for each UAV, how to determine the order in which the destinations should be monitored?
3. Can we find a partition and an order for each UAV such that the total distance travelled by the UAVs is minimum or the total risk encountered is minimum?

If there is only one vehicle, the problem of finding an optimal sequence such that each destination is visited once and the total distance travelled

S. Rathinam and R. Sengupta: *Algorithms for Routing Problems Involving UAVs*, Studies in Computational Intelligence (SCI) **70**, 147–172 (2007)
www.springerlink.com

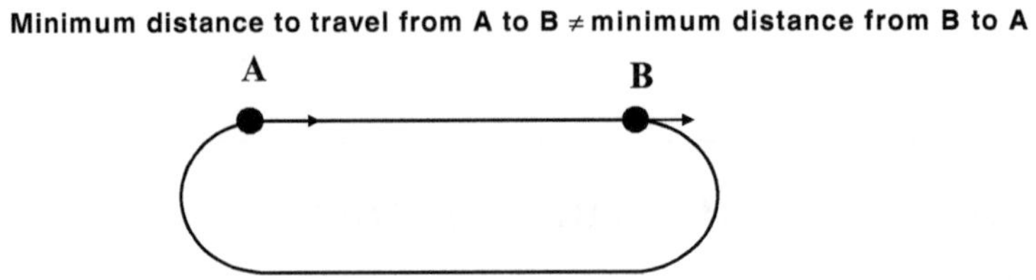

Fig. 1. An example that illustrates the asymmetry in resource allocation problems involving UAVs

by the vehicle is minimum is called the Travelling Salesman Problem (TSP). TSP is known to be NP-Hard [1],[2]. That is, there are no algorithms that are currently available in the literature that can solve the TSP optimally in polynomial time[3]. There are other variants of the standard TSP that can be useful for applications involving UAVs. For example, in multiple UAV problems, vehicles might start their missions from a single depot or from multiple depots.

The presence of kinematic constraints for the UAV complicates these resource allocation problems further. A typical constraint that is used in planning problems involving UAVs is the yaw rate constraint. The yaw rate constraint models the inability of a UAV to turn at any arbitrary yaw rate. This yaw rate constraint introduces asymmetry in the distances travelled between two points. For example, a UAV starting at destination A with heading ψ_A and arriving at destination B with heading ψ_B may not equal the length of its optimal path starting at destination B with heading ψ_B and arriving at destination A with heading ψ_A. An illustration of this asymmetry for an example is shown in figure 1.

This chapter formulates three sets of resource allocation problems that are useful in the context of UAVs and presents algorithms that have been developed in the literature to address each of them. To simplify the presentation, the chapter formulates each resource allocation problem, discusses the relevant literature and presents algorithms to solve the same in separate sections.

2 Single Vehicle Resource Allocation Problem in the Absence of Kinematic Constraints

2.1 Problem Formulation

Let one UAV be required to visit n destinations. Let V be the set of vertices that correspond to the location of the UAV and the destinations, with the first vertex V_1 representing the UAV and $V_2, \ldots, V_{n+1}$ representing the

[3] An algorithm can solve a problem in polynomial time if the number of steps required to run the algorithm is a polynomial function of the input size of the problem.

destinations. Let $E = V \times V$ denote the set of all edges (pairs of vertices) and let $c : E \rightarrow \Re_+$ denote the cost function with $c(V_i, V_j)$ (or simply, c_{ij}) representing the cost of travelling from vertex V_i to vertex V_j. Costs are symmetric, that is, $c_{ij} = c_{ji}$. A tour of the UAV is an ordered set of $n+2$ elements of the form $\{V_1, V_{i_1}, \ldots, V_{i_n}, V_1\}$, where $V_{i_l},\ l = 1, \ldots, n$ corresponds to n distinct destinations being visited in that sequence. The overall cost, $C(TOUR)$, associated with the tour of the UAV is defined as $C(TOUR) = c_{1,i_1} + \sum_{k=1}^{n-1} c_{i_k,i_{k+1}} + c_{i_n,1}$. Given the graph $G = (V, E)$, the single vehicle problem (**SVP**) is to find a tour for the UAV that minimizes the overall cost.

2.2 Relevant Literature

The formulated problem is essentially the single Travelling Salesman Problem (TSP) as referred to in the operations research literature. For a general cost function (i.e. c_{ij}), it has been proved that there exists no algorithm with a constant approximation factor unless P=NP [1]. An approximation factor $\beta(P, A)$ of using an algorithm A to solve the problem P (objective is minimize some cost function) is defined as

$$\beta(P, A) = \sup_I \left(\frac{C(I, A)}{C_o(I)}\right), \tag{1}$$

where I is a problem instance, $C(I, A)$ is the cost of the solution by applying algorithm A to the instance I and $C_o(I)$ is the cost of the optimal solution of I. In simple terms, the algorithm A produces an approximate solution to every instance I of the problem P, whose cost is within $\beta(P, A)$ times the optimal solution of I. Constant factor approximation algorithms are useful in the sense that they give an **upper bound** to the cost of the resulting solution that is independent of the size of the problem. If the cost function satisfies triangle inequality and is symmetric, constant factor approximation algorithms are available. If i, j, k denote the destinations to be visited then satisfying triangle inequality means that $c_{ij} \leq c_{ik} + c_{kj}$. The following are the two main approximation algorithms available for the single TSP:

- 2 approximation algorithm [2].
- 1.5 approximation algorithm by Christofides [3].

When the destinations lie on a Euclidean plane, the cost function has additional properties that was exploited by Arora in [4]. Given any $\epsilon > 0$, Arora's algorithm finds a solution with an approximation factor of $1 + \epsilon$ in time $n^{O(\frac{1}{\epsilon})}$.

As far as the **lower bounds** are concerned, Held and Karp's result [5], [6] is the best known result for the TSP. An advantage of deriving good lower bounds is that they can be incorporated in branch and bound solvers used to solve the TSP to obtain faster results. Also, if one can find lower bounds that are close to the optimal solution in an efficient way, then the quality of using an algorithm can be found out by comparing the solution produced by the algorithm directly with the lower bound than with the optimal solution which may require a large time to compute. The experimental results in [7] show that even for large size problems, Held-Karp's lower bound gets within 1-2% of the optimal solution. An important feature of Held-Karp's algorithm is that the results are very close to the optimal solution for any general cost function (i.e. cost function doesn't have to satisfy triangle inequality). Hence, in the context of UAVs, this might be ideal as the cost function could be determined by the risk of travelling between any two destinations and hence, may not satisfy triangle inequality.

In this section we review three algorithms, namely the 2-Approx algorithm, the 1.5-Approx algorithm and the Held-Karp's lower bounding algorithm that can be used to address the **SVP**.

2.3 Algorithms

Before we present the algorithms, we define some of the terms commonly used in the TSP literature. Let $\{i, j\}$ indicate the edge joining vertex i and vertex j. A subgraph G' of G is defined as $G' := (V, E')$ where $E' \subseteq E$. The cost of a subgraph is defined as the sum of the cost of all the edges present in E'. A spanning tree is a subgraph of G that spans all the vertices in V and does not contain a cycle. A minimum spanning tree of G is a spanning tree of G that has minimum cost. The degree of a vertex v in graph $G = (V, E)$ indicates the total number of edges present in E incident on v. An Eulerian graph is a graph where each vertex has an even degree. A perfect matching of a graph G corresponds to a subgraph G' of G that has each vertex having a degree equal to 1. An Eulerian walk of G is a path that visits each edge in G exactly once and each vertex in G atleast once.

2-Approx Algorithm

Assuming the costs are symmetric and satisfy triangle inequality, the approximation algorithm *2-Approx* [2] is as follows:

1. Find the Minimum Spanning Tree (MST) of the graph G. An example illustrating this step of the algorithm is given in Fig. 2.
2. Double every edge in the MST to get an Eulerian graph (Fig. 3).
3. Find an Eulerian walk on this Eulerian graph (Fig. 4).
4. Find the tour such that the vehicle visits the vertices of G in the order of their first appearance in the Eulerian walk (Fig. 5).

The following theorem in [2] shows that *2-Approx* has an approximation factor of 2.

Theorem 1. *The algorithm 2-Approx solves the **SVP** with an approximation factor of 2 in $O(n^2)$ steps when the costs are symmetric and satisfy triangle inequality.*

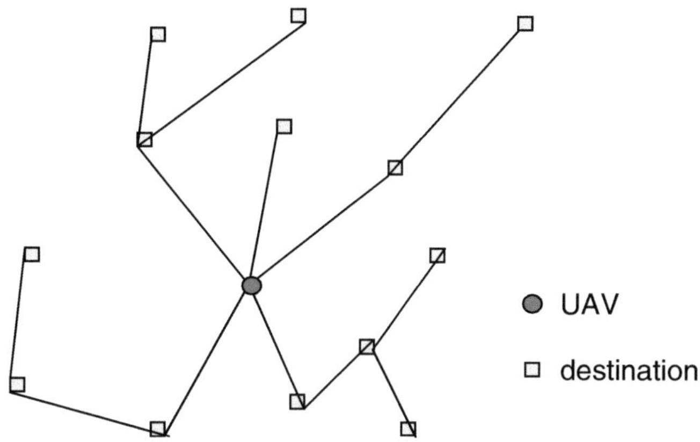

Fig. 2. Find the minimum spanning tree

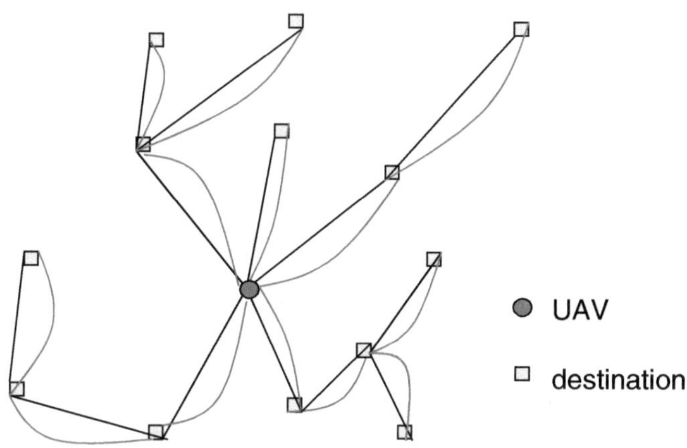

Fig. 3. Double the edges in the minimum spanning tree to construct an Eulerian graph

1.5-Approx Algorithm

Christofides in [3] reduced the approximation factor from 2 to 1.5 by coming up a better way of constructing the Eulerian graph from the minimum spanning tree. This gave rise to the following *1.5-Approx* algorithm for **SVP**:

1. Find the Minimum Spanning Tree (MST) of the graph G. This step is same as the step 1 presented in the *2-Approx* algorithm (Fig. 2).
2. Find the minimum cost perfect matching (PM) on all the odd degree vertices in the MST (Fig. 6).
3. Add all the edges in MST and PM to get an Eulerian graph (Fig. 7).
4. Find an Eulerian walk on this Eulerian graph (Fig. 8).

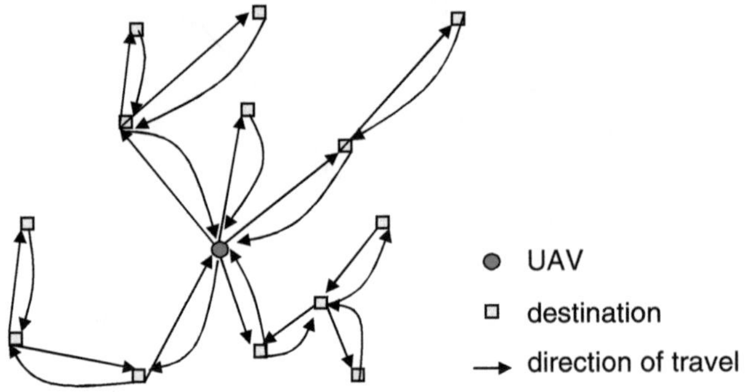

Fig. 4. Find an Eulerian walk

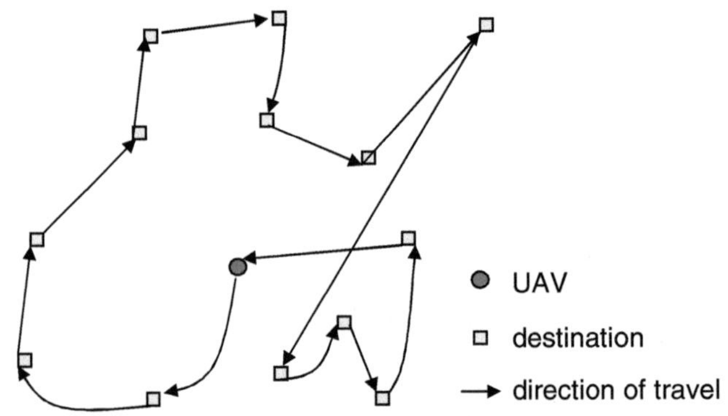

Fig. 5. Find a tour from the Eulerian walk

5. Find the tour such that the vehicle visits the vertices of G in the order of their first appearance in the Eulerian walk (Fig. 9).

The following theorem in [3] shows that **SVP** has an approximation factor of 1.5.

Theorem 2. *The algorithm 1.5-Approx solves the* ***SVP*** *with an approximation factor of 1.5 in* $O(n^{2.5})$ *steps when the costs are symmetric and satisfy triangle inequality.*

Held-Karp's Lower Bound

Held and Karp [5] derived a lower bound to **SVP** by first noting the fact that every tour is a 1-tree. A 1-tree is a tree on vertices $V = \{V_2, \ldots V_{n+1}\}$ with two additional edges incident on vertex V_1. So if the minimum cost 1-tree on the set V turns out to be a tour, it also solves the **SVP**. The basic idea behind the Held-Karp algorithm is the observation that the optimal solution of a **SVP** does not alter by changing the costs of the edges from c_{ij} to $c_{ij} + \pi_i + \pi_j$,

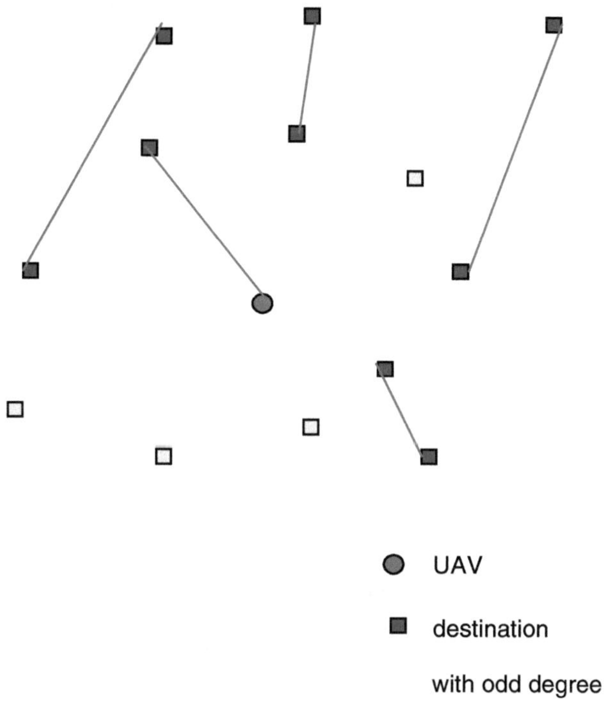

Fig. 6. Find the minimum cost perfect matching (PM) on the odd degree vertices of MST

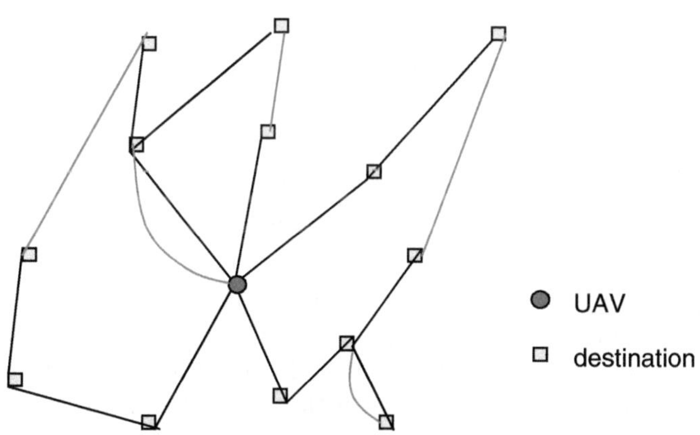

Fig. 7. Add the edges from MST with the edges in PM

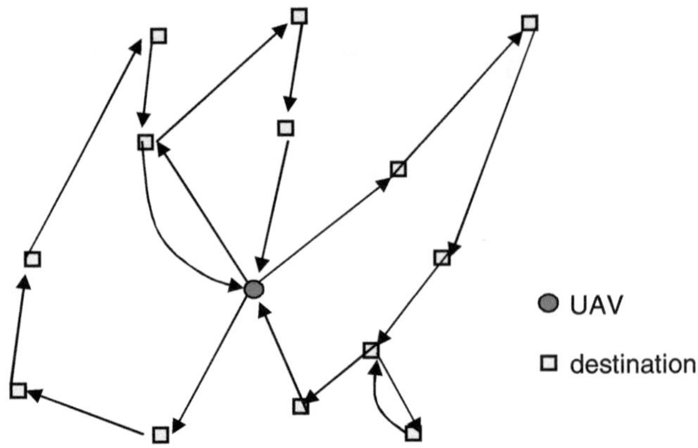

Fig. 8. Find an Eulerian walk

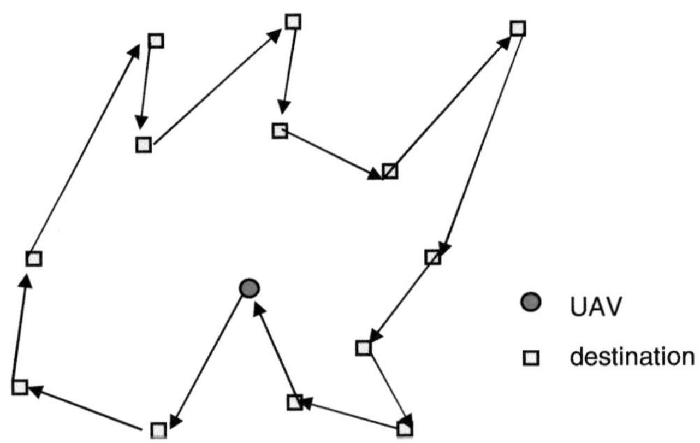

Fig. 9. Find a tour from the Eulerian walk

where as, the optimal solution of the minimum cost 1-tree may change. π_i can be treated as weights on each vertex $i \in V$. The reason why the optimal solution for a **SVP** doesn't change is because for any tour x, $\sum_{\{i,j\}\in x}(c_{ij}+\pi_i+\pi_j) = \sum_{\{i,j\}\in x} c_{ij} + 2\sum_{i\in V} \pi_i$. Therefore, $\arg\min_x\{\sum_{\{i,j\}\in x}(c_{ij} + \pi_i + \pi_j) : x\in \mathrm{T}\} = \arg\min_x\{\sum_{\{i,j\}\in x} c_{ij} : x \in \mathrm{T}\}$, where T is the set of all tours in V. But if y denotes a 1-tree, then, $\sum_{\{i,j\}\in y}(c_{ij}+\pi_i+\pi_j) = \sum_{\{i,j\}\in y} c_{ij} + \sum_{i\in V} \pi_i d_{iy}$, where d_{iy} is the degree of vertex i in y. Hence, the additional cost added depends on the degree of each vertex in the 1-tree. Using the fact that every tour is a 1-tree, we have,

$$\min_{y\in \mathrm{Q}} \sum_{\{i,j\}\in y} c_{ij} + \sum_{i\in V} \pi_i d_{iy} \leq \min_{x\in \mathrm{T}} \sum_{\{i,j\}\in x} c_{ij} + 2\sum_{i\in V} \pi_i, \quad (2)$$

where Q is the set of all 1-trees in V. Therefore, for any given vector π,

$$\min_{y \in \mathrm{Q}} \sum_{\{i,j\} \in y} c_{ij} + \sum_{i \in V} \pi_i (d_{iy} - 2) \leq \min_{x \in \mathrm{T}} \sum_{\{i,j\} \in x} c_{ij}. \tag{3}$$

Since the above equation is true for any π, we get the following result:

Theorem 3.

$$\max_{\pi} \min_{y \in \mathrm{Q}} \sum_{\{i,j\} \in y} c_{ij} + \sum_{i \in V} \pi_i (d_{iy} - 2) \leq \min_{x \in \mathrm{T}} \sum_{\{i,j\} \in x} c_{ij}. \tag{4}$$

The left hand side in the above result provides a lower bound to the **SVP**. Let $w(\pi) = \min_{y \in \mathrm{Q}} \sum_{\{i,j\} \in y} c_{ij} + \sum_{i \in V} \pi_i (d_{iy} - 2)$. For any fixed π, calculating $w(\pi)$ is that of finding an optimal 1-tree. An optimal 1-tree can be easily solved using the Prim's algorithm [2]. Note that the function $w(\pi)$ is concave in π. This lends itself to a gradient ascent algorithm that produces a sequence of lower bounds to the **SVP** as discussed in [5],[6].

3 Multiple Vehicle Resource Allocation Problems in the Absence of Kinematic Constraints

The resource allocation problems considered in this section involves multiple UAV's where vehicles could start from a single depot or from multiple depots. The general problem discussed in this section is as follows: Given a set of UAVs and destinations, find tours for each UAV such that (1) each destination is visited once by only one UAV (2) the sum of the tour cost of all the UAVs is minimum. As mentioned in the introduction, there are several variants of this multiple vehicle problem. In this section, we present three such variants and discuss approaches to solve them. To avoid using redundant variables in the problem formulation, each variant is formulated separately under each subsection.

3.1 Literature Review

The Multiple Travelling Salesmen Problem (MTSP) has two distinct cases - one case where all vehicles start at a root vertex (referred to as Single Depot MTSP) and an other where vehicles may start at different locations (referred to as Multiple Depot MTSP). Please refer to the recent paper by Bektas [8] for an extensive review of MTSP's. Bellmore and Hong [9] consider a Single Depot MTSP where each vehicle is available for service at a specific cost and the edge costs need not satisfy triangle inequality. Since the objective is to reduce the total cost travelled by the vehicles, there could be situations when the optimal solution will not necessitate using all the vehicles. Bellmore and

Hong [9] provide a way of transforming this single depot MTSP to a standard TSP for the asymmetric case and Rao [10] discuss the symmetric version of the same problem. GuoXing [11] also provides a transformation of a variant of an asymmetric, Multiple Depot MTSP to an Asymmetric TSP, wherein most applicable literature for the standard asymmetric TSP can be put to good use. Recently, Rathinam et al. [12] provided a 2−approx algorithm for Multiple Depot MTSP when the edge costs are symmetric and satisfy triangle inequality. In their work, each vehicle start and end at different locations. Also, Darbha [13] discuss a generalized version of the multiple depot MTSP's where there is an upper bound on the number of vehicles that can be used. The following subsections discuss three variants of the multiple vehicle TSP presented in Rao [10], Rathinam et al. [12] and Darbha [13].

3.2 Single Depot, Multiple TSP(SDTSP)

Problem Formulation

Let there be n destinations and m UAVs. V consists of the vertex V_0 representing the depot along with vertices $V_1, \ldots, V_n$ that represent the destinations. There are m UAV's, $u_0, u_1 \ldots u_{m-1}$, present in the depot (vertex V_0). Let $E = V \times V$ denote the set of all edges (pairs of vertices). A edge joining vertices V_i and V_j is represented as (V_i, V_j). Each edge (V_i, V_j) has a cost denoted by $c(V_i, V_j)$ (or simply, c_{ij}). A tour is an ordered set, $TOUR_i$, of at least $r+2$, $r \geq 1$ elements of the form $\{V_0, V_{i_1}, \ldots, V_{i_r}, V_0\}$, where V_{i_l}, $l = 1, \ldots, r$ corresponds to r distinct destinations being visited in that sequence by UAV u_i. There is a cost, $C(TOUR_i)$, associated with a tour for the UAV u_i and is defined as $C(TOUR_i) = c_{0,i_1} + \sum_{k=1}^{r-1} c_{i_k, i_{k+1}} + c_{i_r,0}$. Also, there is a fixed price C_i of using the UAV u_i. Without loss of generality, we assume that $C_0 \leq C_1 \ldots \leq C_{m-1}$. If $\mathcal{S}_p$ is the set of p UAVs chosen to visit the destinations, the overall cost is defined as $\sum_{i \in \mathcal{S}_p} [C(TOUR_i) + C_i]$. Given the graph $G = (V, E)$ the problem is to choose p $(1 \leq p \leq m)$ vehicles so that each destination is visited by only one UAV and the overall cost is a minimum among all possible choices of p and their corresponding tours.

Transformation of SDTSP to a Single TSP

Rao [10] presents an approach to solve **SDTSP** by transforming **SDTSP** to an equivalent single TSP. By doing this, most of the available heuristics for the single TSP can be used to get solutions for the **SDTSP**. It turns out in practice, this method of transforming the given **SDTSP** to a single TSP does not yield good results as the number of the vehicles increases [14]. Nevertheless, this approach gives an insight as to how multiple vehicle problems can be dealt with.

The basic idea is to construct a new graph $G' = (V', E')$ and the corresponding cost function such that finding a single optimal tour on graph G' is equivalent to solving the **SDTSP**. Graph $G' = (V', E')$ is constructed as follows:

- Add additional $m-1$ vertices to V represented by $V_{-1}, V_{-2}...V_{-(m-1)}$. The new set of vertices $V' := V \bigcup \{V_{-1}, V_{-2}...V_{-(m-1)}\}$.
- E' contains
  1. every edge present in E.
  2. an edge (V_{-i}, V_j) if (V_0, V_j) is present in E, $\forall i \in \{1, 2..(m-1)\}$ and $\forall j \in \{1..n\}$.
  3. an edge $(V_{-i}, V_{-(i-1)})$, $\forall i \in \{1..(m-1)\}$.
- The new cost function $c' : E' \rightarrow \Re_+$ is defined as follows:
  1. $c'(V_i, V_j) = c(V_i, V_j)$, $\forall i = \{1, 2..n\}$, $\forall j = \{1, 2..n\}$ and edge $(V_i, V_j) \in E$.
  2. $c'(V_{-i}, V_j) = c(V_0, V_j) + \frac{1}{2}C_i$, $\forall i = \{0, 1, ..(m-1)\}$, $\forall j = \{1, 2..n\}$ and edge $(V_0, V_j) \in E$.
  3. $c'(V_{-i}, V_{-i+1}) = \frac{1}{2}(C_{i-1} - C_i)$, $\forall i \in \{1..(m-1)\}$.

An example of this transformation is shown in Fig. 10 and Fig. 11. The main result in Rao [10] that helps us solve the **SDTSP** is stated in the following theorem.

Theorem 4. *Solving the* ***SDTSP*** *on graph G is equivalent to solving a single TSP on the transformed graph G'.*

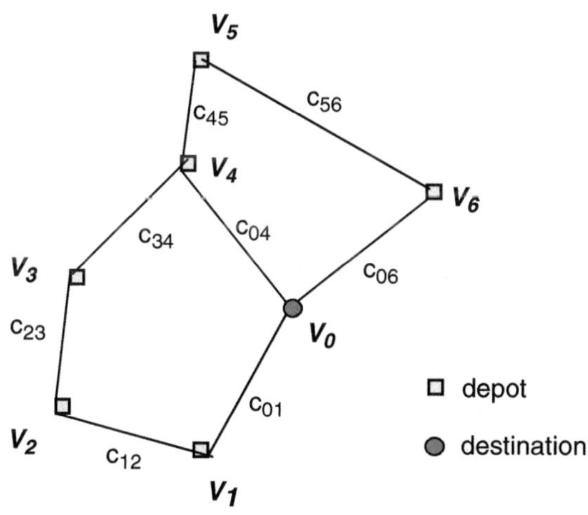

Fig. 10. An example of a graph G with 3 vehicles present at the depot

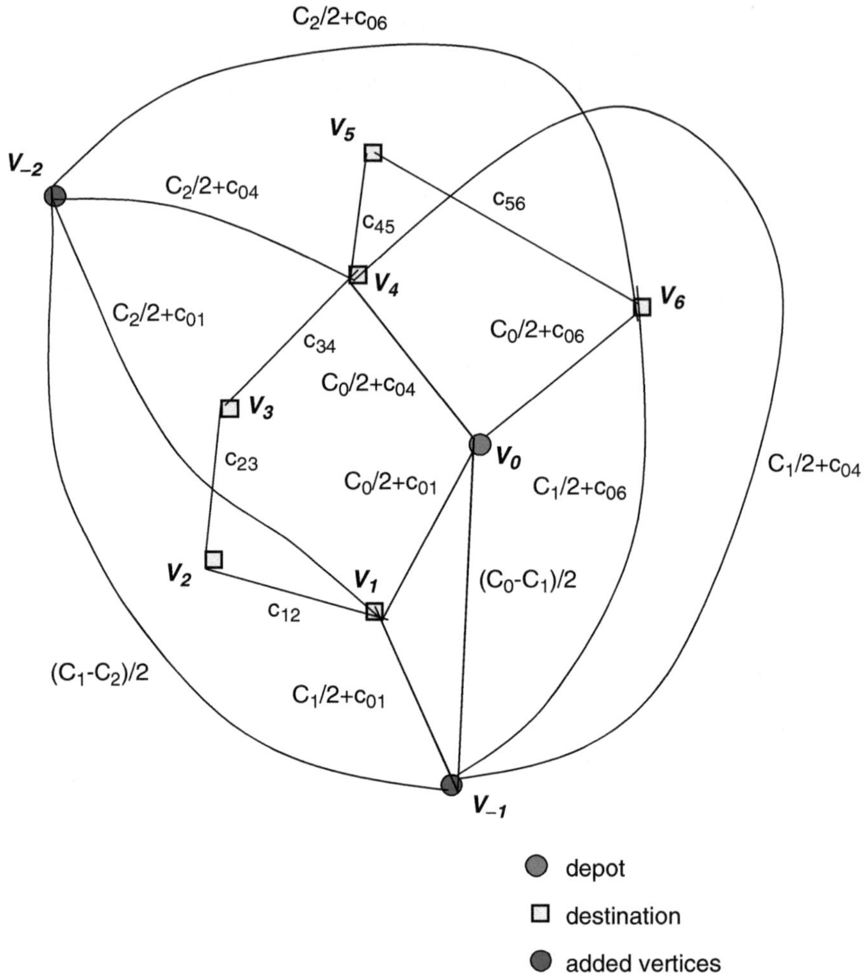

Fig. 11. Transformed graph G'

3.3 Multiple Depot, Multiple TSP (MDMTSP)

Let there be n destinations and m UAVs. Let V be the set of vertices that correspond to the destinations, the starting and the terminal location of the UAVs. The first m vertices of V namely, $V_1, \ldots, V_m$, represents the starting locations of the UAVs (i.e., the vertex V_i corresponds to the starting location of the ith vehicle). The next n vertices in V, $V_{m+1}, \ldots, V_{m+n}$, represents the destinations. Finally, vertices $V_{m+n+1}, \ldots, V_{2m+n}$ in V represents the possible terminal locations of the UAVs. Let $E = V \times V$ denote the set of all edges (pairs of vertices) and let $c : E \rightarrow \Re_+$ denote the cost function with $c(V_i, V_j)$ (or simply, c_{ij}) representing the cost of travelling from vertex

V_i to vertex V_j. We consider costs that are symmetric and satisfy triangle inequality. A path is an ordered set, PATH_i, of at least $r+2$, $r \geq 1$ elements of the form $\{V_i, V_{i_1}, \ldots, V_{i_r}, V_{i_f}\}$, where V_{i_l}, $l = 1, \ldots, r$ corresponds to r distinct destinations being visited in that sequence by the ith UAV and V_{i_f} is a terminal location. Any two paths PATH_i and PATH_j are such that $\text{PATH}_\text{i} \bigcap \text{PATH}_\text{j} = \Phi$. There is a cost, $C(\text{PATH}_\text{i})$, associated with a path for the ith UAV and is defined as $C(\text{PATH}_\text{i}) = c_{i,i_1} + \sum_{k=1}^{r-1} c_{i_k,i_{k+1}} + c_{i_r,i_f}$. Let each UAV be allowed to choose any one of the given terminal locations present in $V_{m+n+1}, \ldots, V_{2m+n}$ not visited by other UAVs. Given the graph $G = (V, E)$, find m UAV paths such that each destination is visited by only one UAV and the overall cost defined as $\sum_{i=1}^{m} C(\text{PATH}_\text{i})$ is minimum.

Approximation Algorithm for MDMTSP

Before, we present the approximation algorithm we give the definition of a constrained forest as discussed in [12]. A constrained forest is a subgraph of G with m disjoint trees such that each tree spans exactly one vertex from $\{V_1, \ldots, V_m\}$, exactly one vertex from $\{V_{m+n+1}, \ldots, V_{2m+n}\}$ and a subset of vertices from $\{V_{m+1}, \ldots, V_{m+n}\}$. (i.e. each tree must consist of exactly one starting vertex and one terminal vertex). The approximation algorithm *CF* [12] that solves the **MDMTSP** is as follows:

1. Find the minimum cost constrained forest. The output of this step for an example with five vehicles is shown in Fig. 12.
2. For each tree corresponding to a vehicle, double its edges to construct its Eulerian graph (Fig. 13).
3. Then construct a path for each vehicle based on its Eulerian graph (Fig. 14). This step essentially uses the same algorithm implemented for the tour computation in the single TSP (section 2.3).

The following theorem in [12] shows algorithm *CF* has an approximation factor of 2.

Theorem 5. *The algorithm CF solves the* ***MDMTSP*** *with an approximation factor of 2 in* $O((n+2m)^6)$ *steps when the costs are symmetric and satisfy triangle inequality.*

3.4 Generalized Multiple Depot Multiple TSP (GMTSP)

Problem Formulation

Let there be n destinations and m UAVs. Let V be the set of vertices that correspond to the location of UAVs and the destinations, with the first m

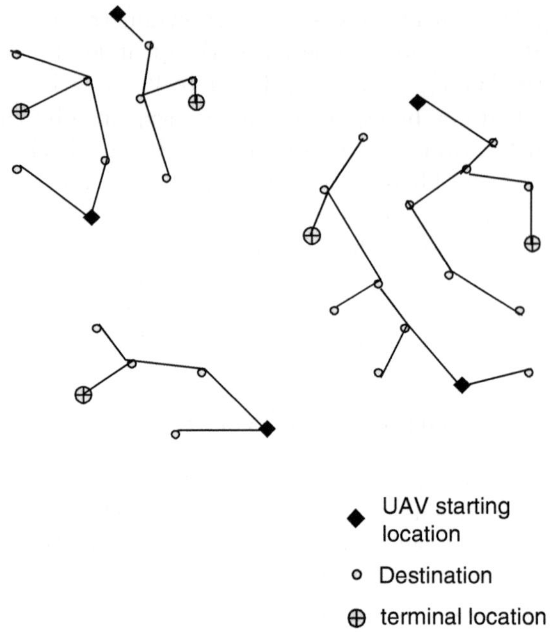

Fig. 12. Step 1 of algorithm *CF* for **MDMTSP**: Find the optimal constrained forest

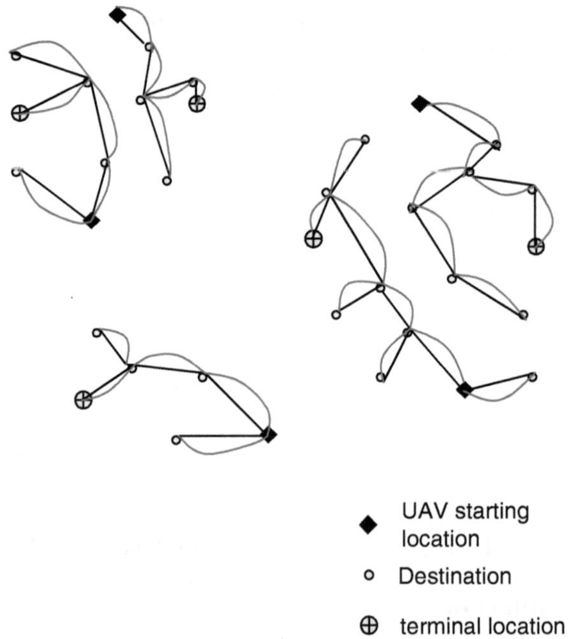

Fig. 13. Step 2 of algorithm *CF* for **MDMTSP**: Double the edges in each tree to get a Eulerian graph for each vehicle

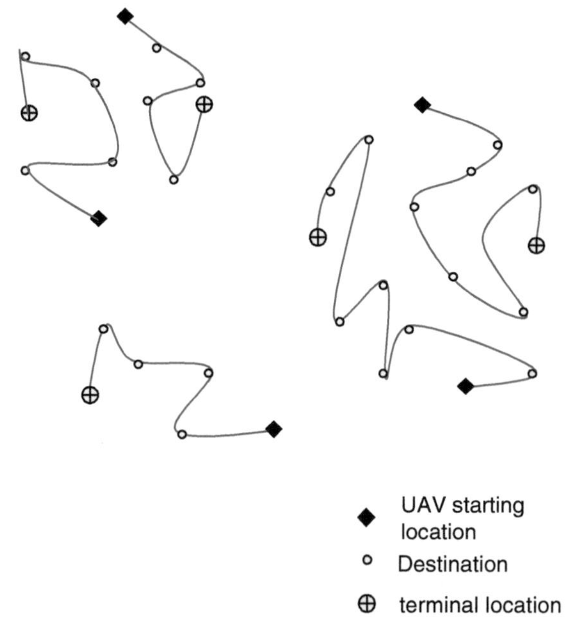

Fig. 14. Step 3 of algorithm *CF* for **MDMTSP**: Construct a path out of each Eulerian graph

vertices $V_1, \ldots, V_m$ representing the UAVs (i.e., the vertex V_i corresponds to the ith UAV) and $V_{m+1}, \ldots, V_{m+n}$ representing the destinations. Let $E = V \times V$ denote the set of all edges (pairs of vertices) and let $c : E \rightarrow \Re_+$ denote the cost function with $c(V_i, V_j)$ (or simply, c_{ij}) representing the cost of travelling from vertex V_i to vertex V_j. We consider costs that are symmetric, i.e. $c_{ij} = c_{ji}$ and satisfy triangle inequality. A tour is an ordered set, $TOUR_i$, of at least $r + 2$, $r \geq 1$ elements of the form $\{V_i, V_{i_1}, \ldots, V_{i_r}, V_i\}$, where V_{i_l}, $l = 1, \ldots, r$ corresponds to r distinct destinations being visited in that sequence by the ith UAV. There is a cost, $C(TOUR_i)$, associated with a tour for the ith UAV and is defined as $C(TOUR_i) = c_{i,i_1} + \sum_{k=1}^{r-1} c_{i_k,i_{k+1}} + c_{i_r,i}$. If $\mathcal{S}_p$ is the set of p vehicles chosen to visit the destinations, the overall cost is defined as $\sum_{i \in \mathcal{S}_p} C(TOUR_i)$. Given the graph $G = (V, E)$, and a number $p \leq m$, choose at most p UAVs so that each destination is visited by at least one UAV and the overall cost is a minimum among all possible choice of p or fewer UAVs and their corresponding tours.

Approximation Algorithm for GMTSP

The approximation algorithm *CT* [13] that solves the **GMTSP** is given as follows:

1. Construct a graph $\tilde{G}$ as follows: Add a new vertex (called as the root) denoted by r. Connect r to all the vertices denoting the UAVs through zero

cost edges. Remove the edges between any pair of vertices representing the UAVs.
2. Construct a constrained Minimum Spanning Tree on $\tilde{G}$ such that the sum of the degrees of the vertices denoting the UAVs to be at most $m + p$.
3. By dropping all the edges between the root vertex and each of the vertices representing the UAVs in the constrained MST found from step 2, one will get a forest consisting of at most p non-trivial trees (a non-trivial tree is one which consists of atleast one edge) that spans all destinations with exactly one UAV in each tree and at least $m - p$ vehicles that are not incident on any edge.
4. We then double the edges of the non-trivial trees and construct a tour for each of the vehicles by following the exact procedure outlined in the 2-approximation algorithm for single TSP in section 2.3.

The following theorem in [13] shows this algorithm CT has an approximation factor of 2.

Theorem 6. *The algorithm CT solves the **MVMDP** with an approximation factor of 2 in $O((n+m)^4)$ steps when the costs are symmetric and satisfy triangle inequality.*

4 Resource Allocation Problems in the Presence of Kinematic Constraints

4.1 Problem Formulation

Let $(x(v_i,t), y(v_i,t), \theta(v_i,t))$ denote the position and the heading of UAV v_i at time t. Let each UAV start at an initial heading $\theta(v_i,0) = \alpha_i$. Similarly, let $(x(d_j,t), y(d_j,t))$ denote the position of destination d_j at time t. Since the destinations are assumed to be stationary, let $(\bar{x}(d_j), \bar{y}(d_j)) = (x(d_j,t), y(d_j,t)) \ \forall\ t$. Given a set of UAVs $\{v_1, v_2, ...v_m\}$ and destinations $\{d_1, d_2, ...d_n\}$, the problem is to

- assign a sequence of destinations P_i to each UAV to visit such that $\{d_1, d_2...d_n\} = \{\bigcup_i P_i\}$ and $\{P_i\} \bigcap \{P_j\} = \emptyset$ if $i \neq j$.
- assign to each UAV v_i, a path through the sequence P_i such that the path of each UAV v_i satisfies the following kinematic constraints:

$$\frac{dx(v_i,t)}{dt} = v_o \cos(\theta(v_i,t)),$$
$$\frac{dy(v_i,t)}{dt} = v_o \sin(\theta(v_i,t)),$$
$$\frac{d\theta(v_i,t)}{dt} = \Omega \text{ where } \Omega \in [-\omega, +\omega], \tag{5}$$

where, v_o denotes the speed, ω represents the bound on the yaw rate and $r = \frac{v_o}{\omega}$ is the minimum turning radius of each UAV.

Let the sequence P_i for UAV v_i be $d_{i_1}, ...d_{i_k}$. Assigning a path for UAV v_i through its sequence P_i of destinations also implies assigning the angles of approach β_{d_i} at each destination and assigning the angle of return β_{v_i} at which the UAV comes back to its initial position $(x(v_i, 0), y(v_i, 0))$. For example, the ith UAV moves from $(x(v_i, 0), y(v_i, 0), \alpha_i)$ to $(\bar{x}(d_{i_1}), \bar{y}(d_{i_1}), \beta(d_{i_1}))$, and then from $(\bar{x}(d_{i_1}), \bar{y}(d_{i_1}), \beta(d_{i_1}))$ to $(\bar{x}(d_{i_2}), \bar{y}(d_{i_2}), \beta(d_{i_2}))$ and so on. After reaching d_{i_k}, it comes back to its initial position $(x(v_i, 0), y(v_i, 0))$ at an angle β_{v_i}.

The objective is to minimize $\sum_{i=1}^{n} Cost(P_i)$, where $Cost(P_i)$ is the total distance travelled by the ith UAV.

The above problem is called as the **RAP(m)**, i.e, Resource Allocation Problem for m UAVs.

4.2 Literature Review

Significant interest in the potential of realizing a mission in battle field environments using a collection of small autonomous UAVs was the main motivation that lead to the formulation of problems such as **RAP(m)**. Resource allocation problems concerning UAVs has received considerable attention in the last 7 years [15], [16], [17], [18], [19],[20], [21], [22], [23]. A more general version of **RAP(m)** with each destination requiring multiple tasks was formulated in [24]. Yang et al. [25] consider path planning for an UAV with kinematic constraints given fixed initial and final positions in the presence of obstacles. The UAV in their work is required to visit a destination and then reach a final position avoiding threats and other obstacles. This is related to **RAP(1)** in the absence of obstacles when there is one destination on the tour. The single vehicle problem (**RAP(1)**) has been addressed by several authors [26], [27], [29], [30]. In [26], Savla et al. bound the distance of the UAV path between any points (x_1, y_1, θ_1) and (x_2, y_1, θ_2) in terms of the Euclidean distance between the corresponding points. Also, using this result, they propose an algorithm which bounds the total distance travelled by the vehicle in terms of the Euclidean distance tour. Ny et al. [27] provide an algorithm with an approximation factor of $(1+\max\{\frac{8\pi r}{D_{min}}, \frac{14}{3}\}) \log n$, where D_{min} is the minimum Euclidean distance between any two locations. They approximate **RAP(1)** as an asymmetric TSP and use the bound of $\log n$ by Frieze et al. [28] to get the approximation factor. In [29], Rathinam et al. provide an algorithm for **RAP(1)** with an approximation factor of 4.56 by assuming that $D_{min} \geq 2r$. The main difference between the result in [29] and [27] is that Rathinam et al. approximate the **RAP(1)** as as symmetric TSP and hence the approximation factor is independent of n. Tang et al. [30] also provide a heuristic for **RAP(1)**that uses an approximate gradient method to determine the path of the UAV. However, there are no bounds presented in [30].

The paper that is most relevant to the multiple vehicle problem (**RAP(m)**) is the work by Tang et al. [30]. In [30], Tang et al. provide heuristics for multiple vehicles tracking moving destinations using clustering and gradient techniques. Even though [30] consider moving destinations, their main results are for stationary destinations which is essentially the **RAP(m)**. Also heuristics for more general versions of **RAP(m)** are presented in [31] [32], but there are no bounds. Rathinam et al. [29] provide a algorithm for **RAP(m)** with an approximation factor of 6.07 by assuming that $D_{min} \geq 2r$. In the following subsections, we review two algorithms, one by Savla et al. [26] for the single vehicle case and an other by Rathinam et al. [29] for the multiple vehicle case.

Remark: Before we discuss the algorithms, we present the result by L.E. Dubins [33] which forms the motivation for the paths chosen in the algorithms. L.E. Dubins [33] gives the optimal path the vehicle must travel between any two points subject to the path constraints given by equations 5. Henceforth, any curved segment of radius r along which the vehicle executes a clockwise (counterclockwise) rotational motion is denoted by $R(L)$, and the segment along which the vehicle travels straight is denoted by S. Thus the path in figure 15 is an RSL path. Dubin's result states that the path joining the two points (x_1, y_1, θ_1) and (x_2, y_2, θ_2) that has minimal length subject to constraints in 5, is one of RSR, RSL, LSR, LSL, RLR and LRL. Such an optimal path between any two points that has minimum length subject to constraints in 5 would be called a *Dubin's* path in this chapter.

4.3 Alternating Algorithm for the Single UAV Case

Let the number of destination points be $(n \geq 2)$.

1. Compute the optimal single TSP tour ignoring the kinematic constraints of the vehicles (i.e. find the optimal single TSP tour based on the Euclidean distances between all the points). Let the sequence of the destinations in the calculated tour be denoted by $d_{i_1}, ...d_{i_n}$.

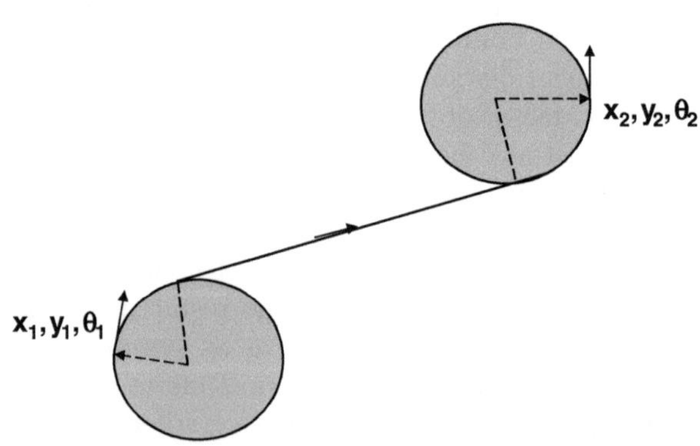

Fig. 15. Shortest path - {*clockwise, straight, counter clockwise*}

2. Since the sequence of the destinations is known, the path of the UAV can be determined by fixing the heading angles at each of the destinations. The heading angles are now fixed as follows:
   a) Let $j = 1$.
   b) If j is odd and $j \leq n - 1$, fix β_{i_j} to be the orientation of the line segment joining d_{i_j} to $d_{i_{j+1}}$, i.e $\beta(d_{i_j}) := \arctan[\frac{\bar{y}(d_{i_{j+1}})-\bar{y}(d_{i_j})}{\bar{x}(d_{i_{j+1}})-\bar{x}(d_{i_j})}]$.
   c) If j is odd and $j = n$, fix β_{i_j} to be the orientation of the line segment joining d_{i_n} to the initial position of the vehicle, i.e $\beta(d_{i_j}) := \arctan[\frac{y(v_1,0)-\bar{y}(d_{i_n})}{x(v_1,0)-\bar{x}(d_{i_n})}]$.
   d) if j is even, fix $\beta(d_{i_j}) := \beta(d_{i_{j-1}})$.
   e) if $j = n$ fix the return angle of the UAV to its initial position, β_{v_1}, equal to $\beta(d_{i_n})$ and stop. Else, if $j < n$, assign $j \Longrightarrow j + 1$ and go to step (b).
3. Now construct Dubin's path from $(x(v_i, 0), y(v_i, 0), \alpha_i)$ to $(\bar{x}(d_{i_1}), \bar{y}(d_{i_1}), \beta(d_{i_1}))$ and then from $(\bar{x}(d_{i_1}), \bar{y}(d_{i_1}), \beta(d_{i_1}))$ to $(\bar{x}(d_{i_2}), \bar{y}(d_{i_2}), \beta(d_{i_2}))$ and so on. For the last leg of the tour that joins d_{i_n} to the initial vehicle location, construct a Dubin's path from $(\bar{x}(d_{i_n}), \bar{y}(d_{i_n}), \beta(d_{i_n}))$ to $(x(v_i, 0), y(v_i, 0), \beta_{v_1})$.

An example of the alternating algorithm is shown in Fig. 16. The main result in [26] bounds the length of the Dubin's path $D(p_1, p_2)$ that joins $p_1 = (x_1, y_1, \theta_1)$ to $p_2 = (x_2, y_2, \theta_2)$ in terms of the Euclidean distance $E(p_1, p_2)$ between the points, where $E(p_1, p_2) := \sqrt{(x_1 - x_2)^2 + (y_1 - y_2)^2}$. This result is stated in the following theorem.

Theorem 7. *$D(p_1, p_2) \leq E(p_1, p_2) + \kappa\pi r$ where $\kappa \in [2.657, 2.658]$ and r is the minimum turning radius of the UAV.*

4.4 Approximation Algorithm for the Multiple UAV Case

Rathinam et al. [29] assume that the Euclidean distances between any two destinations and the Euclidean distance between the initial position of each UAV and a destination is greater than twice the minimum turning radius of the UAV. This is a reasonable assumption in the context of unmanned aerial UAVs which carry sensors that have footprints that are greater than $2r$. This implies that $\sqrt{(\bar{x}(d_j) - \bar{x}(d_k))^2 + (\bar{y}(d_j) - \bar{y}(d_k))^2} \geq 2r$ and $\sqrt{(x(v_i, 0) - \bar{x}(d_j))^2 + (y(v_i, 0) - \bar{y}(d_j))^2} \geq 2r$, $\forall j \neq k$, $\forall j, k \in \{1, 2..n\}, \forall i \in \{1, 2..m\}$.

First, we give a simple algorithm **S** for the UAV v_1 to find a path to travel from positions $(x(v_1), y(v_1), \alpha_1)$ to $(\bar{x}(d_j), \bar{y}(d_j))$. Note that the final approach angle at the position $(\bar{x}(d_j), \bar{y}(d_j))$ is free to be chosen. Algorithm **S** is as follows:

1. Find the distances of two possible paths the UAV could take: RS and LS.
2. Choose the path that has the minimum distance.

Once, this path is followed, the UAV reaches the position $(\bar{x}(d_j), \bar{y}(d_j))$ at some final angle θ and this angle is chosen as the heading at the final position.

The algorithm MVA for the **RAP(m)** is as follows:

1. Construct a complete graph with vertices being all the UAVs and destinations. Assign the Euclidean distance as the cost to each edge that joins a UAV to a destination and a destination to a destination. Assign zero cost to an edge that joins any two UAVs.
2. Find the minimum spanning tree of the graph using Prim's algorithm [2]. This minimum spanning tree will contain exactly $m-1$ zero cost edges where m is the number of UAVs (Fig. 17).
3. Remove the zero cost edges to get a tree for each UAV (Fig. 18).
4. For each tree corresponding to a UAV, double its edges to construct a Eulerian graph (Fig. 19). Then construct a tour for each UAV based on the Eulerian graph. A tour for each UAV is a sequence of destinations for it to visit (Fig. 20). (This step is similar to tour construction for the single TSP discussed in section 2.3).
5. Use the above sequence and construct paths using algorithm **S** between any two consecutive locations. For example, use algorithm **S** to construct a path from $(x(v_1), y(v_1), \alpha_1)$ to $(\bar{x}(d_1), \bar{y}(d_1))$. Say, the UAV reaches the

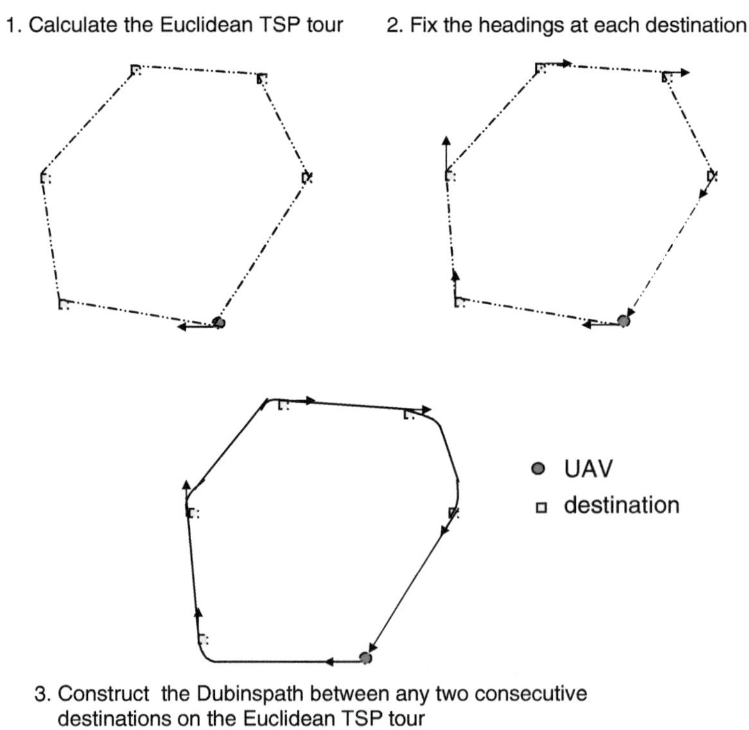

Fig. 16. Alternating Algorithm for the **RAP(1)**

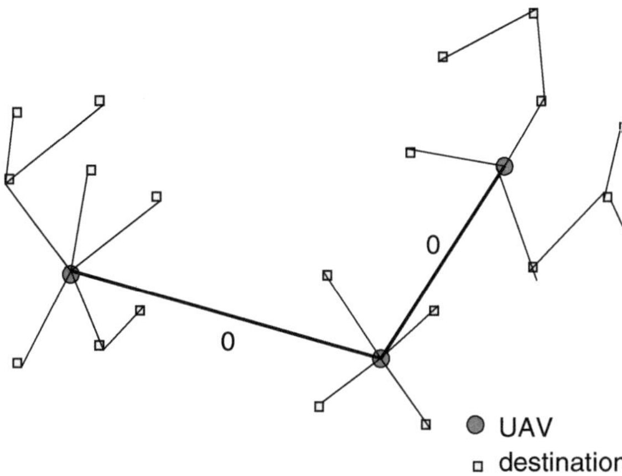

Fig. 17. Calculate the minimum spanning tree (MST). In this example, there are 3 UAVs, hence MST will have 2 zero cost edges

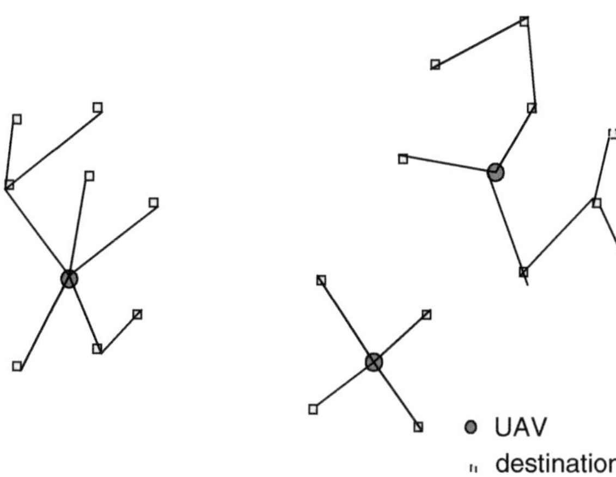

Fig. 18. Remove the zero cost edges from MST to yield a tree for each UAV

destination d_1 at an angle θ. Again, use algorithm **S** to construct a path from $(\bar{x}(d_1), \bar{y}(d_1), \theta)$ to $(\bar{x}(d_2), \bar{y}(d_2))$ and so on. (Fig. 21).

The above algorithm has an approximation factor of 6.07 [29]. This is stated in the following theorem.

Theorem 8. *AlgorithmMVA with the assumptions on the minimum Euclidean distance solves the **RAP(m)** with an approximation factor equal to* $2(\pi + 1 - \tan^{-1}(2)) \approx 6.07$ *in* $O((n+m)^2)$ *steps.*

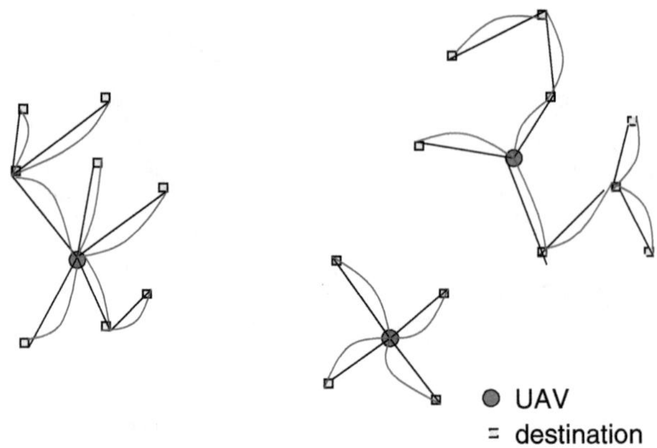

Fig. 19. After removing the zero cost edges, double the edges of the MST to get a Eulerian graph for each UAV

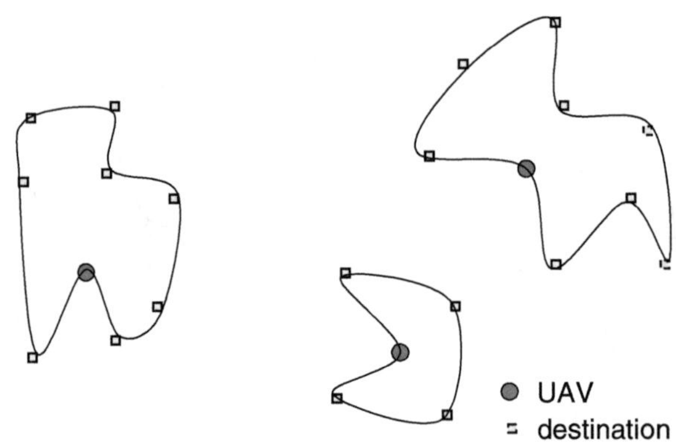

Fig. 20. Compute a tour based on the Eulerian graph for each UAV

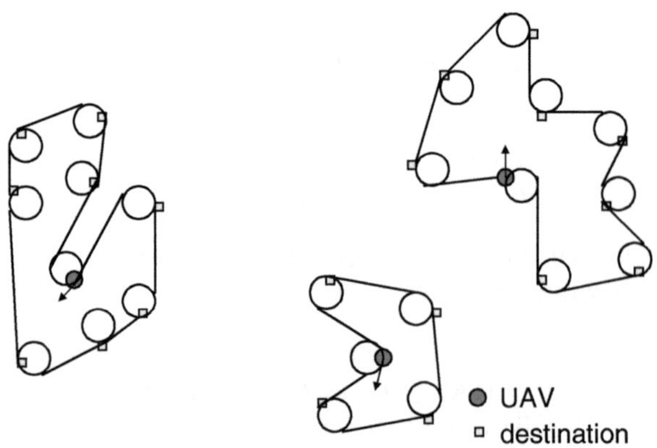

Fig. 21. Use the sequence got from the tour and construct paths using the **S** algorithm between the corresponding locations

5 Summary and Open Problems

This chapter formulated a set of resource allocation problems that are motivated by the applications involving Unmanned Aerial Vehicles. Since UAVs have fuel constraints in them and the distance travelled by the vehicles depend upon its fuel capacity, the problems focussed on the objective of minimizing the total distance travelled. Since these problems are variants or generalizations of the Travelling Salesman Problem that is NP-Hard, approximation algorithms were presented to solve the same. The kinematics of the UAVs further complicate these resource allocation problems and methods that have been presented in this chapter combine results from the TSP and the optimal control literature. The following part of the section discusses some of the key issues that have not been addressed in this chapter and the related open problems in the context of UAV applications:

- *Approximation algorithms with lesser bounding factors:*

  This chapter reviewed algorithms with an approximation factor of 2 for different variants of multiple depot routing problems. It is not clear whether the Christofides algorithm can be extended to the multiple depot case. The main difficulty in deriving lesser approximation factors is due to the hardness in obtaining a suitable partition of the destination vertices. Another result that is worth mentioning here is a complexity result for the bottleneck variants of the multiple depot problem. In [35], it is stated that it is hard to derive an algorithm with an approximation factor less than 2 unless P=NP for bottleneck variants. It is unclear whether a similar result can be derived for the multiple depot problems presented in this chapter.

- *Distributed algorithms:*

  The algorithm for the multi depot problem given in this paper involved finding a minimum spanning tree of all the vertices. It is known that minimum spanning tree computations can be distributed and auction style algorithms can be developed for these problems as shown in [34]. But it seems that there is a tradeoff between obtaining a tighter approximation factor versus distributed computation. It is intuitive that it would be even harder to obtain distributed algorithms with approximation factors less than 2. Recent results in [34] suggest some approaches for these routing problems based on auctions. Further studies on distributed, routing algorithms are suggested in the context of UAV applications.

- *Computational results involving UAVs:*

  The main difference between the routing problems involving UAVs and the TSP variants is that UAVs have additional kinematic and dynamic constraints. Though there are several theoretical results for routing problems involving UAVs currently in the literature, there have been no computational results that compare the performance of different heuristics for these problems. Even though algorithms with approximation factors are helpful, there might be simple heuristics that could perform well in practice. The main difficulty of these routing problems involving UAVs is that there are no existing methods to calculate the optimal cost. However lower bounds based on Euclidean distances can be easily derived using the algorithms presented in this paper. A study comparing the performance of different heuristics for a given number of depots and destinations would be very useful.

- *Heterogeneous vehicles:*

  All the problems considered in this chapter assumed a homogeneous collection of vehicles. Many applications involving UAVs might require vehicles with different capabilities to act in a cooperative manner. A simple case would be when the vehicles have a different minimum turning radius. It is unclear even whether algorithms with approximation factors of 2 are possible for these problems.

- *Adding and deleting destination points:*

  In military applications, it would be common to have tasks removed or added as the mission progresses. A simple scenario would be when certain destination points are deleted or added frequently. A naive approach to deal with such scenarios would be to recompute solutions whenever the destinations change. But this might require a large computation time. A very useful research direction would be to derive algorithms that can adapt itself to changing scenarios. In particular, the following question is the one to ask: Can one devise a routing algorithm for all the vehicles that does not recompute the entire solution from scratch but rather uses old information in building new solutions?

References

1. Vazirani, V.V., 2001. Approximation algorithms, Springer
2. Papadimitriou, C.H., Steiglitz, K., 1998. Combinatorial optimization: algorithms and complexity, Dover publications
3. Christofides, N., 1976. Worst-case analysis of a new heuristic for the travelling salesman problem. In: J.F. Traub (Editor), Algorithms and Complexity: New Directions and Recent Results, Academic Press, pp. 441
4. Arora, S., 1996. Polynomial-time approximation schemes for Euclidean TSP and other geometric problems. Proceedings of the 37th Annual Symposium on the Foundations of Computer Science, pp. 2–11
5. Held, M., Karp, R.M., 1970. The traveling salesman problem and minimum spanning trees. Operations Research 18, pp. 1138–1162
6. Held, M., Karp, R.M., 1971. The travelling salesman problem and minimum spanning trees: Part II. Mathematical Programming 18, pp. 6–25
7. Gutin, G., Punnen, A.P. (Editors), 2002. The travelling salesman problem and its variations. Kluwer Academic Publishers

8. Bektas, T., 2006. The Multiple Traveling Salesman Problem: an Overview of Formulations and Solution Procedures. OMEGA: The International Journal of Management Science, 34(3), 209–219
9. Bellmore, M., Hong, S., 1977. A note on the symmetric multiple travelling salesman problem with fixed charges. Operations Research 25, pp. 871–874
10. Rao, M.R., 1980. A note on multiple travelling salesmen problem. Operations Research 28(3), pp. 628–632
11. GuoXing, Y., 1995. Transformation of multidepot multisalesmen problem to the standard traveling salesman problem. European Journal of Operations Research 81, pp. 557–560
12. Rathinam, S. and Sengupta, R., 2006. Lower and upper bounds for a symmetric, multiple depot, multiple travelling salesman problem. Submitted to IEEE conference on Decision and Control
13. Darbha, S., 2005. Combinatorial motion planning of reed-shepp vehicles, Final Report, American Society for Engineering Education (ASEE)\ Airforce Office of Scientific Research(AFOSR), Summer Faculty Program, Air Force Research Laboratory, Eglin, Florida
14. Gavish, B., Srikanth, K., 1986. An optimal solution method for the multiple travelling salesman problem. Operations Research 34(5), pp. 698–717
15. Chandler, P.R., Pachter, 1998. m., Research issues in autonomous control of tactical UAVs. American Control Conference, pp. 394–398
16. Chandler, P.R., Rasmussen, S.R., Pachter, M., 2000. UAV cooperative path planning. Proceedings of the GNC, pp.1255–1265
17. Chandler, P.R., Pachter, M., 2001. Hierarchical control of autonomous control of tactical UAVs. Proceedings of GNC, pp. 632-642
18. Chandler, P.R., Rasmussen, S.R., Pachter, M., 2001. UAV cooperative control. American Control Conference
19. Schumacher, C., Chandler, P.R., Rasmussen, S.R., 2001. Task allocation for wide area search munitions via network flow optimization. AIAA Guidance, Navigation, and Control Conference and Exhibit, Montreal, Canada
20. Chandler, P.R., Pachter, M., Swaroop, D., Fowler, J.M., Howlett, J.K., Rasmussen, S.R., Schumacher, C., Nygard, K., 2002. Complexity in UAV cooperative control. Proceedings of the American Control Conference, Anchorage, Arkansas
21. Maddula, T., Minai, A.A., Polycarpou, M.M., 2002. Multi-target assignment and path planning for groups of UAVs. S. Butenko, R. Murphey, and P. Pardalos (Eds.), Kluwer Academic Publishers
22. Richards, A., Bellingham, J., Tillerson, M., How, J. P., 2002. Co-ordination and control of multiple UAVs. AIAA Guidance, Navigation, and Control Conference
23. Alighanbari, M., Kuwata, Y., How, J.P., 2003. Coordination and control of multiple UAVs with timing constraints and loitering. Proceeding of the IEEE American Control Conference
24. Darbha, S., 2001. Teaming Strategies for a resource allocation and coordination problem in the cooperative control of UAVs. AFRL Summer Faculty Report, Dayton, Ohio
25. Yang, G., Kapila, V., 2002. Optimal path planning for unmanned air vehicles with kinematic and tactical constraints. Proceedings of the 41st IEEE Conference Decision and Control 2, pp. 1301–1306

26. Savla, K., Frazzoli, E., Bullo, F., 2005. On the point-to-point and traveling salesperson problems for Dubin's vehicle. American Control Conference, Portland, Oregan
27. Ny, J.L., Feron, E., 2005. An approximation algorithm for the curvature constrained traveling salesman problem. Proceedings of the 43rd Annual Allerton Conference on Communications, Control and Computing
28. Frieze, A., Galbiati, G., Maffioli, F., 1982. On the worst-case performance of some algorithms for the asymmetric traveling salesman problem. Networks 12, pp. 23–39
29. Rathinam, S., Sengupta, R., Swaroop, D., 2005. A resource allocation algorithm for multi vehicle systems with non-holonomic constraints. Accepted in IEEE Transactions on Automation Science and Engineering
30. Tang, Z., Ozguner, U., 2005. Motion planning for multi-target surveillance with mobile sensor agents. IEEE Transactions of Robotics
31. Beard, R., Mclain, T., Goodrich, M., Anderson, E., 2002. Coordinated target assignment and intercept for unmanned air vehicles. IEEE Transactions on Robotics and Automation 18(6), pp. 911–922
32. Mclain, T., Beard, R., 2003. Cooperative path planning for timing critical missions. Proceedings of the American Control Conference, Denver, Colorado
33. Dubins, L.E., 1957. On curves of minimal length with a constraint on average curvature, and with prescribed initial and terminal positions and tangents. American Journal of Mathematics 79(3), pp. 487–516
34. Lagoudakis, M. G., Markakis, E., Kempe, D. , Keskinocak, P., Kleywegt, A., Koenig, S., Tovey, C., Meyerson, A., and Jain, S., June 2005. Auction-Based Multi-Robot Routing. Proceedings of Robotics: Science and Systems I, Cambridge, USA
35. Hochbaum, S., July 1996. Approximation Algorithms for NP-Hard Problems

State Estimation for Micro Air Vehicles

Randal W. Beard

Department of Electrical and Computer Engineering
Brigham Young University, Provo, Utah
beard@ee.byu.edu

Abstract. Autopilots for small UAVs are generally equipped with low fidelity sensors that make state estimation challenging. In addition, the sensor suite does not include units that measure angle-of-attack and side-slip angles. The achievable flight performance is directly related to the quality of the state estimates. Unfortunately, the computational resources on-board a small UAV are generally limited and preclude large state Kalman filters that estimate all of the states and sensor biases. In this chapter we describe simple models for the sensors typically found on-board small UAVs. We also describes a simple cascaded approach to state estimation that has been extensively flight tested using the Kestrel autopilot produced by Procerus Technologies. Our intention is to provide a tutorial of continuous-discrete Kalman filtering with application to state estimation for small UAVs.

High fidelity estimates of the position, velocity, attitude, and angular rates are critical for successful guidance and control of intelligent UAVs. The achievable fidelity of the state estimates depends upon the quality of the sensors on-board the UAV. Unfortunately, high quality sensors are usually heavy and expensive. This is particularly true for sensors that directly measure the attitude of the UAV. In this chapter we focus on the problem of state estimation using light weight, inexpensive, low quality sensors. In doing so, our target platforms are small and micro air vehicles with limited payload capacity.

In recent years, several autopilots for small UAVs have appeared on the commercial market. These include the Procerus Kestrel [4], the Cloudcap Piccolo [2], and the Micropilot MP2028 [3]. Each of these autopilots use the following sensors:

- rate gyros,
- accelerometers,
- pressure sensors, and
- GPS.

We will assume throughout this chapter that these are the only sensors that are available for state estimation.

R.W. Beard: *State Estimation for Micro Air Vehicles*, Studies in Computational Intelligence (SCI) **70**, 173–199 (2007)
www.springerlink.com

The limited payload capacity of small UAVs not only restricts the type and quality of the sensors, it also limits the computational resources that can be placed on-board the UAV. For example, the Procerus Kestrel autopilot has an 8-bit Rabbit microcontroller with 512K of memory. Therefore, Kalman filters that estimate all of the states as well as the sensor biases are not feasible. The objective of this chapter is to describe simple attitude estimation techniques for small UAVs that require limited computational resources.

The chapter is organized as follows. In Section 1 we define and briefly describe the states that need to be estimated. In Section 2 we describe the sensors that are generally available on small UAVs and develop mathematical models of their behavior. Section 3 briefly describes the simulation environment that is used to demonstrate the algorithms described in this chapter. Section 4 describes simple state estimation techniques that use digital low pass filters and sensor model inversion. In Section 5 we provide a brief review of the continuous-discrete Kalman filter. Finally, Section 6 describes the application of the continuous-discrete extended Kalman filter to roll, pitch, position, and heading estimation.

1 UAV State Variables

Aircraft have three degrees of translational motion and three degrees of rotational motion. Therefore, there are twelve state variables as listed below:

$$
\begin{aligned}
p_n &= \text{the inertial north (latitude) position of the UAV,} \\
p_e &= \text{the inertial east (longitude) position of the UAV,} \\
h &= \text{the altitude of the UAV,} \\
u &= \text{the body frame velocity measured out the nose,} \\
v &= \text{the body frame velocity measured out the right wing,} \\
w &= \text{the body frame velocity measured through the belly,} \\
\phi &= \text{the roll angle,} \\
\theta &= \text{the pitch angle,} \\
\psi &= \text{the yaw angle,} \\
p &= \text{the roll rate,} \\
q &= \text{the pitch rate,} \\
r &= \text{the yaw rate.}
\end{aligned}
$$

The state variables are shown schematically in Figure 1. As an alternative to expressing the velocity vector as $(u, v, w)^T$, it can be expressed in terms of the airspeed V_a, the angle-of-attack α, and the side-slip angle β. The transformation between the two representations is given by [22]

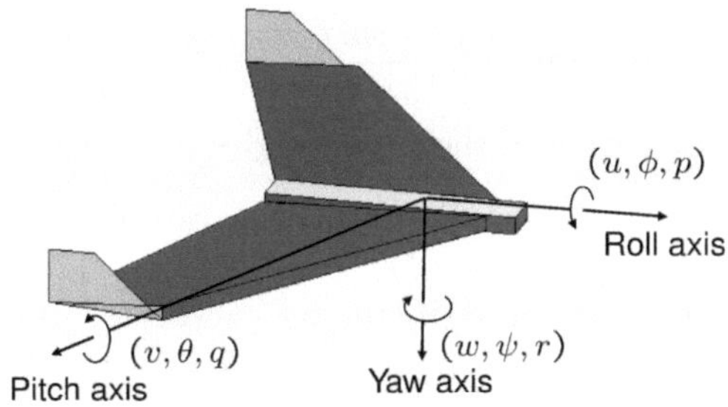

Fig. 1. This figures depicts some of the UAV state variables. The forward velocity u and the roll rate p are defined along the roll axis which points out the nose of the UAV. The side slip velocity v and the pitch rate q are defined along the pitch axis which points out the right wing of the UAV. The downward velocity w and the yaw rate r are defined with respect to the yaw axis which points out the belly of the UAV. The Euler angles are defined by first yawing ψ about the yaw axis, pitching θ about the transformed pitch axis, and finally rolling ϕ about the transformed roll axis

$$\begin{pmatrix} u \\ v \\ w \end{pmatrix} = V_a \begin{pmatrix} \cos\alpha\cos\beta \\ \sin\beta \\ \sin\alpha\cos\beta \end{pmatrix}. \tag{1}$$

$$\begin{aligned} V_a &= \sqrt{u^2 + v^2 + w^2} \\ \alpha &= \tan^{-1}\left(\frac{w}{u}\right) \\ \beta &= \tan^{-1}\left(\frac{v}{\sqrt{u^2 + w^2}}\right). \end{aligned} \tag{2}$$

There are several other quantities that are also of interest for guidance and control of UAVs including the flight path angle γ, the course angle χ, and the ground velocity V_g. The flight path angle defines the inertial climb angle of the UAV and is given by

$$\gamma = \theta - \alpha\cos\phi - \beta\sin\phi.$$

Note that in wings level flight, this formula reduces to the standard equation $\gamma = \theta - \alpha$. The course angle defines the inertial heading of the UAV which may be different than the yaw angle ψ due to wind. If $(w_n, w_e)^T$ is the wind vector in the inertial frame, then we have the following relationships

$$V_g \begin{pmatrix} \cos\chi \\ \sin\chi \end{pmatrix} = V_a \begin{pmatrix} \cos\psi\cos\gamma \\ \sin\psi\cos\gamma \end{pmatrix} + \begin{pmatrix} w_n \\ w_e \end{pmatrix}$$

$$V_g = \sqrt{V_a^2\cos^2\gamma + 2V_a\cos\gamma\sqrt{w_n^2 + w_e^2}\cos\left(\psi - \tan^{-1}\left(\frac{w_e}{w_n}\right)\right) + w_n^2 + w_e^2}$$

$$\chi = \tan^{-1}\left(\frac{V_a \sin\psi \cos\gamma + w_e}{V_a \cos\psi \cos\gamma + w_n}\right).$$

The kinematic evolution of the Euler angles are given by [22]

$$\begin{pmatrix} \dot{\phi} \\ \dot{\theta} \\ \dot{\psi} \end{pmatrix} = \begin{pmatrix} 1 & \sin(\phi)\tan(\theta) & \cos(\phi)\tan(\theta) \\ 0 & \cos(\phi) & -\sin(\phi) \\ 0 & \sin(\phi)\sec(\theta) & \cos(\phi)\sec(\theta) \end{pmatrix} \begin{pmatrix} p \\ q \\ r \end{pmatrix}, \tag{3}$$

and the navigational equations of motion are given by

$$\dot{r}_n = V_a \cos\psi \cos\gamma + w_n = V_g \cos\chi \cos\gamma \tag{4}$$

$$\dot{r}_e = V_a \sin\psi \cos\gamma + w_e = V_g \sin\chi \cos\gamma \tag{5}$$

$$\dot{h} = V_a \sin\gamma. \tag{6}$$

2 Sensor Models

This section derives mathematical models for sensors typically found on small and micro UAVs. In particular, we discuss rate gyros, accelerometers, pressure sensors, and GPS sensors.

2.1 Rate Gyros

A MEMS rate gyro contains a small vibrating lever. When the lever undergoes an angular rotation, Coriolis effects change the frequency of the vibration, thus detecting the rotation. A brief description of the physics of rate gyros can be found in Ref [9, 15, 23].

The output of the rate gyro is given by

$$y_{\text{gyro}} = k_{\text{gyro}}\omega + \beta_{\text{gyro}}(T) + \eta_{\text{gyro}},$$

where y_{gyro} is in Volts, k_{gyro} is a gain, ω is the angular rate in radians per second, β_{gyro} is a temperature dependent bias term, and η_{gyro} is a zero mean Gaussian process with known variance. The bias term $\beta_{\text{gyro}}(T)$ is a function of the temperature T and can be effectively determined by use of a temperature chamber before flight.

If three rate gyros are aligned along the x, y, and z axes of the UAV, then the rate gyros measure the angular body rates p, q, and r as follows:

$$y_{\text{gyro},x} = k_{\text{gyro},x}p + \beta_{\text{gyro},x}(T) + \eta_{\text{gyro},x}$$

$$y_{\text{gyro},y} = k_{\text{gyro},y}q + \beta_{\text{gyro},y}(T) + \eta_{\text{gyro},y}$$

$$y_{\text{gyro},z} = k_{\text{gyro},z}r + \beta_{\text{gyro},z}(T) + \eta_{\text{gyro},z}.$$

We will assume that $k_{\text{gyro},*}$, $\beta_{\text{gyro},*}(T)$, and the covariance of $\eta_{\text{gyro},*}$ have been determined *a priori* and are known in-flight. MEMS gyros are analog devices that are sampled by the on-board processer. We will assume that the sample rate is given by T_s. As an example, the Procerus Kestrel autopilot samples its rate gyros at approximately 120 Hz.

2.2 Accelerometers

A MEMS accelerometer contains a small plate attached to torsion levers. The plate rotates under acceleration which changes the capacitance between the plate and the surrounding walls. The change in capacitance is proportional to the linear acceleration [1, 23].

The output of the accelerometers is given by

$$y_{\text{acc}} = k_{\text{acc}} a + \beta_{\text{acc}}(T) + \eta_{\text{acc}},$$

where y_{acc} is in Volts, k_{acc} is a gain, a is the acceleration in meters per second squared, β_{acc} is a temperature dependent bias term, and η_{acc} is zero mean Gaussian noise with known variance.

Accelerometers measure the specific force in the body frame of the vehicle. A physically intuitive explanation is given in [22, p. 13–15]. An additional explanation is given in [19, p. 27]. Mathematically we have

$$\begin{aligned} \begin{pmatrix} a_x \\ a_y \\ a_z \end{pmatrix} &= \frac{1}{m}(\mathbf{F} - \mathbf{F}_{\text{gravity}}) \\ &= \dot{\mathbf{v}} + \boldsymbol{\omega} \times \mathbf{v} - \frac{1}{m}\mathbf{F}_{\text{gravity}}. \end{aligned}$$

In component form we have

$$\begin{aligned} a_x &= \dot{u} + qw - rv + g\sin\theta \\ a_y &= \dot{v} + ru - pw - g\cos\theta\sin\phi \\ a_z &= \dot{w} + pv - qu - g\cos\theta\cos\phi. \end{aligned}$$

The output of an accelerometer is usually in units of [g], therefore $k_{\text{acc}} = 1/g$. The output of the accelerometers are therefore given by

$$y_{\text{acc},x} = \frac{\dot{u} + qw - rv + g\sin\theta}{g} + \beta_{\text{acc},x}(T) + \eta_{\text{acc},x}$$

$$y_{\text{acc},y} = \frac{\dot{v} + ru - pw - g\cos\theta\sin\phi}{g} + \beta_{\text{acc},y}(T) + \eta_{\text{acc},y} \tag{7}$$

$$y_{\text{acc},z} = \frac{\dot{w} + pv - qu - g\cos\theta\cos\phi}{g} + \beta_{\text{acc},z}(T) + \eta_{\text{acc},z}. \tag{8}$$

As with the rate gyros, we will assume that the biases and noise statistics are known and available in-flight. MEMS accelerometers are analog devices that are sampled by the on-board processer. We will assume that the sample rate is given by T_s.

2.3 Pressure Sensors

Small autopilots typically have two pressure sensors: a static pressure sensor which is used to measure altitude, and a dynamic pressure sensor which is used to measure airspeed. These sensors will be discussed in the following two sections.

Altitude Sensor

Pressure is a measure of force per unit area or

$$P = \frac{F}{A},$$

where P is the pressure, F is the force, and A is the area. The static pressure at a particular altitude is determined by the force exerted by a column of air at that altitude:

$$P = \frac{m_{\text{column}} g}{A},$$

where m_{column} is the mass of the column of air, g is the gravitational constant, and A is the area upon which the column is exerting pressure. The density of air is the mass per unit volume. Since the volume is given by the area times the height we get

$$P = \rho h g,$$

where ρ is the density of air and h is the altitude [8, 11].

Therefore, the output of the static pressure sensor is given by

$$y_{\text{static pres}} = \rho g h + \beta_{\text{static pres}} + \eta_{\text{static pres}},$$

where $\beta_{\text{static pres}}$ is a slowly varying bias and $\eta_{\text{static pres}}$ is a zero mean Gaussian process. To remove the bias, we collect multiple measurements of the pressure on the ground and average to remove the Gaussian noise to obtain

$$\bar{y}_{\text{static pres}}(h_{\text{ground}}) = \rho g h_{\text{ground}} + \beta_{\text{static pres}}.$$

If we know the altitude of the ground station above sea level and the density of the surrounding air, then the bias $\beta_{\text{static pres}}$ can be determined. If, on the other hand, we are interested in the height above the ground station then we can subtract the calibrated ground measurement to obtain

$$\begin{aligned} y_{\text{static pres}}(\Delta h) &\stackrel{\triangle}{=} y_{\text{static pres}}(h) - \bar{y}_{\text{static pres}}(h_{\text{ground}}) \\ &= \rho g(h - h_{\text{ground}}) + \eta_{\text{static pres}}(t), \\ &= \rho g \Delta h + \eta_{\text{static pres}}(t), \end{aligned}$$

where Δh is the height above the ground station.

Air Speed Sensor

When the UAV is in motion, the atmosphere exerts dynamic pressure on the UAV in the direction of airflow. The dynamic pressure is given by [8]

$$P_I = \frac{1}{2}\rho V_a^2,$$

where V_a is the airspeed of the UAV. Bernoulli's theorem states that [8]

$$P_s = P_I + P_O,$$

where P_s is the total pressure, and P_O is the static pressure.

Therefore, the output of the differential pressure sensor is

$$\begin{aligned} y_{\text{diff pres}} &= P_s - P_O + \eta_{\text{diff pres}} \\ &= \frac{1}{2}\rho V_a^2 + \eta_{\text{diff pres}}(t), \end{aligned}$$

where $\eta_{\text{diff pres}}$ is a zero mean Gaussian process with known variance.

The static and differential pressure sensors are analog devices that are sampled by the on-board processer. We will assume that the sample rate is given by T_s.

2.4 GPS

There are several sources of GPS error. Table 1 lists the sources of error and the respective error budget. The data was obtained from http://www.montana.edu/places/gps/lres357/slides/GPSaccuracy.ppt.

The current weather affects the speed of light in the atmosphere. However, this inaccuracy should be relatively constant for a given day. We will model the effect of the atmosphere by a random variable drawn from a Gaussian distribution with a standard deviation equal to 5 meters.

The geometry of the Satellites viewed by the receiver is used to triangulate the location of the GPS receiver. Triangulation is much more effective in the horizontal plane than in the vertical direction. The satellite geometry is slowly changing in time. Therefore we will measure the effect of satellite geometry as a sinusoid with amplitude equal to $2.5\sqrt{2}$ (RMS=2.5), with a constant but unknown frequency ω_{geometry} and a phase that is a random variable drawn from a uniform distribution over $[-\pi, \pi]$.

We will assume that the clock drift is relatively constant over time. Therefore, we will model the clock drift by a constant random variable drawn from a Gaussian distribution with standard deviation of 1.5 meters.

Effect	Ave. Horizontal Error	Ave. Vertical Error
Atmosphere	5.5 meters	5.5 meters
Satellite Geometry	2.5 meters	15 meters
(Ephemeris) data		
Satellite clock drift	1.5 meters	1.5 meters
Multipath	0.6 meters	0.6 meters
Measurement noise	0.3 meters	0.3 meters

Table 1. This table lists average error estimates for commercial grade GPS units. Atmosphere, satellite geometry, clock drift, and multipath produce a near constant bias term. The measurement noise is modeled as an additive Gaussian process

Multipath is a function of the position of the UAV. Therefore we will assume that the error is a sinusoidal signal with a magnitude of $0.6\sqrt{2}$, a frequency equal to $\omega_{\text{multipath}}$ and a random phase drawn from a uniform distribution over $[-\pi, \pi]$.

We will model the measurement noise as a zero mean Gaussian process with a variance equal to 0.3 meters. The model for the GPS signal is therefore given by

$$\begin{aligned}
y_{\text{GPS},n}(t) &= p_n + \nu_{n,\text{atmosphere}} + \nu_{\text{clock}} + \eta_{n,\text{measurement}}(t) \\
&\quad + 2.5\sqrt{2}\sin(\omega_{\text{geometry}} t + \nu_{n,\text{geometry}}) \\
&\quad + 0.6\sqrt{2}\sin(\omega_{\text{multipath}} t + \nu_{n,\text{multipath}}) \\
y_{\text{GPS},e}(t) &= p_e + \nu_{e,\text{atmosphere}} + \nu_{e,\text{clock}} + \eta_{e,\text{measurement}}(t) \\
&\quad + 2.5\sqrt{2}\sin(\omega_{\text{geometry}} t + \nu_{e,\text{geometry}}) \\
&\quad + 0.6\sqrt{2}\sin(\omega_{\text{multipath}} t + \nu_{e,\text{multipath}}) \\
y_{\text{GPS},h}(t) &= h + \nu_{h,\text{atmosphere}} + \nu_{h,\text{clock}} + \eta_{h,\text{measurement}}(t), \\
&\quad + 15\sqrt{2}\sin(\omega_{\text{geometry}} t + \nu_{h,\text{geometry}}) \\
&\quad + 0.6\sqrt{2}\sin(\omega_{\text{multipath}} t + \nu_{h,\text{multipath}}),
\end{aligned}$$

where p_n, p_e, and h are the actual earth coordinates and altitude above sea level respectively. The GPS receiver also computes estimated ground speed and heading from the measurements listed above. Accordingly, we have

$$y_{\text{GPS},V_g} = \sqrt{\left(\frac{y_{\text{GPS},n}(t+T_s) - y_{\text{GPS},n}(t)}{T_s}\right)^2 + \left(\frac{y_{\text{GPS},e}(t+T_s) - y_{\text{GPS},e}(t)}{T_s}\right)^2}$$

$$y_{\text{GPS,course}} = \tan^{-1}\left(\frac{y_{\text{GPS},e}(t+T_s) - y_{\text{GPS},e}(t)}{y_{\text{GPS},n}(t+T_s) - y_{\text{GPS},n}(t)}\right).$$

The update rate of a GPS receiver is typically on the order of $T_{\text{GPS}} = 1$ second. However, the update rate can vary between $0.1 - 2$ seconds, depending on the GPS receiver.

3 Simulation Environment

We will illustrate the quality of the state estimation techniques proposed in this chapter via simulation. This section briefly describes the simulation environment which is a six degree-of-freedom nonlinear flight simulator called Aviones, developed at Brigham Young University using C/C++, and which runs on the Microsoft Windows operating system. The sensor models described in the previous section were implemented in Aviones using the parameters shown in Table 2. We have assumed that sensor biases are estimated before flight and are therefore not included in the simulator, with the exception of GPS, where it is not possible to estimate the biases.

Parameter	Value	Units
$\sigma_{\text{gyro},x}$	0.005	rad/sec
$\sigma_{\text{gyro},y}$	0.005	rad/sec
$\sigma_{\text{gyro},z}$	0.005	rad/sec
$\sigma_{\text{acc},x}$	0.005	m/sec^2
$\sigma_{\text{acc},y}$	0.005	m/sec^2
$\sigma_{\text{acc},z}$	0.005	m/sec^2
$\sigma_{\text{static pres}}$	0.4	meters
$\sigma_{\text{diff pres}}$	0.4	meters/sec
$\sigma_{\text{mag},x}$	500	nanotesla
$\sigma_{\text{mag},y}$	500	nanotesla
$\sigma_{\text{mag},z}$	500	nanotesla
$\sigma_{\text{GPS},n}$	0.5	meters
$\sigma_{\text{GPS},e}$	0.5	meters
$\sigma_{\text{GPS},h}$	0.5	meters
$\bar{\nu}_{\text{atmosphere}}$	5.5	meters
$\bar{\nu}_{\text{clock}}$	1.5	meters
$\bar{\nu}_{\text{geometry}}$	2.5	meters
$\bar{\nu}_{\text{multipath}}$	0.6	meters

Table 2. Sensor parameters used in the Aviones flight simulator. σ_* denote the variance of a zero mean Gaussian process. ν_* denotes a random variable drawn uniformly from the set $[0, \bar{\nu}_*]$

The state estimate plots shown in this chapter are all associated with a similar flight trajectory which was dictated by the following autopilot commands (using full state feedback).

- **Throughout maneuver:**
  Hold airspeed at 10.0 m/s.
- $0 \leq t \leq 2.5$ **seconds:**
  Hold a pitch angle of 20 degrees.
  Hold a roll angle of 30 degrees.
- $2.5 \leq t \leq 5.0$ **seconds:**
  Hold a pitch angle of -20 degrees.
  Hold a roll angle of 0 degrees.
- $5.0 \leq t \leq 8.0$ **seconds:**
  Hold a pitch angle of 20 degrees.
  Hold a roll angle of -30 degrees.
- $8.0 \leq t \leq 10.0$ **seconds:**
  Hold a pitch angle of -20 degrees.
  Hold a roll angle of 0 degrees.
- $8.0 \leq t \leq 10.0$ **seconds:**
  Hold a pitch angle of -20 degrees.
  Hold a roll angle of 0 degrees.
- $10.0 \leq t \leq 13.0$ **seconds:**
  Hold a pitch angle of 20 degrees.
  Hold a roll angle of 30 degrees.

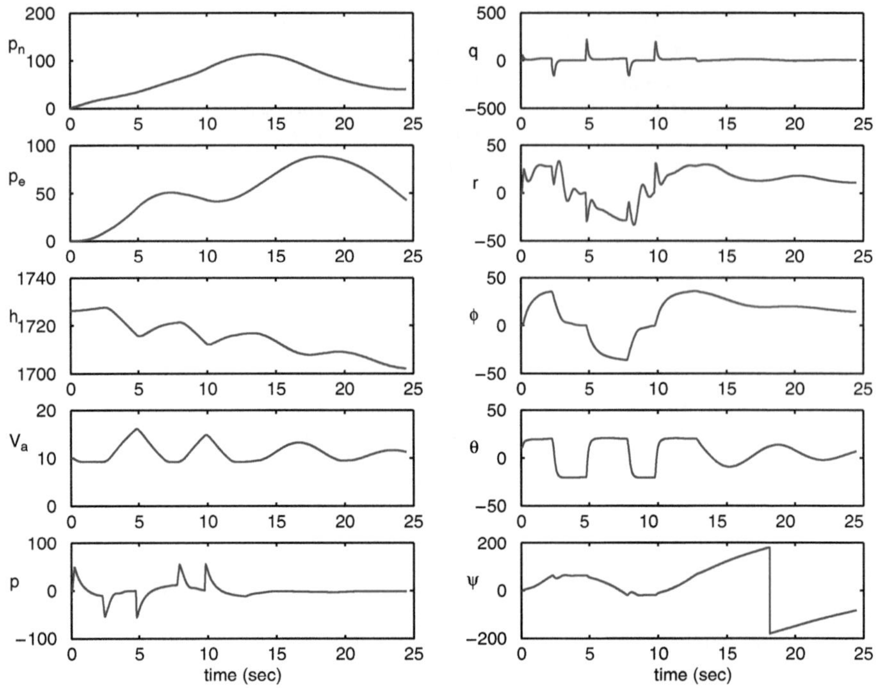

Fig. 2. Actual states during the simulated test maneuver used throughout the article. The positions p_n and p_e are in units of meters from home base, h is in units of meters above sea level, V_a is in meters/sec, p, q, and r are in units of degrees/sec, and ϕ, θ, and ψ are in units of degrees

- $13.0 \leq t < 30.0$ **seconds:**
  Hold a pitch angle of 0 degrees.
  Hold a roll angle of 0 degrees.

  A plot of the state variables during this maneuver is shown in Figure 2.

4 State Estimation via Model Inversion

The objective of this section is to demonstrate that computationally simple state estimation models can be derived by inverting the sensor models. As we shall demonstrate, the quality of the estimates produced by this method is, unfortunately, relatively poor for some of the states.

4.1 Low Pass Filters

All of the state estimation schemes require low-pass filtering of the sensor signals. For completeness, we will briefly discuss digital implementation of a first order low-pass filter.

The Laplace transform representation of a simple unity DC gain low-pass filter is given by

$$Y(s) = C(s)U(s) \triangleq \frac{a}{s+a}U(s),$$

were $u(t)$ is the input of the filter and $y(t)$ is the output. Taking the inverse Laplace transform we obtain

$$\dot{y} = -ay + au. \tag{9}$$

By introducing an integrating factor, it is straightforward to show that the solution to this differential equation is given by

$$y(t+T) = e^{-aT}y(t) + a\int_0^T e^{-a(T-\tau)}u(\tau)\,d\tau.$$

Assuming that $u(t)$ is constant between sample periods results in the expression

$$\begin{aligned} y(t+T) &= e^{-aT}y(t) + a\int_0^T e^{-a(T-\tau)}\,d\tau u(t) \\ &= e^{-aT}y(t) + (1-e^{-aT})u(t). \end{aligned} \tag{10}$$

Note that this equation has a nice physical interpretation: the new value of y (filtered value) is a weighted average of the old value of y and u (unfiltered value). We will use the notation $C(s)\{\cdot\}$ to represent the low-pass filter operator. Therefore $\hat{x} = C(s)\{x\}$ is the low-pass filtered version of x.

4.2 State Estimation by Inverting the Sensor Model

In this section we will derive the simplest possible state estimation scheme based on inverting the sensor models. While this method is effective for angular rates, altitude, and airspeed, it is not effective for estimating the position and Euler angles.

Position and Heading

The position variables p_n, p_e and the course heading χ can be estimated by low-pass filtering the GPS signals:

$$\hat{p}_n = C(s)\{y_{\text{GPS,n}}\} \tag{11}$$

$$\hat{p}_e = C(s)\{y_{\text{GPS,e}}\} \tag{12}$$

$$\hat{\chi} = C(s)\{y_{\text{GPS,course}}\}. \tag{13}$$

Figure 3 shows the actual and estimated states using this scheme. Note that since the measurements are received at 1 Hz, the estimates have a sampled data characteristic that includes significant delay.

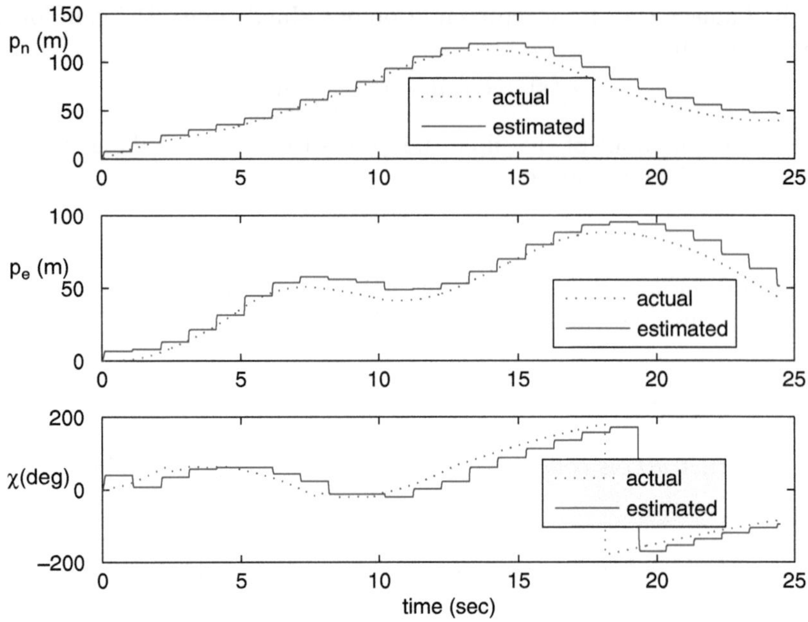

Fig. 3. Actual and estimated values of p_n, p_e, and h after low pass filtering the GPS sensor. The actual and estimated values of χ are wrapped so that they lie between ± 180 degrees

Angular Rates

Similarly, the angular rates p, q, and r can be estimated by low-pass filtering the rate gyro signals:

$$\hat{p} = C(s)\{y_{\text{gyro,x}}\}/k_{\text{gyro,x}} \tag{14}$$

$$\hat{q} = C(s)\{y_{\text{gyro,y}}\}/k_{\text{gyro,y}} \tag{15}$$

$$\hat{r} = C(s)\{y_{\text{gyro,z}}\}/k_{\text{gyro,z}}. \tag{16}$$

Figure 4 shows the actual and estimated states using this scheme. Note that low pass filtering the rate gyros results in acceptable estimates of p, q, and r.

Altitude

GPS is not accurate enough to estimate the altitude. Therefore, we will use the absolute pressure sensor. Recall that

$$y_{\text{static pressure}} = \rho g(h - h_{\text{ground}}) + \eta_{\text{static pressure}}.$$

Therefore, a simple estimation scheme is

$$\hat{h} = h_{\text{ground}} + \frac{C(s)\{y_{\text{static pressure}}\}}{\rho g}. \tag{17}$$

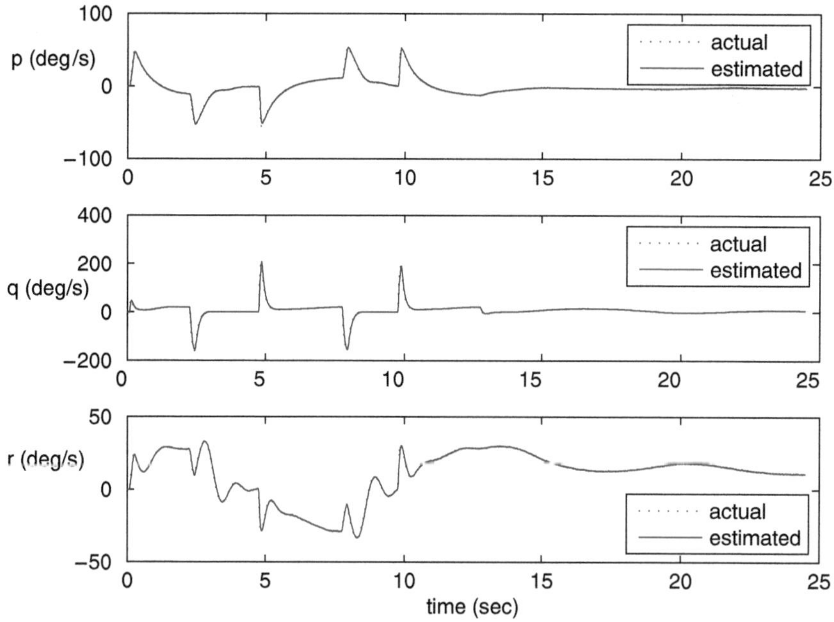

Fig. 4. Actual and estimated values of the angular rates p, q, and r after low pass filtering the rate gyros

Airspeed

Recall that

$$y_{\text{diff pres}} = \frac{1}{2}\rho V_a^2 + \eta_{\text{diff pres}}.$$

Therefore, a simple estimation scheme is

$$\hat{V}_a = \sqrt{\frac{2}{\rho} C(s)\{y_{\text{diff pres}}\}}. \tag{18}$$

Figure 5 shows the actual and estimated altitude and airspeed using this scheme. Again note that inverting the sensor models results in acceptable estimates of altitude and airspeed.

Roll and Pitch Angles

Roll and pitch angles are the most difficult variables to estimate well on small UAVs. A simple scheme, that works in unaccelerated flight, can be derived as follows. Recalling that

$$y_{\text{accel,x}} = \frac{\dot{u} + qw - rv + g\sin\theta}{g} + \eta_{\text{accel,x}}$$

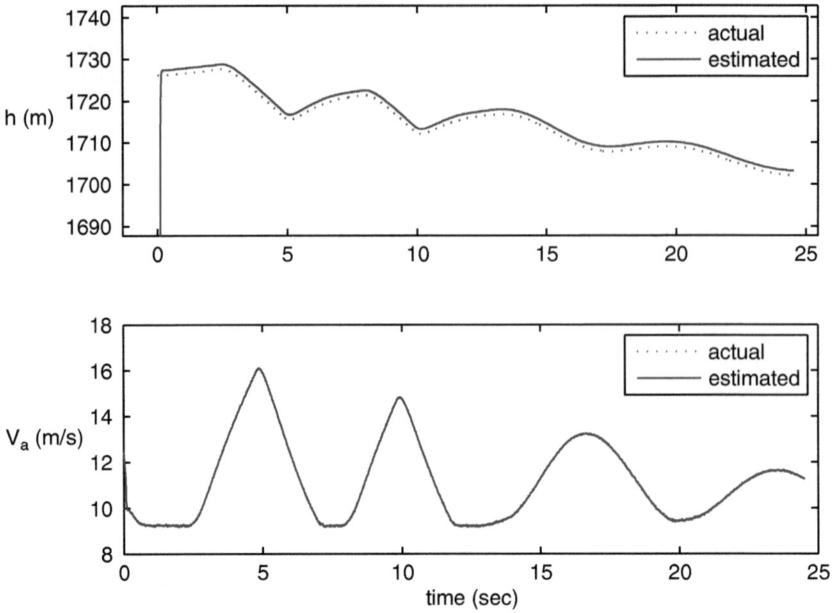

Fig. 5. Actual and estimated values of h and V_a after low pass filtering the pressure sensors and inverting their models

$$
\begin{aligned}
y_{\text{accel,y}} &= \frac{\dot{v} + ru - pw - g\cos\theta\sin\phi}{g} + \eta_{\text{accel,y}} \\
y_{\text{accel,z}} &= \frac{\dot{w} + pv - qu - g\cos\theta\cos\phi}{g} + \eta_{\text{accel,z}}.
\end{aligned}
$$

and that in unaccelerated flight $\dot{u} = \dot{v} = \dot{w} = p = q = r = 0$, we get that

$$
\begin{aligned}
C(s)\{y_{\text{accel,x}}\} &= \sin\theta \\
C(s)\{y_{\text{accel,y}}\} &= -\cos\theta\sin\phi \\
C(s)\{y_{\text{accel,z}}\} &= -\cos\theta\cos\phi.
\end{aligned}
$$

Solving for ϕ and θ we get

$$
\hat{\phi}_{\text{accel}} = \tan^{-1}\left(\frac{C(s)\{y_{\text{accel,y}}\}}{C(s)\{y_{\text{accel,z}}\}}\right) \tag{19}
$$

$$
\hat{\theta}_{\text{accel}} = \tan^{-1}\left(\frac{C(s)\{y_{\text{accel,x}}\}}{\sqrt{C(s)\{y_{\text{accel,y}}\}^2 + C(s)\{y_{\text{accel,z}}\}^2}}\right). \tag{20}
$$

Figure 6 shows the actual and estimated roll and pitch angles during the sample trajectory using this scheme. Note that the sample trajectory severely violates the unaccelerated flight assumptions. Clearly, model inversion does

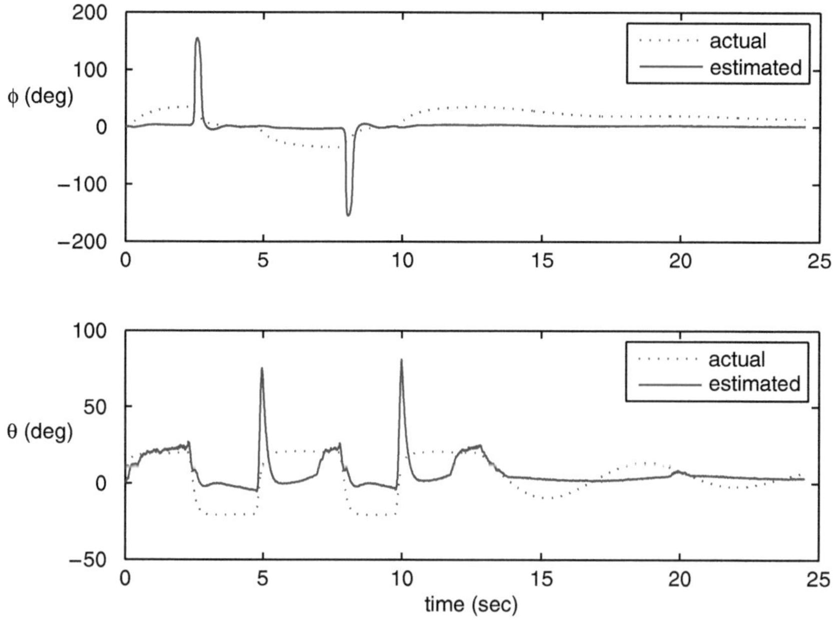

Fig. 6. Actual and estimated values of roll angle ϕ and pitch angle θ using simple model inversion

not work well for attitude estimation during accelerated flight. Another idea is to combine model inversion with the integral of roll and pitch as estimated by the rate gyros.

Recalling that

$$\begin{aligned}\dot{\phi} &= p + q\sin\phi\tan\theta + r\cos\phi\tan\theta \\ \dot{\theta} &= q\cos\phi - r\sin\phi\end{aligned}$$

and assuming that $\phi \approx 0$ and $\theta \approx 0$ we get

$$\begin{aligned}\dot{\phi} &= p \\ \dot{\theta} &= q.\end{aligned}$$

Therefore we can integrate these equations to obtain an additional estimate of ϕ and θ:

$$\begin{aligned}\hat{\phi}_{\text{int}} &= \int_{-\infty}^{t} p(\tau)\,d\tau \\ \hat{\theta}_{\text{int}} &= \int_{-\infty}^{t} q(\tau)\,d\tau.\end{aligned}$$

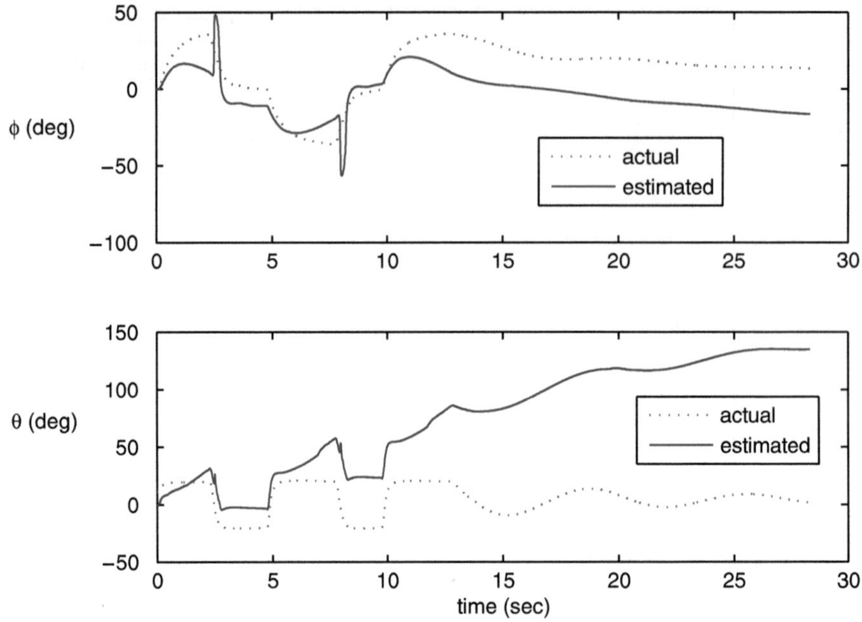

Fig. 7. Actual and estimated values of roll angle ϕ and pitch angle θ combining model inversion with the integral of the rate gyros

Combining the estimate from the integrator and the accelerometers we obtain

$$\hat{\phi} = \kappa\hat{\phi}_{\text{int}} + (1-\kappa)\hat{\phi}_{\text{accel}}$$

$$\hat{\theta} = \kappa\hat{\theta}_{\text{int}} + (1-\kappa)\hat{\theta}_{\text{accel}},$$

where $\kappa \in (0,1)$.

Figure 7 shows the actual and estimated roll and pitch angles using this scheme. It can be observed that the integration of the rate gyros causes a drift in the estimate of ϕ and θ.

While low pass filtering and model inversion work well for estimates of p, q, r, V_a and h, we need more sophisticated techniques to adequately estimate p_n, p_e, χ, ϕ, and θ. In Section 5 we will review the basics of Kalman filter theory. In Section 6 we use two extended Kalman filters to obtain estimates for p_n, p_e, χ, ϕ, and θ.

5 The Continuous-Discrete Kalman Filter

The objective of this section is to give a brief review of Kalman filter theory. There are many excellent references on Kalman filtering including [12, 13, 14, 16, 5]. We will provide a brief derivation and then focus on the application of the Kalman filter to UAV state estimation.

5.1 Dynamic Observer Theory

As a first step in deriving the Kalman filter, we briefly review dynamic observer theory. Consider the linear time-invariant system modeled by the equations

$$\dot{x} = Ax + Bu$$
$$y = Cx.$$

A continuous-time observer for this system is given by the equation

$$\dot{\hat{x}} = \underbrace{A\hat{x} + Bu}_{\text{copy of the model}} + \underbrace{L\,(y - C\hat{x})}_{\text{correction due to sensor reading}}, \tag{21}$$

where $\hat{x}$ is the estimated value of x. Letting $\tilde{x} = x - \hat{x}$ we get that

$$\dot{\tilde{x}} = (A - LC)\tilde{x}$$

which implies that the observation error decays exponentially to zero if L is chosen such that the matrix $A - LC$ is Hurwitz [20].

In practice, the sensors are usually sampled and processed in digital hardware at a sample rate T_s. How should the observer equation shown in Eq. (21) be modified to account for sampled sensor readings? The typical approach is to propagate the system model between samples using the equation

$$\dot{\hat{x}} = A\hat{x} + Bu \tag{22}$$

and then to update the estimate when a measurement is received using the equation

$$\hat{x}^+ = \hat{x}^- + L(y(t_k) - C\hat{x}^-), \tag{23}$$

where t_k is the instant in time that the measurement is received and $\hat{x}^-$ is the state estimate produced by Eq. (22) at time t_k. Equation (22) is then re-instantiated with initial conditions given by $\hat{x}^+$. The continuous-discrete observer is summarized in Table 3 [16]. The observation process is shown graphically in Figure 8. Note that a fixed sample rate is not required. The continuous-discrete observer can be implemented using Algorithm 1 which is listed below.

5.2 Essentials from Probability Theory

Let $X = (x_1, \ldots, x_n)^T$ be a vector whose elements are random variables. The mean, or expected value of X is denoted by

$$\boldsymbol{\mu} = \begin{pmatrix} \mu_1 \\ \vdots \\ \mu_n \end{pmatrix} = \begin{pmatrix} E\{x_1\} \\ \vdots \\ E\{x_n\} \end{pmatrix} = E\{X\},$$

System model:
$\dot{x} = Ax + Bu$
$y(t_k) = Cx(t_k)$
Initial Condition $x(0)$.
Assumptions:
Knowledge of A, B, C, $u(t)$.
No measurement noise.
In between measurements ($t \in [t_{k-1}, t_k)$):
Propagate $\dot{\hat{x}} = A\hat{x} + Bu$.
Initial condition is $\hat{x}^+(t_{k-1})$.
Label the estimate at time t_k as $\hat{x}^-(t_k)$.
At sensor measurement ($t = t_k$):
$\hat{x}^+(t_k) = \hat{x}^-(t_k) + L\left(y(t_k) - C\hat{x}^-(t_k)\right)$.

Table 3. Continuous-discrete observer for linear time-invariant systems

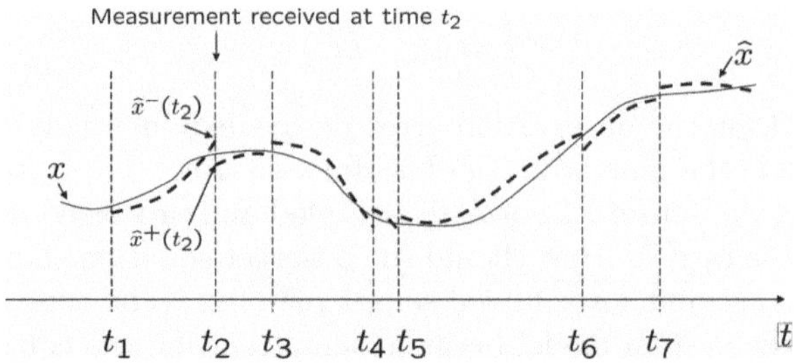

Fig. 8. This figure shows qualitatively the evolution of the state estimate. The solid line represents the actual state variable and the dashed line represents the state estimate. Measurements are received at discrete times denoted by t_i. Between measurements, the state estimate is computed by propagating the state model. At the measurements, the estimate is updated via a weighted average of the current estimate and the measurement

Algorithm 1 Continuous-Discrete Observer

1: Initialize: $\hat{x} = 0$.
2: Pick an output sample rate T_{out} which is much less than the sample rates of the sensors.
3: At each sample time T_{out}:
4: **for** $i = 1$ to N **do** {Propagate the state equation.}
5: $\quad \hat{x} = \hat{x} + \left(\frac{T_{out}}{N}\right)(A\hat{x} + Bu)$
6: **end for**
7: **if** A measurement has been received from sensor i **then** {Measurement Update}
8: $\quad \hat{x} = \hat{x} + L_i(y_i - C_i\hat{x})$
9: **end if**

where

$$E\{x_i\} = \int \xi f_i(\xi)\, d\xi,$$

and $f(\cdot)$ is the probability density function for x_i. Given any pair of components x_i and x_j of X, we denote their covariance as

$$cov(x_i, x_j) = \Sigma_{ij} = E\{(x_i - \mu_i)(x_j - \mu_j)\}.$$

The covariance of any component with itself is the variance, i.e.,

$$var(x_i) = cov(x_i, x_i) = \Sigma_{ii} = E\{(x_i - \mu_i)(x_i - \mu_i)\}.$$

The standard deviation of x_i is the square root of the variance:

$$stdev(x_i) = \sigma_i = \sqrt{\Sigma_{ii}}.$$

The covariances associated with a random vector X can be grouped into a matrix known as the covariance matrix:

$$\Sigma = \begin{pmatrix} \Sigma_{11} & \Sigma_{12} & \cdots & \Sigma_{1n} \\ \Sigma_{21} & \Sigma_{22} & \cdots & \Sigma_{2n} \\ \vdots & & \ddots & \vdots \\ \Sigma_{n1} & \Sigma_{n2} & \cdots & \Sigma_{nn} \end{pmatrix} = E\{(X - \boldsymbol{\mu})(X - \boldsymbol{\mu})^T\} = E\{XX^T\} - \boldsymbol{\mu}\boldsymbol{\mu}^T.$$

Note that $\Sigma = \Sigma^T$ so that Σ is both symmetric and positive semi-definite, which implies that its eigenvalues are real and nonnegative.

The probability density function for a Gaussian random vector is given by

$$f_X(X) = \frac{1}{\sqrt{2\pi \det \Sigma}} \exp\left[-\frac{1}{2}(X - \boldsymbol{\mu})^T \Sigma^{-1} (X - \boldsymbol{\mu})\right],$$

in which case we write

$$X \sim \mathcal{N}(\boldsymbol{\mu}, \Sigma),$$

and say that X is normally distributed with mean $\boldsymbol{\mu}$ and covariance Σ. Figure 9 shows the level curves for a 2D Gaussian random variable with different covariance matrices.

5.3 Continuous-Discrete Kalman Filter

In this section we assume the following state model:

$$\begin{aligned} \dot{x} &= Ax + Bu + G\xi \\ y_k &= Cx_k + \eta_k, \end{aligned} \tag{24}$$

where $y_k = y(t_k)$ is the kth sample of y, $x_k = x(t_k)$ is the kth sample of x, η_k is the measurement noise at time t_k, ξ is a zero-mean Gaussian process with covariance Q, and η_k is a zero-mean Gaussian random variable with covariance R. Note that the sample rate does not need to be be fixed. The covariance R can usually be estimated from sensor calibration, but the covariance Q

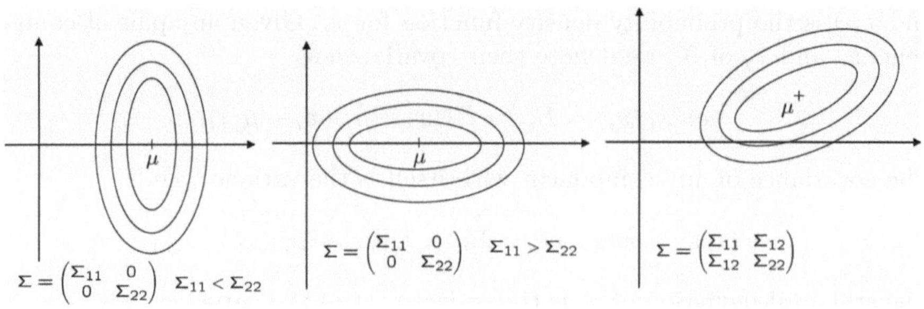

Fig. 9. Level curves for the pdf of a 2D Gaussian random variable. On the left is the pdf when the covariance matrix is diagonal with $\Sigma_{11} < \Sigma_{22}$. In the middle is a pdf when $\Sigma_{22} < \Sigma_{11}$. On the right is a pdf for general $\Sigma = \Sigma^T > 0$. The eigenvalues and eigenvectors of Σ define the major and minor axes of the level curves of the pdf

is generally unknown and therefore becomes a system gain that can be tuned to improve the performance of the observer.

We will use the observer given by Eqs. (22) and (23). Define the estimation error as $\tilde{x} = x - \hat{x}$. The covariance of the estimation error is given by

$$P(t) = E\{\tilde{x}(t)\tilde{x}(t)^T\}.$$

Note that $P(t)$ is symmetric and positive semi-definite, therefore its eigenvalues are real and non-negative. Also small eigenvalues of $P(t)$ imply small variance, which implies low average estimation error. Therefore, we would like to choose L to minimize the eigenvalues of $P(t)$. Recall that

$$tr(P) = \sum_{i=1}^{n} \lambda_i,$$

where $tr(P)$ is the trace of P and λ_i are the eigenvalues. Therefore, minimizing $tr(P)$ minimizes the estimation error covariance. Our objective is to pick the estimation gain L in Table 3 to minimize $\mathrm{tr}(P(t))$.

Between Measurements.

Differentiating $\tilde{x}$ we get

$$\begin{aligned}\dot{\tilde{x}} &= \dot{x} - \dot{\hat{x}} \\ &= Ax + Bu + G\xi - A\hat{x} - Bu \\ &= A\tilde{x} + G\xi,\end{aligned}$$

which implies that

$$\tilde{x}(t) = e^{At}\tilde{x}_0 + \int_0^t e^{A(t-\tau)} G\xi(\tau)\, d\tau.$$

We can compute the evolution for P as

$$\begin{aligned}\dot{P} &= \frac{d}{dt}E\{\tilde{x}\tilde{x}^T\}\\ &= E\{\dot{\tilde{x}}\tilde{x}^T + \tilde{x}\dot{\tilde{x}}^T\}\\ &= E\left\{A\tilde{x}\tilde{x}^T + G\xi\tilde{x}^T + \tilde{x}\tilde{x}^T A^T + \tilde{x}\xi^T G^T\right\}\\ &= AP + PA^T + GE\{\xi\tilde{x}^T\}^T + E\{\tilde{x}\xi^T\}G^T,\end{aligned}$$

where

$$\begin{aligned}E\{\xi\tilde{x}^T\} &= E\left\{\xi(t)\tilde{x}_0 e^{A^T t} + \int_0^t \xi(t)\xi^T(\tau)G^T e^{A^T(t-\tau)}\,d\tau\right\}\\ &= \frac{1}{2}QG^T,\end{aligned}$$

which implies that

$$\dot{P} = AP + PA^T + GQG^T.$$

At Measurements.

At a measurement we have that

$$\begin{aligned}\tilde{x}^+ &= x - \hat{x}^+\\ &= x - \hat{x}^- - L\left(Cx + \eta - C\hat{x}^-\right)\\ &= \tilde{x}^- - LC\tilde{x}^- - L\eta.\end{aligned}$$

Therefore

$$\begin{aligned}P^+ &= E\{\tilde{x}^+\tilde{x}^{+T}\}\\ &= E\left\{\left(\tilde{x}^- - LC\tilde{x}^- - L\eta\right)\left(\tilde{x}^- - LC\tilde{x}^- - L\eta\right)^T\right\}\\ &= E\left\{\tilde{x}^-\tilde{x}^{-T} - \tilde{x}^-\tilde{x}^{-T}C^T L^T - \tilde{x}^-\eta^T L^T\right.\\ &\quad - LC\tilde{x}^-\tilde{x}^{-T} + LC\tilde{x}^-\tilde{x}^{-T}C^T L^T + LC\tilde{x}^-\eta^T L^T\\ &= \left. -L\eta\tilde{x}^{-T} + L\eta\tilde{x}^{-T}C^T L^T + L\eta\eta^T L^T\right\}\\ &= P^- - P^- C^T L^T - LCP^- + LCP^- C^T L^T + LRL^T.\end{aligned} \tag{25}$$

Our objective is to pick L to minimize $\text{tr}(P^+)$. A necessary condition is

$$\begin{aligned}&\frac{\partial}{\partial L}\text{tr}(P^+) = -P^- C^T - P^- C^T + 2LCP^- C^T + 2LR = 0\\ &\implies 2L(R + CP^- C^T) = 2P^- C^T\\ &\implies L = P^- C^T (R + CP^- C^T)^{-1}.\end{aligned}$$

Plugging back into Eq. (25) give

$$\begin{aligned} P^+ &= P^- + P^-C^T(R + CP^-C^T)^{-1}CP^- - P^-C^T(R + CP^-C^T)^{-1}CP^- \\ &\quad + P^-C^T(R + CP^-C^T)^{-1}(CP^-C^T + R)(R + CP^-C^T)^{-1}CP^- \\ &= P^- - P^-C^T(R + CP^-C^T)^{-1}CP^- \\ &= (I - P^-C^T(R + CP^-C^T)^{-1}C)P^- \\ &= (I - LC)P^-. \end{aligned}$$

Extended Kalman Filter.

If instead of the linear state model given in (24), the system is nonlinear, i.e.,

$$\begin{aligned} \dot{x} &= f(x, u) + G\xi \\ y_k &= h(x_k) + \eta_k, \end{aligned} \tag{26}$$

then the system matrices A and C required in the update of the error covariance P are computed as

$$\begin{aligned} A(x) &= \frac{\partial f}{\partial x}(x) \\ C(x) &= \frac{\partial h}{\partial x}(x). \end{aligned}$$

The extended Kalman filter (EKF) for continuous-discrete systems is given by Algorithm 2.

Algorithm 2 Continuous-Discrete Extended Kalman Filter

Initialize: $\hat{x} = 0$.
Pick an output sample rate T_{out} which is much less than the sample rates of the sensors.
At each sample time T_{out}:
for $i = 1$ to N **do** {Propagate the equations.}
  $\hat{x} = \hat{x} + \left(\frac{T_{out}}{N}\right) f(\hat{x}, u)$
  $A = \frac{\partial f}{\partial x}(\hat{x})$
  $P = P + \left(\frac{T_{out}}{N}\right)\left(AP + PA^T + GQG^T\right)$
end for
if A measurement has been received from sensor i **then** {Measurement Update}
  $C_i = \frac{\partial h_i}{\partial x}(\hat{x})$
  $L_i = PC_i^T(R_i + C_iPC_i^T)^{-1}$
  $P = (I - L_iC_i)P$
  $\hat{x} = \hat{x} + L_i(y_i - C_i\hat{x})$.
end if

6 Application of the EKF to UAV State Estimation

In this section we will use the continuous-discrete extended Kalman filter to improve estimates of roll and pitch (Section 6.1) and position and course (Section 6.2).

6.1 Roll and Pitch Estimation

From Eq. 3, the equations of motion for ϕ and θ are given by

$$\begin{aligned}\dot{\phi} &= p + q\sin\phi\tan\theta + r\cos\phi\tan\theta + \xi_\phi \\ \dot{\theta} &= q\cos\phi - r\sin\phi + \xi_\theta,\end{aligned}$$

where we have added the noise terms $\xi_\phi \sim \mathcal{N}(0, Q_\phi)$ and $\xi_\theta \sim \mathcal{N}(0, Q_\theta)$ to model the sensor noise on p, q, and r. We will use the accelerometers as the output equations. From Eq. (7), the output of the accelerometers is given by

$$y_{\text{accel}} = \begin{pmatrix} \frac{\dot{u}+gw-rv}{g} + \sin\theta \\ \frac{\dot{v}+ru-pw}{g} - \cos\theta\sin\phi \\ \frac{\dot{w}+pv-qu}{g} - \cos\theta\cos\phi \end{pmatrix} + \eta_{\text{accel}}. \tag{27}$$

However, since we do not have a method for directly measuring $\dot{u}$, $\dot{v}$, $\dot{w}$, u, v, and w, we will assume that $\dot{u} = \dot{v} = \dot{w} \approx 0$ and we will use Eq. (1) and assume that $\alpha \approx \theta$ and $\beta \approx 0$ to obtain

$$\begin{pmatrix} u \\ v \\ w \end{pmatrix} \approx V_a \begin{pmatrix} \cos\theta \\ 0 \\ \sin\theta \end{pmatrix}.$$

Substituting into Eq. (27) gives

$$y_{\text{accel}} = \begin{pmatrix} \frac{qV_a\sin\theta}{g} + \sin\theta \\ \frac{rV_a\cos\theta - pV_a\sin\theta}{g} - \cos\theta\sin\phi \\ \frac{-qV_a\cos\theta}{g} - \cos\theta\cos\phi \end{pmatrix} + \eta_{\text{accel}}.$$

Letting $x = (\phi, \theta)^T$, $u = (p, q, r, V_a)^T$, $\xi = (\xi_\phi, \xi_\theta)^T$, and $\eta = (\eta_\phi, \eta_\theta)^T$, we get the nonlinear state equation

$$\begin{aligned}\dot{x} &= f(x, u) + \xi \\ y &= h(x, u) + \eta,\end{aligned}$$

where

$$f(x, u) = \begin{pmatrix} p + q\sin\phi\tan\theta + r\cos\phi\tan\theta \\ q\cos\phi - r\sin\phi \end{pmatrix}$$

$$h(x,u) = \begin{pmatrix} \frac{qV_a \sin\theta}{g} + \sin\theta \\ \frac{rV_a\cos\theta - pV_a\sin\theta}{g} - \cos\theta\sin\phi \\ \frac{-qV_a\cos\theta}{g} - \cos\theta\cos\phi \end{pmatrix}.$$

Implementation of the extended Kalman filter requires the Jacobians

$$\frac{\partial f}{\partial x} = \begin{pmatrix} q\cos\phi\tan\theta - r\sin\phi\tan\theta & \frac{q\sin\phi - r\cos\phi}{\cos^2\theta} \\ -q\sin\phi - r\cos\phi & 0 \end{pmatrix}$$

$$\frac{\partial h}{\partial x} = \begin{pmatrix} 0 & \frac{qV_a}{g}\cos\theta + \cos\theta \\ -\cos\phi\cos\theta & -\frac{rV_a}{g}\sin\theta - \frac{pV_a}{g}\cos\theta + \sin\phi\sin\theta \\ \sin\phi\cos\theta & \left(\frac{qV_a}{g} + \cos\phi\right)\sin\theta \end{pmatrix}.$$

The state estimation algorithm is given by Algorithm 2.

Figure 10 shows the actual and estimated roll and pitch attitudes obtained by using this scheme, where we note significant improvement over the results shown in Figures 6 and 7. The estimates are still not precise due to the approximation that $\dot{u} = \dot{v} = \dot{w} = \beta = \theta - \alpha = 0$. However, the results are adequate enough to enable non-aggressive MAV maneuvers.

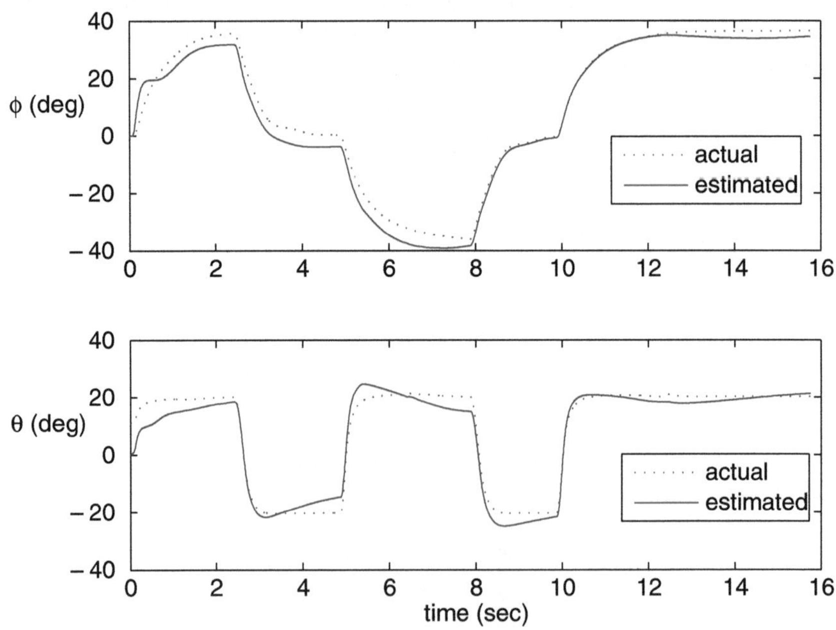

Fig. 10. Actual and estimated values of ϕ and θ using the continuous-discrete extended Kalman filter

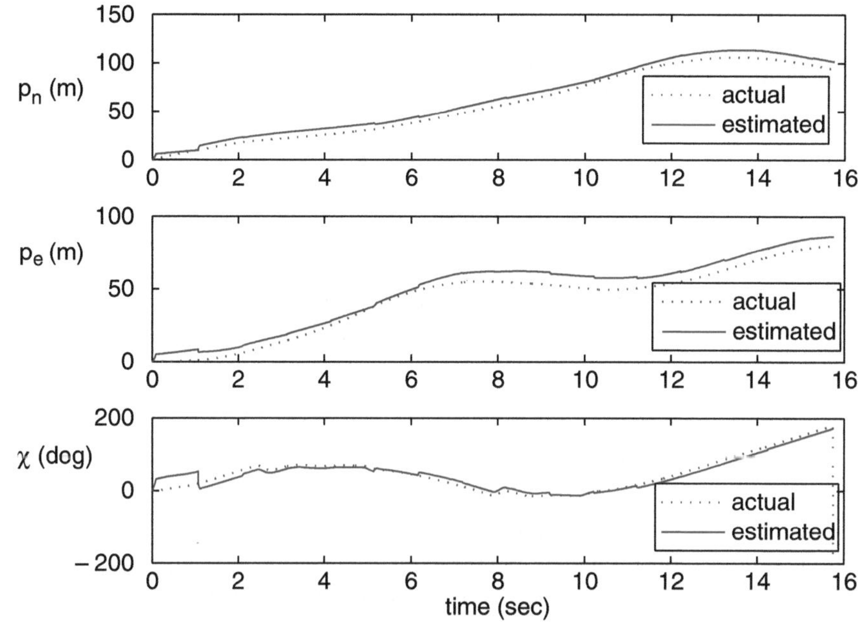

Fig. 11. Actual and estimated values of p_n, p_e, and χ using the continuous-discrete extended Kalman filter

6.2 Position and Course Estimation

The objective in this section is to estimate p_n, p_e, and χ using the GPS sensor. From Eq. (3), the model for χ is given by

$$\dot{\chi} = \dot{\psi} = q\frac{\sin\phi}{\cos\theta} + r\frac{\cos\phi}{\cos\theta}.$$

Using Eqs. (4) and (5) for the evolution of p_n and p_e results in the system model

$$\begin{pmatrix} \dot{p}_N \\ \dot{p}_E \\ \dot{\chi} \end{pmatrix} = \begin{pmatrix} V_g \cos\chi \\ V_g \sin\chi \\ q\frac{\sin\phi}{\cos\theta} + r\frac{\cos\phi}{\cos\theta} \end{pmatrix} + \xi_p$$
$$\triangleq f(x, u) + \xi_p,$$

where $x = (p_n, p_e, \chi)^T$, $u = (V_g, q, r, \phi, \theta)^T$ and $\xi_p \sim \mathcal{N}(0, Q)$.

GPS returns measurements of p_n, p_e, and χ directly. Therefore we will assume the output model

$$y_{\text{GPS}} = \begin{pmatrix} p_n \\ p_e \\ \chi \end{pmatrix} + \eta_p,$$

where $\eta_p \sim \mathcal{N}(0, R)$ and $C = I$, and where we have ignored the GPS bias terms. To implement the extended Kalman filter in Algorithm 2 we need the Jacobian of f which can be calculated as

$$\frac{\partial f}{\partial x} = \begin{pmatrix} 0 & 0 & -V_g \sin\chi \\ 0 & 0 & V_g \cos\chi \\ 0 & 0 & 0 \end{pmatrix}.$$

Figure 10 shows the actual and estimated values for p_n, p_e, and χ obtained by using this scheme. The inaccuracy in the estimates of p_n and p_e is due to the GPS bias terms that have been neglected in the system model. Again, these results are sufficient to enable non-aggressive maneuvers.

7 Summary

Micro air vehicles are increasingly important in both military and civil applications. The design of intelligent vehicle control software pre-supposes accurate state estimation techniques. However, the limited computational resources on board the MAV require computationally simple, yet effective, state estimation algorithms. In this chapter we have derived mathematical models for the sensors commonly deployed on MAVs. We have also proposed simple state estimation techniques that have been successfully used in thousands of hours of actual flight tests using the Procerus Kestrel autopilot (see for example [7, 6, 21, 10, 17, 18]).

Acknowledgments

This work was partially supported under grants AFOSR grants FA9550-04-1-0209 and FA9550-04-C-0032 and by NSF award no. CCF-0428004.

References

1. `http://www.silicondesigns.com/tech.html`.
2. Cloudcap technology. `http://www.cloudcaptech.com`.
3. Micropilot. `http://www.micropilot.com/`.
4. Procerus technologies. `http://procerusuav.com/`.
5. Brian D.O. Anderson and John B. Moore. *Optimal Control: Linear Quadratic Methods*. Prentice Hall, Englewood Cliffs, New Jersey, 1990.
6. D. Blake Barber, Stephen R. Griffiths, Timothy W. McLain, and Randal W. Beard. Autonomous landing of miniature aerial vehicles. In *AIAA Infotech@ Aerospace*, Arlington, Virginia, September 2005. American Institute of Aeronautics and Astronautics. AIAA-2005-6949.

7. Randal Beard, Derek Kingston, Morgan Quigley, Deryl Snyder, Reed Christiansen, Walt Johnson, Timothy McLain, and Mike Goodrich. Autonomous vehicle technologies for small fixed wing UAVs. *AIAA Journal of Aerospace, Computing, Information, and Communication*, 2(1):92–108, January 2005.
8. Robert E. Bicking. Fundamentals of pressure sensor technology. `http://www.sensorsmag.com/articles/1198/fun1198/main.shtml`.
9. Crossbow. Theory of operation of angular rate sensors. `http://www.xbow.com/Support/Support_pdf_files/RateSensorAppNote.pdf`.
10. Stephen Griffiths, Jeff Saunders, Andrew Curtis, Tim McLain, and Randy Beard. Obstacle and terrain avoidance for miniature aerial vehicles. *IEEE Robotics and Automation Magazine*, 13(3):34–43, 2006.
11. David Halliday and Robert Resnick. *Fundamentals of Physics.* John Wiley & Sons, 3rd edition, 1988.
12. Andrew H. Jazwinski. *Stochastic Processes and Filtering Theory*, volume 64 of *Mathematics in Science and Engineering.* Academic Press, Inc., New York, New York, 1970.
13. R.E. Kalman. A new approach to linear filtering and prediction problems. *Transactions ASME Journal of Basic Engineering*, 82:34–35, 1960.
14. R.E. Kalman and R.S. Bucy. New results in linear filtering and prediction theory. *Transaction of the ASME, Journal of Basic Engineering*, 83:95–108, 1961.
15. Robert P. Leland. Lyapunov based adaptive control of a MEMS gyroscope. In *Proceedings of the American Control Conference*, pages 3765–3770, Anchorage, Alaska, May 2002.
16. Frank L. Lewis. *Optimal Estimation: With an Introduction to Stochastic Control Theory.* John Wiley & Sons, New York, New York, 1986.
17. Timothy W. McLain and Randal W. Beard. Unmanned air vehicle testbed for cooperative control experiments. In *American Control Conference*, pages 5327–5331, Boston, MA, June 2004.
18. Derek R. Nelson, D. Blake Barber, Timothy W. McLain, and Randal W. Beard. Vector field path following for miniature air vehicles. *IEEE Transactions on Robotics*, in press.
19. Marc Rauw. *FDC 1.2 - A SIMULINK Toolbox for Flight Dynamics and Control Analysis*, February 1998. Available at `http://www.mathworks.com/`.
20. Wilson J. Rugh. *Linear System Theory.* Prentice Hall, Englewood Cliffs, New Jersey, 2nd edition, 1996.
21. Jeffery B. Saunders, Brandon Call, Andrew Curtis, Randal W. Beard, and Timothy W. McLain. Static and dynamic obstacle avoidance in miniature air vehicles. In *AIAA Infotech@Aerospace*, number AIAA-2005-6950, Arlington, Virginia, September 2005. American Institute of Aeronautics and Astronautics.
22. Brian L. Stevens and Frank L. Lewis. *Aircraft Control and Simulation.* John Wiley & Sons, Inc., Hoboken, New Jersey, 2nd edition, 2003.
23. Navid Yazdi, Farrokh Ayazi, and Khalil Najafi. Micromachined inertial sensors. *Proceedings of the IEEE*, 86(8):1640–1659, August 1998.

Evolutionary Design of a Control Architecture for Soccer-Playing Robots

Steffen Prüter[1], Hagen Burchardt[1], and Ralf Salomon[1]

[1]Institute of Applied Microelectronics and Computer Engineering
University of Rostock
18051 Rostock, Germany
{steffen.prueter, hagen.burchardt, ralf.salomon}@uni-rostock.de

Abstract. Soccer-playing robots provide a good environment for the application of evolutionary algorithms. Among other problems, slipping wheels, changing friction values, and real-world noise are significant problems to be considered. This chapter demonstrates how artificial intelligence techniques such as Kohonen maps, genetic algorithms, and evolutionary-evolved neural networks, can compensate those effects. As soccer robots are physical entities, all adaptation algorithms have to meet real-time constraints.

1 Introduction

The Robot World Cup Initiative (RoboCup) [1] is an international project to advance research on mobile robots and artificial intelligence (AI). The long-time goal is to create a humanoid robot soccer team that is able to compete against the human world-champion team by 2050. RoboCup consists of the following three different fields. RoboCup Junior focuses on teaching children and beginners on how to build and operate simple robots. The RoboCup Rescue section works on robots for disaster and other hostile environments. And RoboCup Soccer is on robot teams that play soccer against each other in different leagues.

RoboCup Soccer itself is further divided into five different leagues, each having its own rules, goals, and robot designs. The simulation league considers only teams of simulated robots. The four-legged league, by contrast, only Sony's physical AIBO robot dogs play each other. Both the small-size and the middle size league focus on self made robots with varying degrees of complexities and capabilities. Finally, the humanoid league is about two-legged robots that behave in a human-like fashion. The division into different fields and leagues allows virtually every research team to participate in and to contribute to the RoboCup initiative within its given financial and human resource limits.

S. Prüter et al.: *Evolutionary Design of a Control Architecture for Soccer-Playing Robots*, Studies in Computational Intelligence (SCI) **70**, 201–222 (2007)
www.springerlink.com

RoboCup soccer is of particular interest for many AI-researchers, because it combines engineering tasks, such as building robot hardware and designing electronic components, with computer science applications, such as localization of objects, finding the robots' positions, and calculating the best path through obstacles. Another interesting challenge emerges from the requirement that all team members have to communicate with each other in order to develop a cooperative behavior. Research on artificial intelligence may help find the optimal solution in all of these areas. Within the field of RoboCup soccer, the small-size league (SSL) allows for pursuing the research mentioned above at a relatively low budget.

Fig. 1 illustrates the general setup used by all teams of the small-size league. Two cameras are mounted approximately four meters above the floor and observe a field of four by five meters in size on which two teams each consisting of five robots play against each other. The cameras send their images to a host PC on which an image processing software determines the ball's as well as robots' positions. Depending on all recognized positions a software component derives the next actions for its own team members such that the team exhibits a cooperative behavior. By utilizing wireless DECT modules, the PC software transmits the derived actions to the robots, which execute them properly and eventually play the ball.

Fig. 2 shows the omnidirectional drive commonly used by most robots of the small-size league. As can be seen, an omnidirectional drive consists of

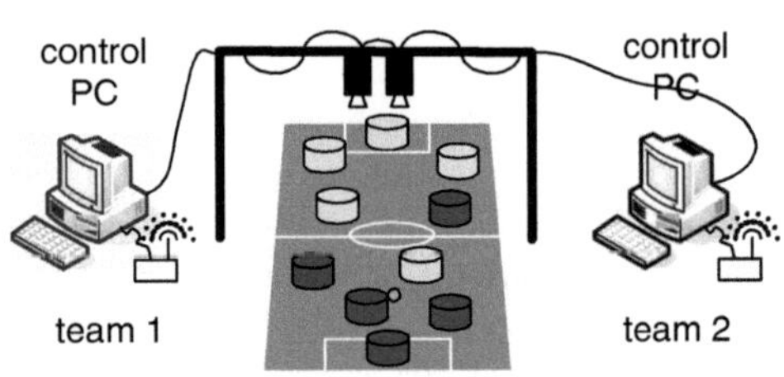

Fig. 1. The physical setup in RoboCup's small-size league

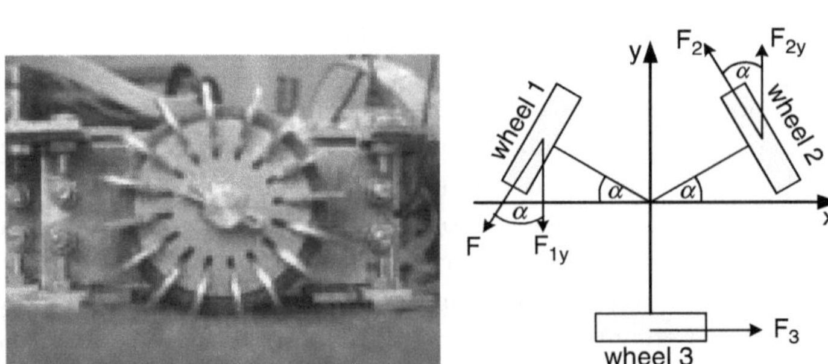

Fig. 2. An omnidirectional drive with its calculation model

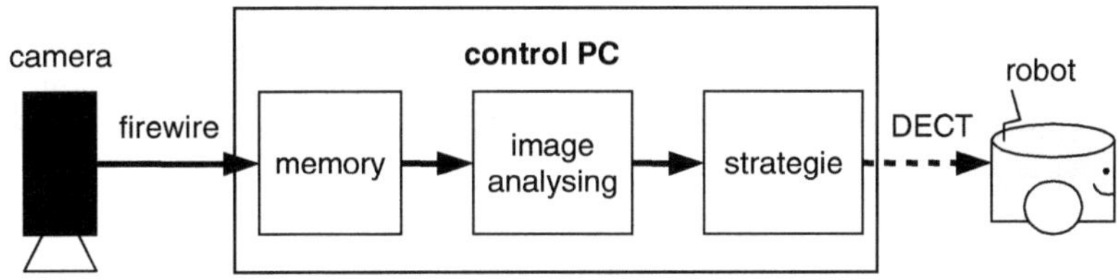

Fig. 3. The image processing system consists of five stages which all contribute to the processing delays, which are also known as latency times

three wheels, which are located at an angle of 120 degrees to one another. This drive has the advantage that a robot can be simultaneously doing both moving forward and spinning around its own central axis. Furthermore, the particular wheels, as shown on the left-hand-side of Fig. 2, yield high grip in the rotation direction, but almost-vanishing friction perpendicular to it. The specific orientation of all three wheels, as illustrated on the right-hand-side of Fig. 2, requires advanced controllers and they exhibit higher friction than standard two-wheel drives. The later drive requires sophisticated servo loops and (PID[1]) controllers [8].

Depending on the carpet and the resulting wheel-to-carpet friction, one or more wheels may slip. As a consequence, the robot leaves its desired moving path. Section 2 shows how Kohonen feature maps [4] can alleviate this problem to a large extent. The results indicate that in comparison to linear algorithms, neural networks yield a better compensation with less effort.

The processing sequence starting at the camera image and ending with the robots executing their action commands suffer from significant time delays, as illustrated in Fig. 3. These time delays have the consequence that when receiving a command, the robot's current position does not correspond to the position shown in the camera image. Consequently, the actions are either inaccurate or may lead to improper behavior in the extreme case. For example, the robot may try to kick the ball even though it is no longer within reach.

These time delays induce two problems: (1) The actual robot position has to be extrapolated on the PC. (2) The robot has to track its current position. Section 3 discusses how by utilizing back-propagation networks [4], the control software, which runs on the host PC, can compensate for those time delays. The experiments indicate that this approach yields significant improvements.

Section 4 discusses how the position correction can be further improved by the robot itself. To this end, the robot employs its own back-propagation network to learn its own specific slip and friction effects. This local, robot specific mechanism complements the global correction done by the neural network as discussed in Section 3. Section 5 demonstrates the implementation of path planning using genetic algorithms. Experiments demonstrate that the robot

[1] PID is the abbreviation of Proportional-Integrate-Differential. For further details, the reader is referred to [8]

hardware is capable of running this task in real-time and that the algorithm adapts to environmental changes such as moving obstacles. Section 6 concludes this chapter with a brief discussion including on outline of possible future research.

2 The Slip Problem

As is well known, robots are moving by spinning their wheels. The resulting direction of the robot depends on the wheels' speeds relative to each other. Usually, PID controllers regulate the motors by comparing the target speed with the tick-count delivered by the attached wheel encoders. However, the PID controllers are not always able to archive this goal when controlling omni-directional drives, because some wheels occasionally slip. As a consequence, the robot deviates from its expected path. This section uses self-organizing Kohonen feature maps [3, 6] to precisely control the wheels.

2.1 Slip and Friction

Slip occurs when accelerating or decelerating a wheel in case the friction between wheel and ground is too low. In case of slipping wheels the driven distance does not match the distance that corresponds to the measured wheel ticks. In other words, the robot has moved a distance shorter (acceleration) or longer (deceleration) than it "thought". Fig. 4 illustrates the effect when wheel 3 is slipping.

Friction is another problem that leads to similar effects. It results from mechanical problems between moving and non-moving parts and also between the robot parts and the floor. In most cases, but not always, servo loops can compensate for those effects. Similarly to the slip problem, friction leads to imprecise positions. In addition, high robot speeds, non-constant friction values, and real-world noise make this problem even worse.

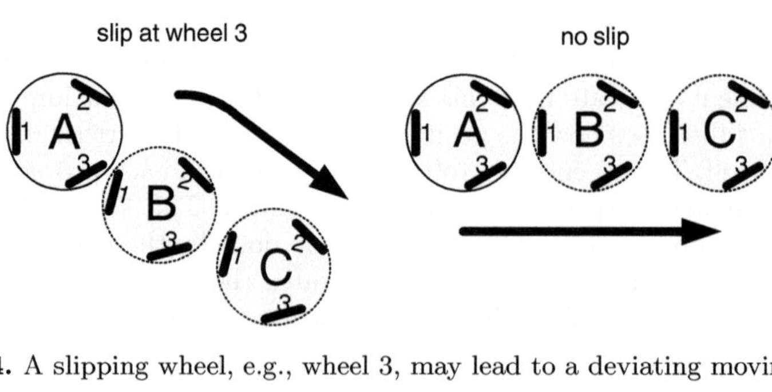

Fig. 4. A slipping wheel, e.g., wheel 3, may lead to a deviating moving path

2.2 Experimental Analysis

The various effects of slip and friction are experimentally measured in two stages. The first stage is dedicated to the determination of the robot's orientation error $\Delta\alpha$. To this end, the robot is located in the center of a circle with 1m in radius. The wheel speeds are set as follows:

$$\begin{aligned} r_1 &= v \cdot \sin(\pi - 60). \\ r_2 &= v \cdot \sin(\pi + 60). \\ r_3 &= v \cdot \sin(\pi + 180) = -r_1 - r_2. \end{aligned} \tag{1}$$

with $r_{i=1\ldots3}$ denoting the rotation speed of wheel i and v denoting the robot's center speed. In an ideal case, these speed settings would make the robot move in a straight line. In the experiment, the robot moves 1 m. After moving a distance with a moderate speed, the robot's orientation offset $\Delta\alpha$ is measured by using the camera and the image processing system. Fig. 5 illustrates this procedure.

Stage two repeats the experiments of the first stage. However, the rear wheel is adjusted by hand such that the robot's orientation does not change while moving. This stage then measures the drift $\Delta\varphi$, as illustrated in Fig. 6.

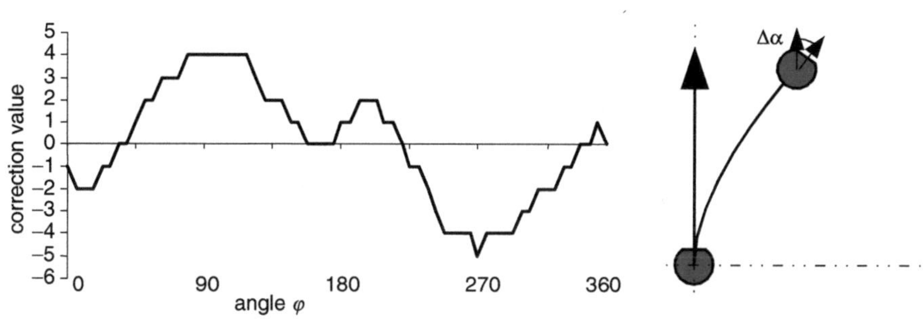

Fig. 5. Turning behavior of the robot due to internal rotation and its correction with an additional correction value of the rear wheel 1 for different angles φ

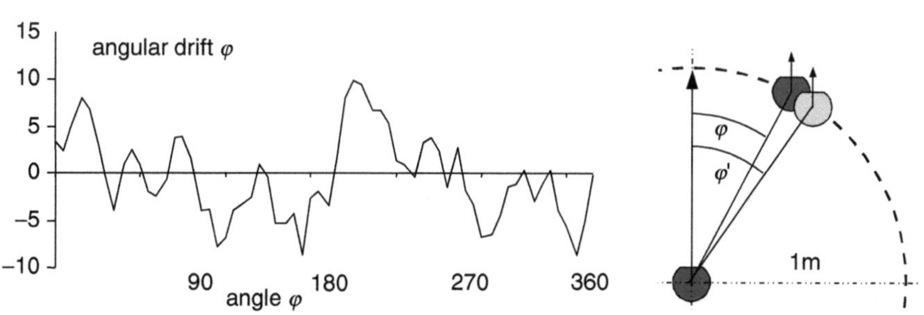

Fig. 6. Drift values $\Delta\varphi$ for different angels φ with rotation compensation

Stages one and two were repeated for twelve different values of φ. The corresponding correction values for wheel 1 and drift values $\Delta\varphi$ are plotted in Fig. 5 and Fig. 6, respectively.

2.3 Self-Organizing Kohonen Feature Maps and Methods

Since similar friction and slip values have similar effects with respect to the moving path, self-organizing Kohonen feature maps [3, 4, 6] are the method of choice for the problem at hand. To train a Kohonen map, the input vectors $\vec{x}_k$ are presented to the network. All nodes calculate the Euclidean distance $d_i = \|\vec{x}_k - \vec{w}_i\|$ of their own weight vector $\vec{w}_i$ to the input vector $\vec{x}_k$. The winner, that is, the node i with the smallest distance d_i, and its neighbors j update their vectors $\vec{w}_i$ and $\vec{w}_j$, respectively, according to the following formula

$$\vec{w}_j = \vec{w}_j + \eta\,(\vec{x}_j - \vec{w}_j)\cdot h\,(i,j)\,. \tag{2}$$

Here $h(i,j)$ denotes a neighborhood function and η denotes a small learning constant. Both the presentation of the input vectors and the updating of the weight vectors continue until the updates are reduced to a small margin. It is important to note that during the learning process, the learning rate η as well as the distance function $h(i,j)$ has to be decreased. After the training is completed, a Kohonen network maps previously unseen input data onto appropriate output values.

As Fig. 7 shows, all experiments have used a one-dimensional Kohonen map. The number of neurons was varied between 1 and 256. Normally, training a Kohonen maps includes finding an optimal distribution of all nodes. Since in this application, all driving directions are equally likely, the neurons were equally distributed over the input range $0 \le \varphi < 360$. Thus, learning could be speed up significantly by initializing all nodes at equidistant input values $\varphi_i \leftarrow i \cdot 360/n$. Here n denotes the number of neurons.

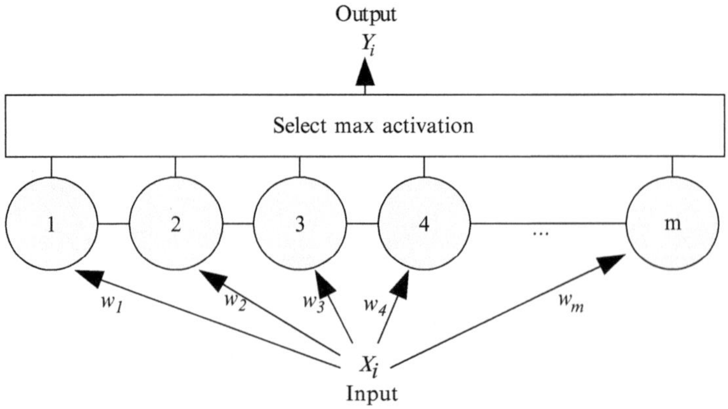

Fig. 7. A one-dimensional Kohonen map used to compensate for slip and friction errors

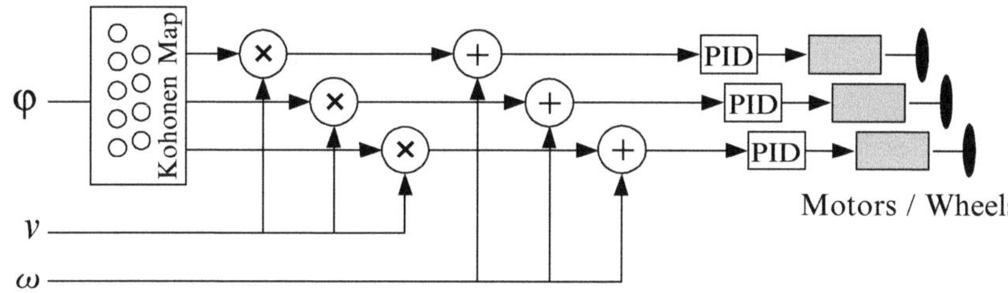

Fig. 8. From a given translational moving direction φ, a Kohonen feature map determines three motor speeds which are multiplied by the desired speed v and updated by an additional desired rotation speed ω

In addition, the neurons were also labeled with the rotation speed of all three wheels. Such architectures are also known as extended Kohonen maps in the literature [4, 6]. For the output value, the network calculates the weighted average over the outputs of the two highest activated units. It should be noted that the activation of the nodes is inversely proportional to the distance d_i.

$$a_i = e^{-d_i} \tag{3}$$

Training was started with a learning rate $\eta = 0.3$. The neighborhood function $h(i, j)$ is 1 for $i = j$, 0.5 for $|i - j| = 1$, and 0 otherwise. After every 30 cycles, the learning rate was divided by 2, and training was stopped after 150 iterations.

After training is finished, the map has been uploaded into the robot. As shown in Fig. 8, the desired direction is applied as input to the map. The two most active units are selected and the motor speeds are interpolated linearly based on the corresponding angles of the units and the input angle. After that, the motor speeds are multiplied by the desired velocity. An additional rotation component ω is added in the last step.

2.4 Results

As shown in Fig. 9, the results of the experimental analysis have indicated that the hand-crafted rotation compensation for α works well over a large range of speeds v.

Therefore, the Kohonen feature maps were only used to compensate for the drift $\Delta\varphi$. Fig. 10 shows the maximum angular drift as a function of the number of nodes. As expected, the error decreases with an increasing number of nodes. With respect to both the computational demands and resulting precision, 32 neurons are considered to be suitable. It should be noted that choosing a power of two greatly simplifies the implementation.

Fig. 11 shows the correction of the drift by means of the Kohonen feature map with 32 neurons. It can be seen that in comparison to Fig. 6, the error has been reduced by a factor of five. It should be mentioned that further improvements are not achievable due to mechanical limitations.

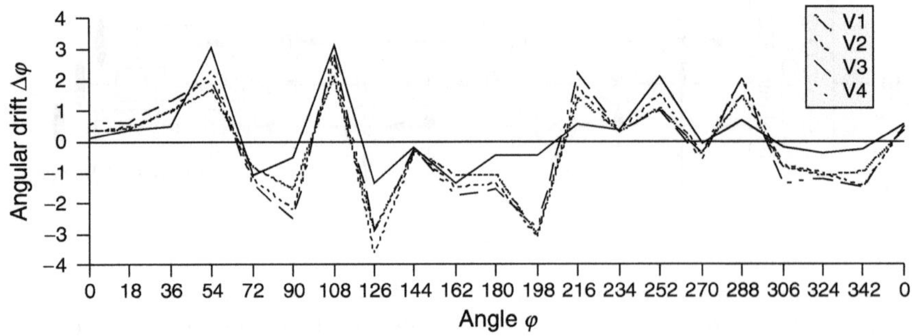

Fig. 9. Robot drift depending on the direction ϕ and the robot speed v

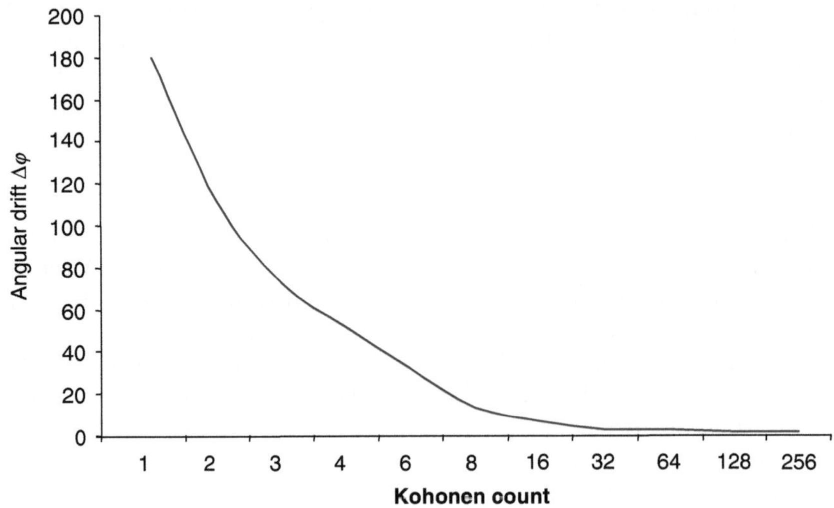

Fig. 10. Angular drift $\Delta\varphi$ of the Kohonen map as a function its number of neurons

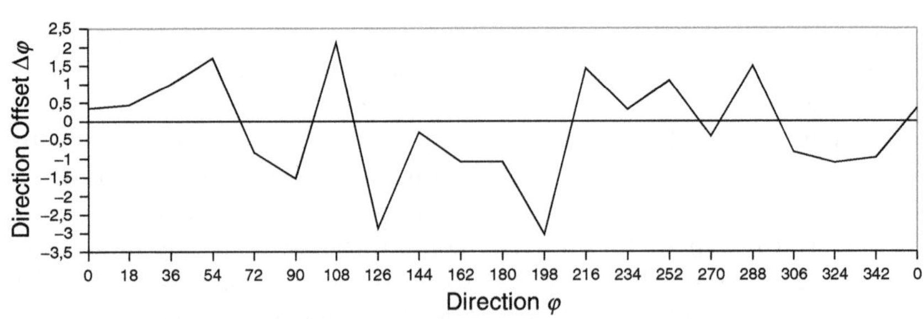

Fig. 11. Robot behavior after direction drift compensation

3 Improved Position Prediction

As has been outlined in the introduction, the latency caused by the image-processing-and-action-generation loop leads to non-matching robot positions. As a measurable effect, the robot starts oscillating, turning around the target position, missing the ball, etc. This section utilizes a three-layer back-propagation network to extrapolate the robot's true position from the camera images.

3.1 Latency Time

RoboCup robots are real-world vehicles rather than simulated objects. Therefore, all algorithms have to account for physical effects, such as inertia and delays, and have to meet real-time constraints. Because of the real-time constraints, exact algorithms would usually require too much a calculation time. Therefore, the designer has to find a good compromise between computational demands and the precision of the results. In other words, fast algorithms with just a sufficient precision are chosen.

As mentioned in the introduction, latency is caused by various components which include the camera's image grabber, the image compression algorithm, the serial transmission over the wire, the image processing software, and the final transmission of the commands to the robots by means of the DECT modules. Even though the system uses the compressed YUV411 image format [7], the image processing software, and the DECT modules are the most significant parts with a total time delay of about 200 ms. For the top-level control software, which is responsible for the coordination of all team members, all time delays appear as a constant-time lag element. The consequences of the latency problem are further illustrated in Fig. 12 and Fig. 13.

Fig. 12 illustrates the various process stages and corresponding robot positions. At time t_0, the camera takes an image with the robot being on the left-hand-side. At the end of the image analysis (with the robot being at

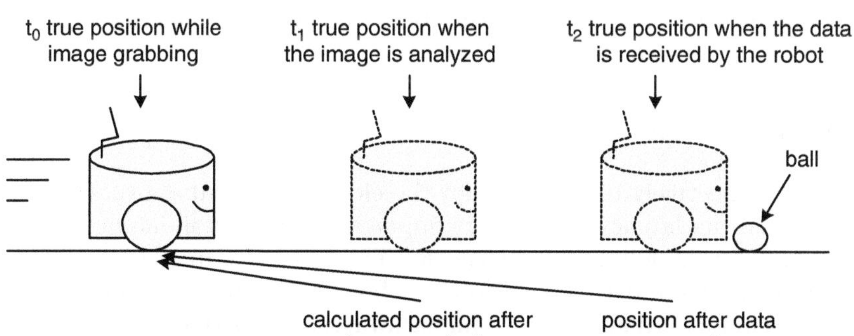

Fig. 12. Due to the latency problem, the robot receives its commands at time t_2, which actually corresponds to the image at time t_0

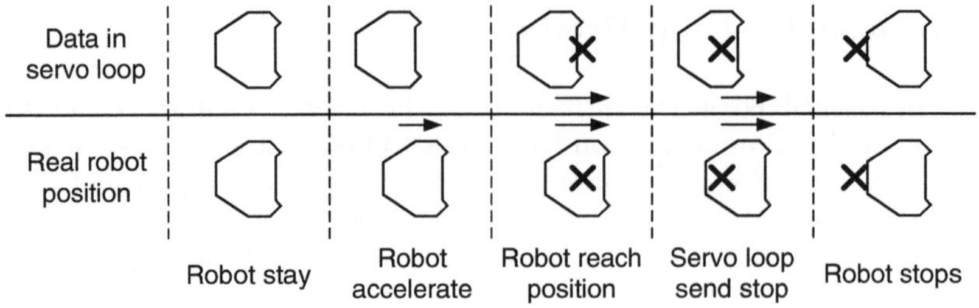

Fig. 13. Problem of stopping the robot at the desired position

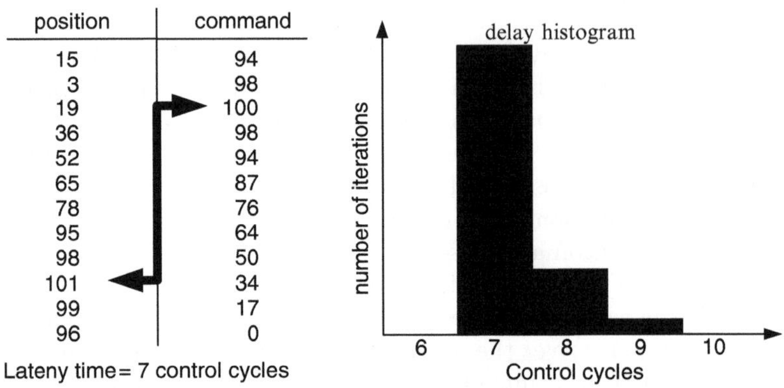

Fig. 14. Detection of the Latency time in the control loop

the old position), the robot has already advanced to the middle position. At time t_2, the derived action commands arrive at the robot, which has further advanced to the position at the right-hand-side. In this example, when being in front of the ball, the robots receive commands which actually belong to a point in time in which the robot was four times its body length away from the ball.

Fig. 13 illustrates how the time delay between image grabbing and receiving commands leads to an oscillating behavior at dedicated target positions (marked by a cross in the figure).

3.2 Experimental Analysis

In order to effectively compensate for the effects discussed above, the knowledge of the exact latency time is very important. The overall latency time was determined by the following experiment: The test software was continuously sending a sinusoidal drive signal to the robot. With this approach, the robot travels 40 cm forward and than 40 cm backwards. The actual robot position as was seen in the image data was then correlated with the control commands. Fig. 14 shows, the duration of the latency time is seven time slots in length, which totals up to 234 ms with 30 frames send by the camera.

For technical reasons, the time delay of the DECT modules is not constant and the jitter is in the order of up to 8 ms. The values given above are averages taken over 100 measurements.

3.3 Back-Propagation Networks and Methods

In general, Kohonen feature maps could be used for addressing the present problem, as was shown in Section 2. In the present case, however, the robot would have to employ a multi-dimensional Kohonen map. For five dimensions with ten nodes each, the network would consist of $10^5 = 100,000$ nodes, which would greatly exceed the robot's computational capabilities.

Multi-layer feed-forward networks are another option, since they are general problem solvers [4] and have low resource requirements. The principal constituents of this network are nodes with input values, an internal activation function, and one output value. A feed-forward network is normally organized in layers, each layer having its own specific number of nodes. The number of nodes in the input and output layers are given by the environment and/or problem description.

The activation of the nodes is propagated layer by layer from input to output. In so doing, each node i calculates its net input $net_i = \sum_i w_{ij} o_j$ as a weighted sum of all nodes j to which it is connected by means of weight w_{ij}. Each node then determines its activation $o_i = f(net_i)$, $f(net_i) = 1/(1 + e^{-net_i})$, with $f(net_i)$ called the logistic function [4].

During training, the algorithm presents all available training patterns, and calculates the total error sum.

$$E(\vec{w}) = \sum_p E_p(\vec{w}) = \frac{1}{2} \sum_p \sum_i (o_i^p - t_i^p)^2. \tag{4}$$

Here p denotes the number of patterns (input/output vector) number and t_i^p denotes the target value for pattern p at output neuron i. After calculating the total error $E(\vec{w})$ the back-propagation algorithm then updates all weights w_i. This is done by performing a gradient-descend step:

$$\vec{w} \leftarrow \vec{w} - \eta \nabla E(\vec{w}), \tag{5}$$

with η denoting a small learning constant, also called the step size. The calculation of the total error sum $E(\vec{w})$ and the subsequent weight update is repeated, until a certain criterion, such as a minimal error sum or stagnation, is met. For further implementation detail, the interested reader is referred to the literature [4].

The experiments in this section were performed with a network having one hidden layer, and a varying number of hidden neurons between $2 \leq h \leq 15$. The learning rate was set to $\eta = 1/2000$ and training was performed over at most 10,000 iterations.

To simplify the task for the neural network, the network adopts a compact coding for the input patterns. This was achieved in the following way. The origin of the coordinate system is set to the robot's current position, and all other vectors are given relative to that one. The network's output is also given in relative coordinates. The input vector consists of the following nine values: six values for the position and orientation of the previous two time steps, and three values for the target position and orientation. The output vector has three values as well. For training and testing, 800 plus 400 patterns were obtained by means of practical experiments.

Fig. 15 shows the average prediction error of a feed-forward network as a function of the number of hidden neurons. It can be seen that three hidden

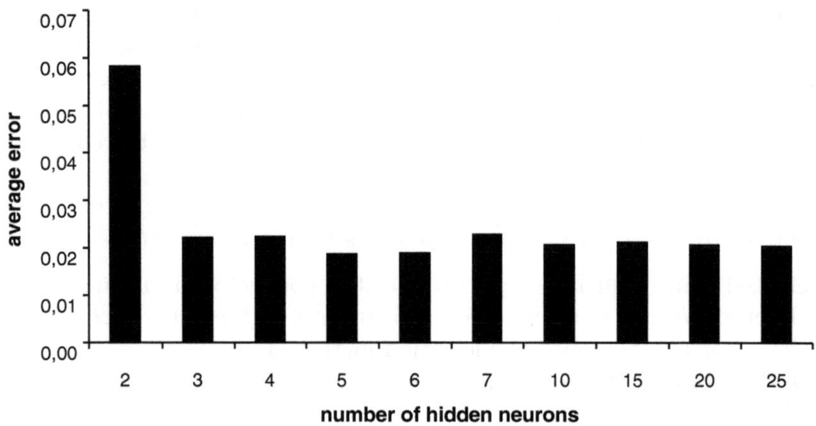

Fig. 15. The average prediction error of a feed-forward network as a function of the number of hidden neurons

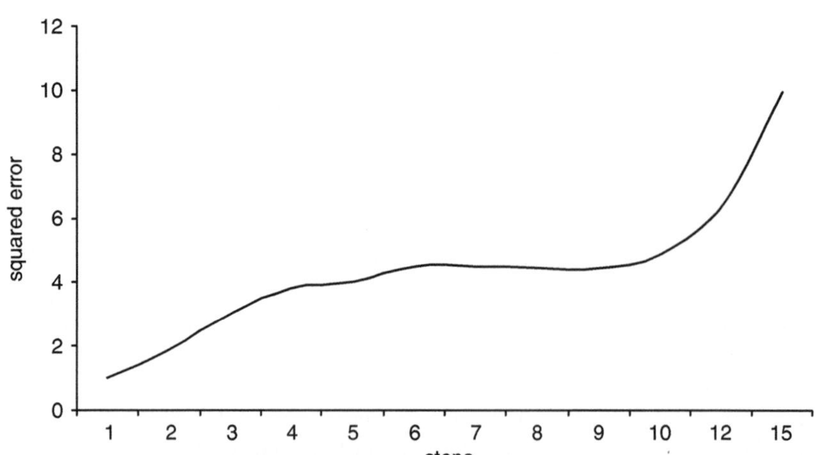

Fig. 16. Square average error as a function of the number of prediction steps into the future

units yield sufficient good results, larger networks do not decrease the network's error.

Since the time delay equals seven camera images the network has to make its prediction for seven time steps in the future.

Fig. 16 plots the networks accuracy when predicting more than one timestamp. It can be seen that the accuracy drastically degrades beyond eleven time steps.

4 Local Position Correction

Another approach to solve the latency problem is to do the compensation on the robot itself. The main advantage of this approach is that the robot's wheel encoders can be used to obtain additional information about the robot's actual behavior. However, since the wheel encoders measure only the wheel rotations, they cannot sense any slip or friction effects directly.

4.1 Increased Position Accuracy by Local Sensors

In the ideal case of slip-free motion, the robot can extrapolate its current position by combining the position delivered by the image processing system, the duration of the entire time delay, and the traveled distance as reported by the wheel encoders. In other words: When slip does not occur, the robot can compensate for all the delays by storing previous and current wheel tick counts. This calculation is illustrated in Fig. 17.

Since the soccer robots are real-world entities, they also have to account for slip and friction, which are among other things, nonlinear and stochastic by nature. The following subsection employs back-propagation networks to account for those effects.

4.2 Embedded Back-Propagation Networks

This section uses the same neural network architectures as have already been discussed in Subsection 3.3. Due to the resource limitations of the robot

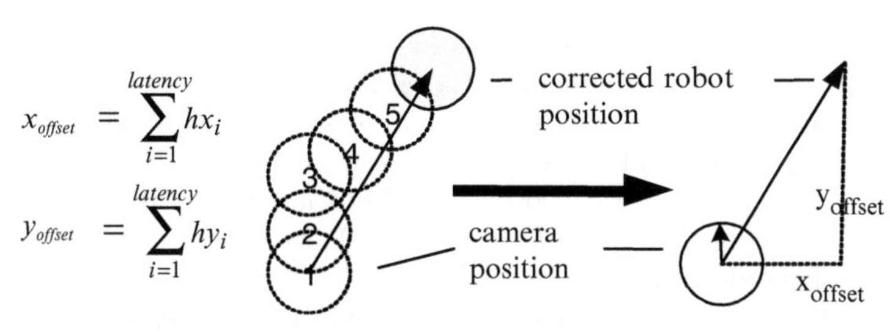

Fig. 17. Extrapolation of the robot's position using the image processing system and the robot's previous tick count

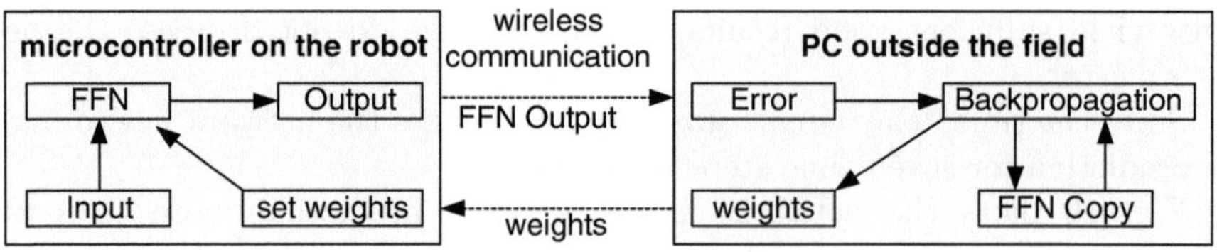

Fig. 18. Separation of the actual feed-forward network (indicated by FFN in the figure) and the back-propagation training algorithm

hardware, the numbers of nodes and connections that the robot can store on its hardware is limited. From a hardware point of view, the memory available on the robot itself is the major constraint. In addition to the actual learning problem, this section is also faced with the challenge of finding a good compromise between the network's complexity and its processing accuracy.

A second constraint to be taken into account concerns the update mechanism of the learning algorithm. It is known that, back-propagation temporarily stores the calculated error counts as well as all the weight changes Δw_{ij} [4]. This leads to a doubling of the memory requirements, which would exhaust the robot's onboard memory size even for moderately sized networks. As a solution for the problem, this section stores those values on the central control PC and communicates the weight changes by means of the wireless communication facility. This separation is illustrated in Fig. 18. Thereby, the neural network can be trained on a PC using the current outputs of the FFN on the robot. A further benefit of the method is that the training can be done during the soccer game, provided that the communication channel has enough capacity for game-control and FFN data. The FFN sends its output values to the PC, which then compares them with the camera data after the latency time t. The PC uses the comparison results to train its network weights without interfering with the robot control. When training is completed and the results are better than the currently used configuration, the new weights are sent to the robot, which start computing the next cycle with these weights.

4.3 Methods

Since the coding of the present problem is not trivial, this section provides a detailed description. In order to avoid a combinatorial explosion, the robot is set at the origin of the coordinate system for every iteration. All other values, such as target position and orientation, are relative to that point. The relative values mentioned above are scaled to be within the range -40 to 40. All angles are directly coded between 0 and 359 degrees. With all these values, the input layer has to have seven nodes.

Fig. 19 illustrates an example configuration. This configuration considers three robot positions labeled "global", "offset", and "target". The first robot

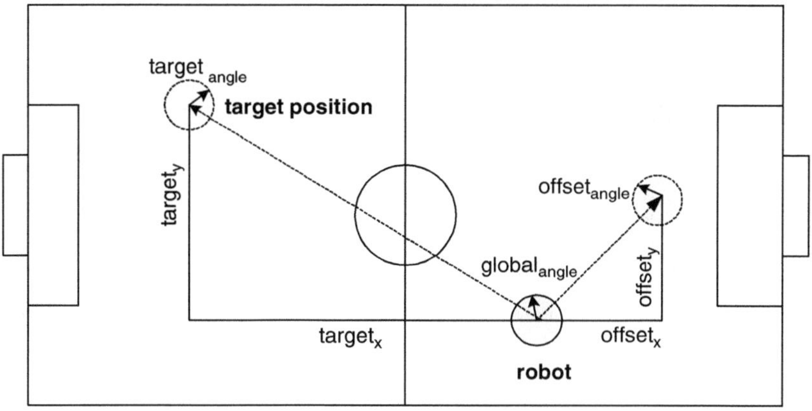

Fig. 19. And example of the configuration for the slip and friction compensation. See text for details

corresponds to the position as provided by the image processing system. The second position called "offset", corresponds to the robot's true position and hence includes the traveled distance during the time delay. The third robot symbolizes the robot's target position. As mentioned previously, the neural network estimates the robot's true positions (labeled by "offset") from the target position, the robot's previous position, and its traveled distances.

All experiments were done using 400 pre-selected training patterns and 800 test patterns. The initial learning rate was set to $\eta = 0.1$. During the course of learning, the learning rate was increased by 2% in case of decreasing error values and decreased by 50% for increasing error values. In 10% of all experiments, the back-propagation became 'stuck' in local optima. These runs were discarded. Learning was terminated, if no improvement was obtained over 100 consecutive iterations.

4.4 Results

Fig. 20 shows the average and maximal error for 3 to 50 hidden neurons organized in one hidden layer. It can be seen that above 20 hidden neurons, the network does not yield any further improvement. This suggests that in order to account for the limited resources available, at most 20 hidden neurons should be used.

Fig. 21 and Fig. 22 summarize some results achieved by networks with two hidden layers. Preliminary experiments have focused on finding a suitable ratio between the hidden neurons in the two hidden layers. Fig. 21 suggests that a ratio 3:1 yield the best results.

Similar to Fig. 20, Fig. 22 shows the error values for two hidden layers with a ratio of 3:1 neurons. The numbers on the x-axis indicate the number

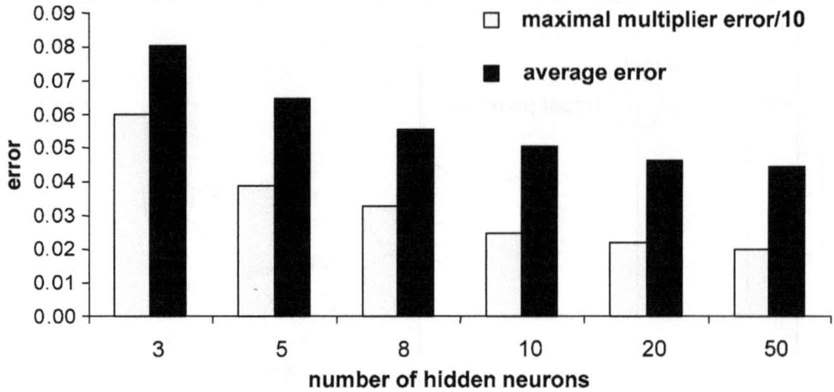

Fig. 20. Average and maximal error of a feed-forward back-propagation network as a function of the number of hidden neurons

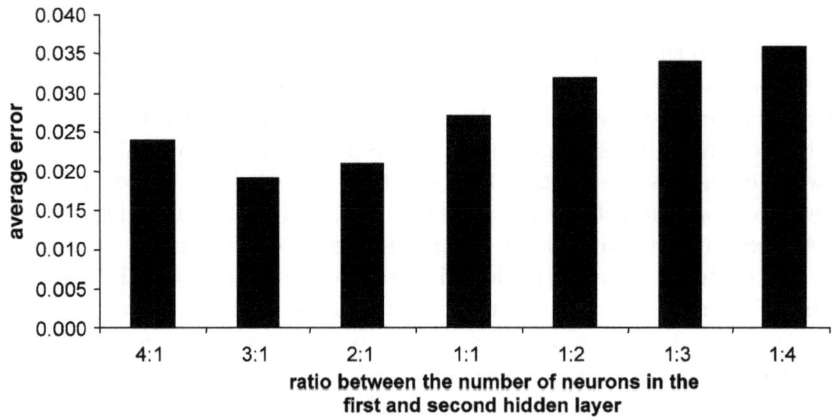

Fig. 21. Average error of a network with two hidden layers as a function of the ratio of the numbers of neurons of two hidden layers

of units in the first and second hidden layer, respectively. From the results, it may be concluded that a network with 45 and 15 neurons in the hidden layers constitutes a good compromise. Furthermore, a comparison of Fig. 20 and Fig. 22 suggest that in this particular application, networks with one-hidden layer perform better than those with two-hidden layers.

When training neural networks, the network's behavior on unseen patterns is of particular interest. Fig. 23 depicts the evolution of both the averaged training and test errors. It is evident that after about 100,000 iterations, the test error stagnates or even increases even though the training error continues decreasing. This behavior is known as Over-Learning in the literature [4].

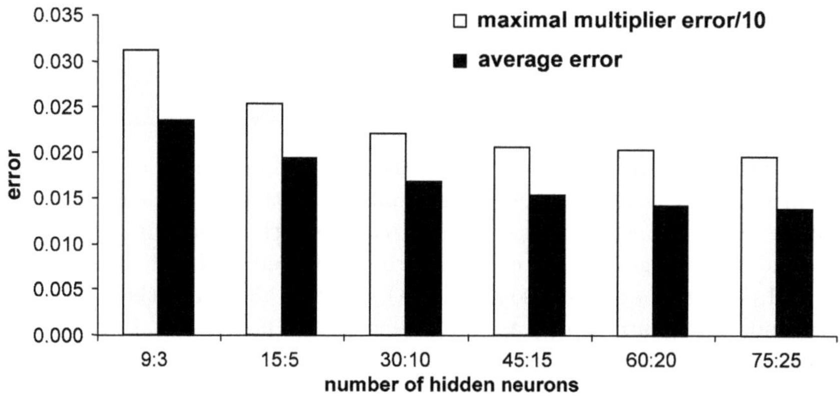

Fig. 22. Average and maximal error for a feed-forward back-propagation network with two hidden layers as a function of the two numbers of hidden neurons

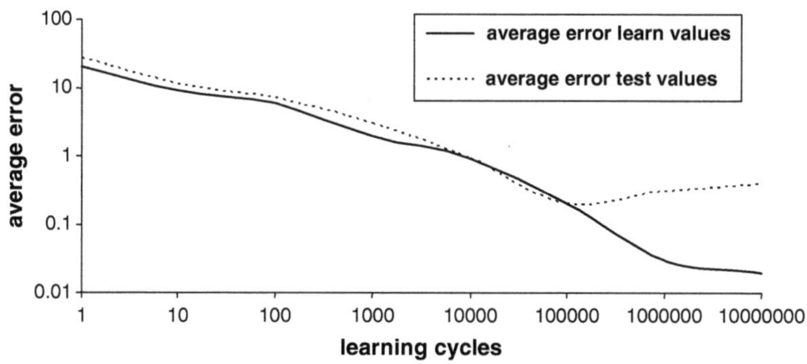

Fig. 23. Typical difference between the training and test error during the course of learning

5 Path Planning using Genetic Algorithms

This section demonstrates how genetic-algorithm-based path planning can be employed on a RoboCup robot. It further demonstrates that a first solution is continuously updated to a changing environment.

The purpose of path planning algorithms is to find a collision free route that satisfies certain optimization parameters between two points. In dynamic environments, a found solution needs to be re-evaluated and updated to environmental changes.

In case of RoboCup, all robots on the field are obstacles. Due to the global camera view, the positions of all robots and hereby all obstacles are known by the robot.

Genetic algorithms use evolutionary methods to find an optimal solution. The solution space is formed by parameters. Possible solutions are represented as individuals of a population. Each gene of an individual represents

Length	x_1	y_1	x_2	y_2	x_3	y_3

Fig. 24. Gene Encoding of an Individual

a parameter. A complete set of genes forms an individual. A new generation is formed by selecting the best individuals from the parent generation and applying evolutionary methods, such as recombination and mutation. After a new generation is generated, each offspring is tested with a fitness function. From all offspring, and in case of $(\mu + \lambda)$-strategy also from the parents, the μ best individuals are chosen as the parents of the next generation. μ usually denotes the number of parents whereas λ is the number of generated children for the next generation.

5.1 Gene Encoding

To apply genetic algorithms to the problem of path planning, the path needs to be encoded into genes. An individual represents a possible path. The path is stored in way points. The start and the destination point of the path are not part of an individual. As the needed number of way points is not known in advance, it is variable. Consequently, the gene length is variable too.

As shown in Fig. 24, each way point is stored in its x and y coordinates as integer values.

The obstacles are relatively small compared to the size of the field and their number cannot exceed nine because each team consists of five robots. This leaves enough room for navigation, three way points between start and end positions are sufficient to find a route. Therefore, the maximal number of way points is set to three.

5.2 Fitness Function

The fitness function is important for the algorithm's stability, because an inadequate function may lead to either stuck at local minima or oscillations around an optimum. Fitness functions are usually constructed by accumulation of weighted evaluation functions. In case of path planning, needed evaluation functions are the path length and a collision avoidance term.

When choosing the representation of the obstacles, it needs to be considered that the calculation is done on the robot. Therefore, the memory footprint is a very important factor.

Each obstacle is stored with its coordinates and its size. This allows for obstacles of any shape. Vectored storing of obstacles provides a higher accuracy and a lower memory consumption but also rises the calculation effort.

The error function consists of the path length and the collision penalty where $path_i$ denotes the length of the sub path, d_i the distance between path and the obstacle center in case the obstacle is hit, r_o the radius of the obstacle, and $c_{penalty}$ a penalty constant. The penalty for hitting an obstacle depends on the distance to its center. The deeper the path is in the obstacle, the higher the penalty should be. Consequently, the fitness raises when the error function lowers.

$$f = \sum_{i=1}^{4} path_i + \sum_{i=0}^{n_{collision}} c_{penalty} \cdot max(0, r_o - d_i) \qquad (6)$$

The collision penalty needs to have a larger influence than a long route. Therefore, $c_{penalty}$ is set to twice the length of the field. Consequently, when the error function has a higher value than twice the field length, no collision free route has been found.

5.3 Evolutionary operations

Evolutionary algorithms find a problem solution by generating new individuals using evolutionary operators. The operators split into two main classes. Crossover operators exchange genes of two individuals, while the mutation operators modify genes of individuals by altering the values of genes. Both classes help to keep the population diverse.

Zheng et al. [15] proposed six mutation operators, which are specially designed for the problem field of path planning. These operators range from modification of one gene over exchange operators to insertion and deletion of way points.

Genetic as well as evolutionary operators can influence the number of way points in the path and thereby the length of the gene.

5.4 Continous calculation

Robots are not static devices. They move around, and their environment and with it the obstacle positions change. Even the destination position of the robot may change. Therefore, the path finding algorithm needs to run during the entire course from the start position to the destination. Due to this reasons, path finding on a robot is a continuing process. On the other hand, the robot does not need to know the best route before it starts driving; a found collision free route is sufficient.

The calculation is done in the main loop of the robot's control program. In the same loop, the data frame is evaluated, and the wheel speeds are calculated. The time between two received data frames is 35 ms. Due to the other tasks that need to be finished in the main loop, the evaluation time for path planning is limited to 20 ms. As the experiments will show, these constraints allow only for the evaluation of one complete generation during every control loop cycle. As mentioned above, the found route does not need to be perfect to start moving. Therefore, the robot does never need to wait longer then four cycles until it can start moving.

5.5 Calculation Time

In this experiment, the time needed to evaluate a population is measured. The parameters vary from 1 to 3 for μ and 10 to 30 for λ. μ is denoting the parent population size while λ is denoting the number of children. The scenario includes four obstacles along the path. For this measurement a plus strategy is used. All times in Table 1 are averaged measurements with a maximal error of 0.9 ms. The timings vary because the randomly chosen genetic operators need different times.

The result indicates that it is possible to use up to 30 offspring in one generation. However, due to variations in calculation speed, it is saver to use only 20 offspring.

5.6 Finding a Path in Dynamic Environments

In real-world scenarios, the obstacles as well as the robot are moving. The movement of the obstacles starts at time step 10 and finishes at time step 30. The robot drives with a speed of 5 pixels per time step. At the beginning, the obstacles are positioned in a way that the robot has enough space between them. In their end position, the robot needs to drive around them.

Fig. 25 shows that until the obstacles start to move, the error function has the same value as the direct distance to the destination. As soon as the obstacle starts to move, the robot is adjusting its path. At time step 22, the distance between both obstacles is smaller than the robot size. At this point,

Table 1. Calculation time for one generation depending on μ and λ

μ	$\lambda = 10$	$\lambda = 20$	$\lambda = 30$
1	5.5 ms	11.2 ms	15.5 ms
2	6.5 ms	14.8 ms	20.7 ms
3	7.2 ms	14.4 ms	20.5 ms

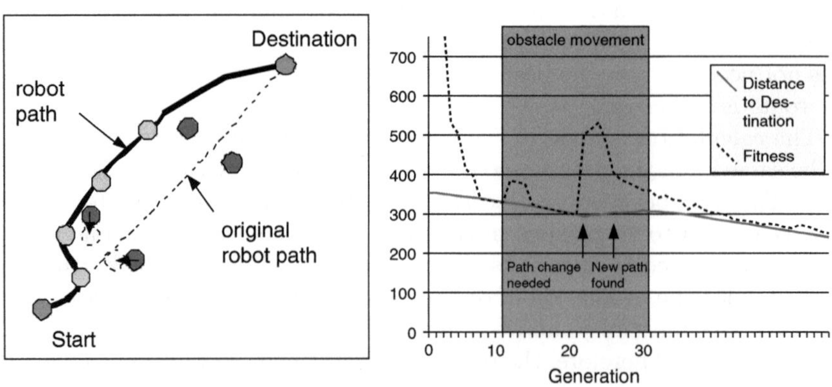

Fig. 25. Path planning and robot movement in a dynamic environment

the fitness function raises by factor of two. The algorithm finds a new route within four time steps.

For this experiment, a (2+20)-strategy was used. Because the fitness function changes when the robot or the obstacles move, found solutions need to be re-calculated in each step. Otherwise, the robot will not change its path as a found solution remains valid.

6 Discussion

This chapter has given a short introduction to the world-wide RoboCup initiative. The focus was on the small-size league, where two teams of five robots play soccer against each other. Since no human control is allowed, the system has to control the robots in an autonomous way. To this end, a control software analyzes images obtained by two cameras and then derives appropriate control commands for all team members.

The omnidirectional drives used by most research teams exhibit certain inaccuracies due to two physical effects called 'slip' and 'friction'. Section 2 has applied Kohonen feature maps to compensate for rotational and directional drift caused by the two effects.

Unfortunately, the image processing system exhibits various time delays at different stages, which leads to erroneous robot behavior. Sections 3 and 4 have incorporated back-propagation networks in order to alleviate this problem by learning techniques which enable precise predictions to be made.

The results presented in this chapter show that neural networks can significantly improve the robot's behavior with respect to accuracy, drift, and response. Additional experiments, which are not discussed in this chapter, have shown that these enhancements lead to an improved team behavior.

The experimental results have also revealed the following deficiencies: Both Kohonen and back-propagation networks require a training phase prior to the actual operation. This limits the networks' online adaptation capabilities. Furthermore, the architectures presented here still require hand-crafted adjustments to some extent. In addition, the resources available on the mobile robots significantly limit the complexity of the employed networks. Finally, the usage of back-propagation networks create the two well-known problems of over-learning and local minima.

Path planning based on evolutionary algorithms on a RoboCup small-size league robot is a possible option. The implementation meets the real-time constraints that are given by the robot's hardware and the environment. The algorithm is capable of finding a path from source to destination and to adapt to environmental changes.

Future research will address the problems discussed above. For this goal, the incorporation of short-cuts into the back-propagation networks seems to be a promising option. The investigation of other learning and self-adaptive principles, such as Hebbian learning [4], seems essential for developing truly self-adaptive control architectures. Another important aspect will be the development of complex controllers which could fit into the low computational resources provided by the robot's onboard hardware.

Acknowledgements

The authors gratefully thank Thorsten Schulz, Guido Moritz, Christian Fabian and Mirko Gerber for helping with all the very time consuming practical time-consuming experiments. Special thanks are due to Prof. Timmermann and Dr. Golatowski for their continuous support.

References

1. http://www.robocup.org
2. A. Gloye, M. Simon, A. Egorova, F. Wiesel, O. Tenchio, M. Schreiber, S. Behnke, and R. Rojas: Predicting away robot control latency, Technical Report B-08-03, FU-Berlin, June 2003.
3. T. Kohonen: Self-Organizing Maps,Springer Series in Information Sciences, Vol. 30, Springer, Berlin, Heidelberg, New York, 1995, 1997, 2001. Third Extended Edition, ISBN 3-540-67921-9, ISSN 0720-678X.
4. R. Rojas: Neural Networks - A Systematic Introduction, Springer-Verlag, Berlin, 1996.
5. Rosenblatt, Frank (1958), The Perceptron: A Probabilistic Model for Information Storage and Organization in the Brain, Cornell Aeronautical Laboratory, Psychological Review, v65, No. 6, pp. 386–408.
6. H. Ritter, K. Schulten: Convergence Properties of Kohonen's Topology Conserving Maps, Biological Cybernetics, Vol. 60, pp 59, 1988
7. J.C. Russ, The Image Processing Handbook, Fourth Edition, CRC Press, 2002, ISBN: 084931142X
8. K.J. Astrom, T. Hagglund, PID Controllers: Theory, Design, and Tuning, International Society for Measurement and Con; 2nd edition, 1995
9. D. Rumelhart, J. Mccelland: Parallel Distributed Processing, MIT Press, 1986
10. D. Rumelhart: The basic ideas in neural net-works, Communications of the ACM 37, 1994 86–92
11. Mohamad H. Hassoun, Fundamentals of artificial neural networks, MIT Press, 1995
12. Marvin L. Minsky and Seymour Papert, Perceptrons (expanded addition), MIT Press, 1988
13. J.C. Alexander and J.H. Maddocks, "On the kinematics of wheeled mobile robots" Autonomous Robot Vehicles, Springer Verlag, pp. 5–24, 1990.
14. Balakrishna, R., and Ghosal, A., "Two dimensional wheeled vehicle kinematics," IEEE Transaction on Robotics and Automation, vol.11, no.l, pp. 126–130, 1995
15. C.W. Zheng, M.Y. Ding, C.P. Zhou, "Cooperative Path Planning for Multiple Air Vehicles Using a Co-evolutionary Algorithm", Proceedings of International Conference on Machine Learning and Cybernetics 2002, Beijing, 1:219–224.

Toward Robot Perception through Omnidirectional Vision

José Gaspar[1], Niall Winters[2], Etienne Grossmann[1], and José Santos-Victor[1] *

[1] Instituto de Sistemas e Robótica
Instituto Superior Técnico
Av. Rovisco Pais, 1
1049-001 Lisboa - Portugal.
(jag,etienne,jasv)@isr.ist.utl.pt

[2] London Knowledge Lab
23-29 Emerald St
London WC1N 3QS, UK.
n.winters@ioe.ac.uk

"My dear Miss Glory, Robots are not people. They are mechanically more perfect than we are, they have an astounding intellectual capacity..."

From the play R.U.R. (Rossum's Universal Robots) *by Karel Capek, 1920.*

1 Introduction

Vision is an extraordinarily powerful sense. The ability to perceive the environment allows for movement to be regulated by the world. Humans do this effortlessly but we still lack an understanding of how perception works. Our approach to gaining an insight into this complex problem is to build artificial visual systems for semi-autonomous robot navigation, supported by human-robot interfaces for destination specification. We examine how robots can use images, which convey only 2D information, in a robust manner to drive its actions in 3D space. Our work provides robots with the perceptual capabilities to undertake everyday navigation tasks, such as *go to the fourth office in the second corridor*. We present a complete navigation system with a focus on building – in line with Marr's theory [57] – mediated perception modalities. We address fundamental design issues associated with this goal; namely sensor design, environmental representations, navigation control and user interaction.

* This work was partially supported by Fundação para a Ciência e a Tecnologia (ISR/IST plurianual funding) through the POS_Conhecimento Program that includes FEDER funds. Etienne Grossmann is presently at Tyzx.com.

J. Gaspar et al.: *Toward Robot Perception through Omnidirectional Vision*, Studies in Computational Intelligence (SCI) **70**, 223–270 (2007)
www.springerlink.com

A critical component of any perceptual system, human or artificial, is the sensing modality used to obtain information about the environment. In the biological world, for example, one striking observation is the diversity of *ocular* geometries. The majority of insects and arthropods benefit from a wide field of view and their eyes have a space- variant resolution. To some extent, the perceptual capabilities of these animals can be explained by their specially adapted eye geometries. Similarly, in this work, we explore the advantages of having large fields of view by using an omnidirectional camera with a 360 degree azimuthal field of view.

Part of the power of our approach comes from the way we construct representations of the world. Our internal environmental representations are tailored to each navigation task, in line with the information perceived from the environment. This is supported by evidence from the biological world, where many animals make alternate use of landmark-based navigation and (approximate) route integration methods [87]. Taking a human example when walking along a city avenue, it is sufficient to know our position to within an accuracy of one block. However, when entering our hall door we require much more precise movements. In a similar manner, when our robot is required to travel long distances, an appearance-based environmental representation is used to perceive the world [89]. This is a long-distance/low-precision navigation modality. For precise tasks, such as docking or door traversal, perception switches from the appearance-based method to one that relies on image features and is highly accurate. We characterize these two modes of operation as: Topological Navigation and Visual Path Following, respectively.

Combining long-distance/low-precision and short-distance/high-accuracy perception modules plays an important role in finding efficient and robust solutions to the robot navigation problem. This distinction is often overlooked, with emphasis being placed on the construction of world models, rather than concentrating on how these models can be used effectively.

In order to effectively navigate using the above representations, the robot needs to be provided with a destination. We have developed human-robot interfaces for this task using (omnidirectional) images for interactive scene modelling. From a perception perspective, our aim is to design an interface where an intuitive link exists between how the user perceives the world and how they control the robot. We achieve this by generating a rich scene description of a remote location. The user is free to rotate and translate this model to specify a particular destination to the robot. Scene modelling, from a single omnidirectional image, is possible with limited user input in the form of co-linearity, co-planarity and orthogonality properties. While humans have an immediate qualitative understanding of the scene encompassing co-planarity and co-linearity properties of a number of points in the scene, robots equipped with an omnidirectional camera can take precise azimuthal and elevation measurements.

1.1 State of the Art

There are many types of omnidirectional vision systems and the most common ones are based on rotating cameras, fish-eye lenses or mirrors [3, 45, 18]. Baker and Nayar listed all the mirror and camera setups having a Single View Point (SVP) [1, 3]. These systems are omnidirectional, have the 360° horizontal field of view, but do not have constant resolution for the most common scene surfaces. Mirror shapes for linearly imaging 3D planes, cylinders or spheres were presented in [32] within a unified approach that encompasses all the previous constant resolution designs [46, 29, 68] and allowed for new ones.

Calibration methods are available for (i) most (static) SVP omnidirectional setups, even where lenses have radial distortion [59] and (ii) for *non-SVP* cameras set-ups, such as those obtained by mounting in a mobile robot multiple cameras, for example [71]. Given that knowledge of the geometry of cameras is frequently used in a back-projection form, [80] proposed a general calibration method for general cameras (including non-SVP) which gives the back-projection line (representing a light-ray) associated with each pixel of the camera. In another vein, precise calibration methods have begun to be developed for pan-tilt-zoom cameras [75]. These active camera set-ups, combining pan-tilt-zoom cameras and a convex mirror, when precisely calibrated, allow for the building of very high resolution omnidirectional scene representations and for zooming to improve resolution, which are both useful characteristics for surveillance tasks. Networking cameras together have also provided a solution in the surveillance domain. However, they pose new and complex calibration challenges resulting from the mixture of various camera types, potentially overlapping fields-of-view, the different requirements of calibration quality and the type of calibration data used (for example, static or dynamic background) [76].

On a final note, when designing catadioptric systems, care must be taken to minimize defocus blur and optical aberrations as the spherical aberration or astigmatism [3, 81]. These phenomena become more severe when minimising the system size, and therefore it is important to develop optical designs and digital image processing techniques that counter-balance the image malformation.

The applications of omnidirectional vision to robotics are vast. Starting with the seminal idea of enhancing the field of view for teleoperation, current challenges in omnidirectional vision include autonomous and cooperative robot-navigation and reconstruction for human and robot interaction [27, 35, 47, 61].

Vision based autonomous navigation relies on various types of information, e.g. scene appearance or geometrical features such as points or lines. When using point features, current research, which combines simultaneous localization and map building, obtains robustness by using sequential Monte-Carlo methods such as particle filters [51, 20]. Using more stable features, such as lines, allows for improved self-localization optimization methods [19]. [10, 54] use sensitivity analysis in order to choose optimal landmark configurations for self-localization. Omnidirectional vision has the advantage of tracking features over a larger azimuth range and therefore can bring additional robustness to navigation.

State of the art automatic scene reconstruction, based on omnidirectional vision, relies on graph cutting methodologies for merging point clouds, acquired at different robot locations [27]. Scene reconstruction is mainly useful for human robot interaction, but can also be used for inter-robot interaction. Current research shows that building robot teams can be framed as a scene independent problem, provided that the robots observe each other and have reliable motion measurements [47, 61]. The robot teams can then share scene models allowing better human to robot-team interaction.

This chapter is structured as follows. In Section 2, we present the modelling and design of omnidirectional cameras, including details of the camera designs we used. In Section 3, we present Topological Navigation and Visual Path Following. We provide details of the different image dewarpings (views) available from our omnidirectional camera: standard, panoramic and bird's–eye views. In addition, we detail geometric scene modelling, model tracking, and appearance-based approaches to navigation. In Section 4, we present our Visual Interface. In all cases, we demonstrate mobile robots navigating autonomously and guided interactively in structured environments. These experiments show that the synergetic design, combining perception modules, navigation modalities and humanrobot interaction, is effective in realworld situations. Finally, in Section 5, we present our conclusions and future research directions.

2 Omnidirectional Vision Sensors: Modelling and Design

In 1843 [58], a patent was issued to Joseph Puchberger of Retz, Austria for the first system that used a rotating camera to obtain omnidirectional images. The original idea for the (static camera) omnidirectional vision sensor was initially proposed by Rees in a US patent dating from 1970 [72]. Rees proposed the use of a hyperbolic mirror to capture an omnidirectional image, which could then be transformed to a (normal) perspective image.

Since those early days, the spectrum of application has broadened to include such diverse areas as tele-operation [84, 91], video conferencing [70], virtual reality [56], surveillance [77], 3D reconstruction [33, 79], structure from motion [13] and autonomous robot navigation [35, 89, 90, 95, 97]. For a survey of previous work, the reader is directed to [94]. A relevant collection of papers, related to omnidirectional vision, can be found in [17] and [41].

Omnidirectional images can be generated by a number of different systems which can be classified into four distinct design groupings: Camera-Only Systems; Multi-Camera – Multi-Mirror Systems; Single Camera – Multi-Mirror Systems, and Single Camera – Single Mirror Systems.

Camera-Only Systems: A popular method used to generate omnidirectional images is the rotation of a standard CCD camera about its vertical axis. The captured information, i.e. perspective images (or vertical line scans) are then stitched together so as to obtain panoramic 360° images. Cao *et al.* [11] describe such a system fitted with a fish-eye lens [60]. Instead of relying upon a single rotating camera, a second camera-only design is to combine cameras pointing in differing directions [28]. Here, images are acquired using inexpensive board cameras and are again stitched together to form panoramas. Finally, Greguss [40] developed a lens, he termed the Panoramic Annular Lens, to capture a panoramic view of the environment.

Multi-Camera – Multi-Mirror Systems: This approach consists of arranging a cluster of cameras in a certain manner along with an equal number of mirrors. Nalwa [63] achieved this by placing four triangular planar mirrors side by side, in the shape of a pyramid, with a camera under each. One significant problem with multi-camera – multi-mirror systems is geometric registering and intensity blending the images together so as to form a seamless panoramic view. This is a difficult problem to solve given that, even with careful alignment, unwanted visible artifacts are often found at image boundaries. These occur not only because of variations between the intrinsic parameters of each camera, but also because of imperfect mirror placement.

Single Camera – Multi-Mirror Systems: The main goal behind the design of single camera – multi-mirror systems is compactness. Single camera – multi-mirror systems are also known as Folded Catadioptric Cameras [66]. A simple example of such a system is that of a planar mirror placed between a light ray travelling from a curved mirror to a camera, thus "folding" the ray. Bruckstein and Richardson [9] presented a design that used two parabolic mirrors, one convex and the other concave. Nayar [66] used a more general design consisting of any two mirrors with a conic-section profile.

Single Camera – Single Mirror Systems:

In recent years, this system design has become very popular; it is the approach we chose for application to visual-based robot navigation. The basic method is to point a CCD camera vertically up, towards a mirror.

There are a number of mirror profiles that can be used to project light rays to the camera. The first, and by far the most popular design, uses a **standard mirror profile**: planar, conical, elliptical, parabolic, hyperbolic or spherical. All of the former, with obvious exception of the planar mirror, can image a 360° view of the environment horizontally and, depending on the type of mirror used approximately 70° to 120°, vertically. Some of the mirror profiles, yield simple projection models. In general, to obtain such a system it is necessary to place the mirror at a precise location relative to the camera. In 1997, Nayar and Baker [64] patented a system combining a parabolic mirror and a telecentric lens, which is well described by a simple model and simultaneously overcomes the requirement of precise assembly. Furthermore, their system is superior in the acquisition of non-blurred images.

The second design involves specifying a **specialised mirror profile** in order to obtain a particular, possibly *task-specific*, view of the environment. In both cases, to image the greatest field-of-view the camera's optical axis is aligned with that of the mirrors'. A detailed analysis of both the standard and specialised mirror designs are given in the following Sections.

2.1 A Unifying Theory for Single Centre of Projection Systems

Recently, Geyer and Daniilidis [37, 38] presented a unified projection model for all omnidirectional cameras *with* a single centre of projection. They showed that these systems (parabolic, hyperbolic, elliptical and perspective[3]) can be modelled by a two-step mapping via the sphere. This mapping of a point in space to the image plane is graphically illustrated in Fig. 1 (left). The two steps of the mapping are as follows:

1. Project a 3D world point, $\mathbf{P} = (x, y, z)$ to a point $\mathbf{P_s}$ on the sphere surface, such that the projection is normal to the sphere surface.
2. Subsequently, project *to* a point on the image plane, $\mathbf{P_i} = (u, v)$ from a point, $\mathbf{O}$ on the vertical axis of the sphere, through the point $\mathbf{P_s}$.

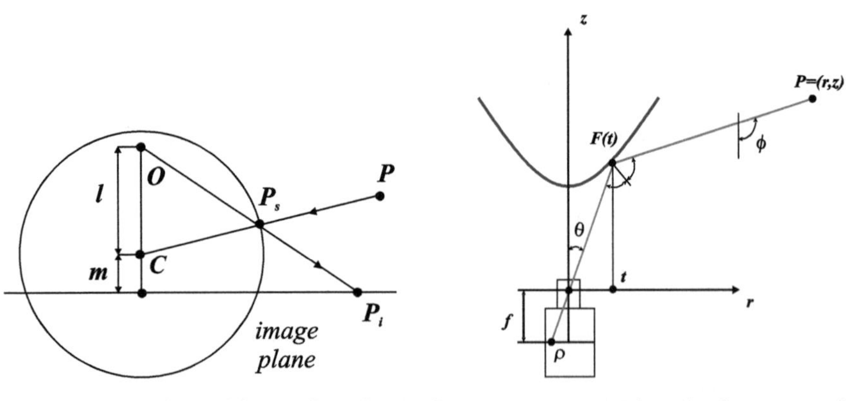

Fig. 1. A Unifying Theory for all catadioptric sensors *with* a single centre of projection (left). Main variables defining the projection model of non-single projection centre systems based on arbitrary mirror profiles, $F(t)$ (right)

[3] A parabolic mirror with an orthographic lens and all of the others with a standard lens. In the case of a perspective camera, the mirror is virtual and planar.

The mapping is mathematically defined by:

$$\begin{bmatrix} u \\ v \end{bmatrix} = \frac{l+m}{l \cdot r - z} \begin{bmatrix} x \\ y \end{bmatrix}, \text{ where } r = \sqrt{x^2+y^2+z^2} \quad (1)$$

As one can clearly see, this is a two-parameter, (l and m) representation, where l represents the distance from the sphere centre, **C** to the projection centre, **O** and m the distance from **O** to the image plane. Modelling the various catadioptric sensors with a single centre of projection is then just a matter of varying the values of l and m in 1. As an example, to model a parabolic mirror, we set $l = 1$ and $m = 0$. Then the image plane passes through the sphere centre, **C** and **O** is located at the north pole of the sphere. In this case, the second projection is the well known stereographic projection. We note here that a standard perspective is obtained when $l = 0$ and $m = 1$. In this case, **O** converges to **C** and the image plane is located at the south pole of the sphere.

2.2 Model for Non-Single Projection Centre Systems

Non-single projection centre systems cannot be represented exactly by the unified projection model. One such case is an omnidirectional camera based on an spherical mirror. The intersections of the projection rays incident to the mirror surface, define a continuous set of points distributed in a volume[2], unlike the unified projection model where they all converge to a single point. In the following, we derive a projection model for non-single projection centre systems.

The image formation process is determined by the trajectory of rays that start from a 3D point, reflect on the mirror surface and finally intersect with the image plane. Considering first order optics [44], the process is simplified to the trajectory of the principal ray. When there is a single projection centre it immediately defines the direction of the principal ray starting at the 3D point. If there is no single projection centre, then we must first find the reflection point at the mirror surface.

In order to find the reflection point, a system of non-linear equations can be derived which directly gives the reflection and projection points. Based on first order optics [44], and in particular on the reflection law, the following equation is obtained:

$$\phi = \theta + 2.atan(F') \quad (2)$$

where θ is the camera's vertical view angle, ϕ is the system's vertical view angle, F denotes the mirror shape (it is a function of the radial coordinate, t) and F' represents the slope of the mirror shape. See Fig. 1 (right).

Equation (2) is valid both for single [37, 1, 96, 82], and non-single projection centre systems [12, 46, 15, 35]. When the mirror shape is known, it provides the projection function. For example, consider the single projection centre system combining a parabolic mirror, $F(t) = t^2/2h$ with an orthographic camera [65], one obtains the projection equation, $\phi = 2atan(t/h)$ relating the (angle to the) 3D point, ϕ and an image point, t.

In order to make the relation between world and image points explicit it is only necessary to replace the angular variables by cartesian coordinates. We do this assuming the pin-hole camera model and calculating the slope of the light ray starting at a generic 3D point (r, z) and hitting the mirror:

$$\theta = atan\left(\frac{t}{F}\right) , \quad \phi = atan\left(-\frac{r-t}{z-F}\right) . \tag{3}$$

The solution of the system of equations (2) and (3) gives the reflection point, (t, F) and the image point $(f.t/F, f)$ where f is the focal length of the lens.

2.3 Design of Standard Mirror Profiles

Omnidirectional camera mirrors can have standard or specialised profiles, $F(t)$. In standard profiles the form of $F(t)$ is known, we need only to find its parameters. In the specialised profiles the form of $F(t)$ is also a degree of freedom to be derived numerically. Before detailing the design methodology, we introduce some useful properties.

Property 1 ***(Maximum vertical view angle)*** *Consider a catadioptric camera with a pin-hole at* $(0, 0)$ *and a mirror profile* $F(t)$*, which is a strictly positive* C_1 *function, with domain* $[0,\ t_M]$ *that has a monotonically increasing derivative. If the slope of the light ray from the mirror to the camera,* t/F *is monotonically increasing then the* maximum vertical view angle, ϕ *is obtained at the mirror rim,* $t = t_M$.

Proof: from Eq. (2) we see that the maximum vertical view angle, ϕ is obtained when t/F and F' are maximums. Since both of these values are monotonically increasing, then the maximum of ϕ is obtained at the maximal t, i.e. $t = t_M$.

□

The maximum vertical view angle allows us to precisely set the system scaling property. Let us define the scaling of the mirror profile (and distance to camera) $F(t)$ by $(t_2, F_2) \doteq \alpha.(t, F)$, where t denotes the mirror radial coordinate. More precisely, we are defining a new mirror shape F_2 function of a new mirror radius coordinate t_2 as:

$$t_2 \doteq \alpha t \quad \wedge \quad F_2(t_2) \doteq \alpha F(t). \tag{4}$$

This scaling preserves the geometrical property:

Property 2 ***(Scaling)*** *Given a catadioptric camera with a pin-hole at* $(0, 0)$ *and a mirror profile* $F(t)$*, which is a* C_1 *function, the* vertical view angle is invariant *to the system scaling defined by Eq. (4).*

Proof: we want to show that the vertical view angles are equal at corresponding image points, $\phi_2(t_2/F_2) = \phi(t/F)$ which, from Eq. (2), is the same as comparing the corresponding derivatives $F_2'(t_2) = F'(t)$ and is demonstrated using the definition of the derivative:

$$F_2'(t_2) = \lim_{\tau_2 \to t_2} \frac{F_2(\tau_2) - F_2(t_2)}{\tau_2 - t_2} = \lim_{\tau \to t} \frac{F_2(\alpha\tau) - F_2(\alpha t)}{\alpha\tau - \alpha t} = \lim_{\tau \to t} \frac{\alpha F(\tau) - \alpha F(t)}{\alpha\tau - \alpha t} = F'(t)$$

□

Simply put, the scaling of the system geometry does not change the local slope at mirror points defined by fixed image points. In particular, the mirror slope at the mirror rim does not change and therefore the vertical view angle of the system does not change.

Notice that despite the vertical view angle remaining constant the observed 3D region actually changes but usually in a negligible manner. As an example, if the system sees an object 1 metre tall and the mirror rim is raised 5 cm due to a scaling, then only those 5 cm become visible on top of the object.

Standard mirror profiles are parametric functions and hence implicitly define the design parameters. Our goal is to specify a large vertical field of view, ϕ given the limited field of view of the lens, θ. In the following we detail the designs of cameras based on spherical and hyperbolic mirrors, which are the most common standard mirror profiles.

Cameras based on spherical and hyperbolic mirrors, respectively, are described by the mirror profile functions:

$$F(t) = L - \sqrt{R^2 - t^2} \quad \text{and} \quad F(t) = L + \frac{a}{b}\sqrt{b^2 + t^2} \tag{5}$$

where R is the spherical mirror radius, (a, b) are the major and minor axis of the hyperbolic mirror and L sets the camera to mirror distance (see Fig. 2). As an example, when $L = 0$ for the hyperbolic mirror, we obtain the omnidirectional camera proposed by Chahl and Srinivasan's [12]. Their design yields a constant gain mirror that linearly maps 3D vertical angles into image radial distances.

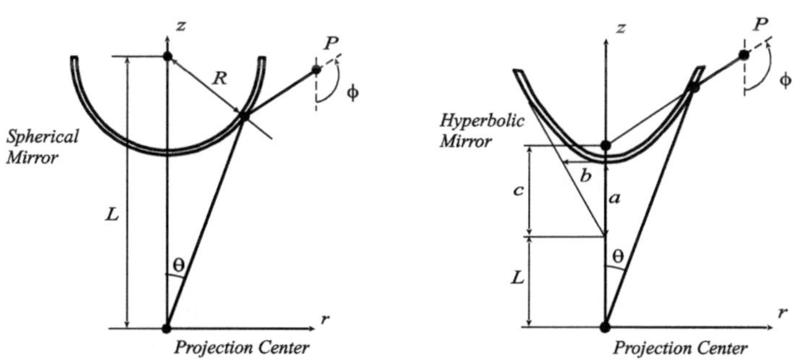

Fig. 2. Catadioptric Omnidirectional Camera based on a spherical (left) or an a hyperbolic mirror (right). In the case of a hyperbolic mirror $L = 0$ or $L = c$ and $c = \sqrt{a^2 + b^2}$

Chahl and Srinivasan's design does not have the single projection centre property, which is obtained placing the camera at one hyperboloid focus, i.e. $L = \sqrt{a^2 + b^2}$, as Baker and Nayar show in [1] (see Fig. 2 (right). In both designs the system is described just by the two hyperboloid parameters, a and b.

In order to design the spherical and hyperbolic mirrors, we start by fixing the focal length of the camera, which directly determines the view field θ. Then the maximum vertical view field of the system, ϕ, is imposed with the reflection law Eq. (2). This gives the slope of the mirror profile at the mirror rim, F'. Stating, without loss of generality, that the mirror rim has unitary radius (i.e. $(1, F(1))$ is a mirror point), we obtain the following non-linear system of equations:

$$\begin{cases} F(1) = 1/\tan\theta \\ F'(1) = \tan(\phi - \theta)/2 \end{cases} . \tag{6}$$

The mirror profile parameters, (L, R) or (a, b), are embedded in $F(t)$, and are therefore found solving the system of equations.

Since there are minimal focusing distances, D_{min} which depend on the particular lens, we have to guarantee that $F(0) \geq D_{min}$. We do this applying the scaling property (Eq. (4)). Given the scale factor $k = D_{min}/F(0)$ the scaling of the spherical and hyperbolic mirrors is applied respectively as $(R, L) \leftarrow (k.R, k.L)$ and $(a, b) \leftarrow (k.a, k.b)$. If the mirror is still too small to be manufactured then an additional scaling up may be applied. The camera self-occlusion becomes progressively less important when scaling up.

Figure 3 shows an omnidirectional camera based on a spherical mirror, built in house for the purpose of conducting navigation experiments. The mirror was designed to have a view field of 10° above the horizon line. The lens has $f = 8mm$ (vertical view field, θ is about $\pm 15^o$ on a $6.4mm \times 4.8mm$ CCD). The minimal distance from the lens to the mirror surface was set to $25cm$. The calculations indicate a spherical mirror radius of $8.9cm$.

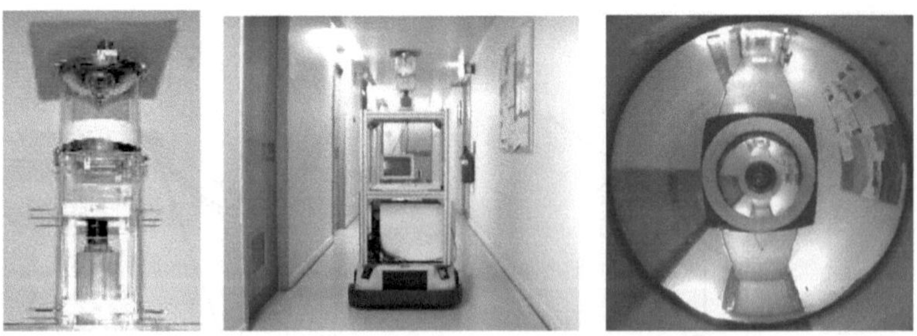

Fig. 3. Omnidirectional camera based on a spherical mirror (left), camera mounted on a Labmate mobile robot (middle) and omnidirectional image (right)

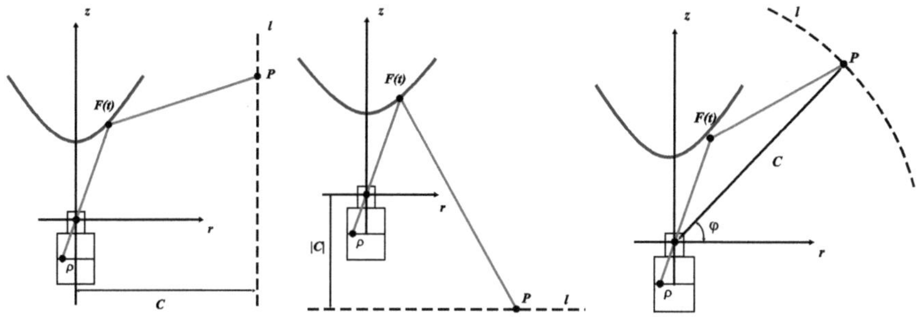

Fig. 4. Constant vertical, horizontal and angular resolutions (respectively left, middle and right schematics). Points on the line l are linearly related to their projections in pixel coordinates, ρ

2.4 Design of Constant Resolution Cameras

Constant Resolution Cameras, are omnidirectional cameras that have the property of linearly mapping 3D measures to imaged distances. The 3D measures can be either elevation angles, vertical or horizontal distances (see Fig. 4). Each linear mapping is achieved by specializing the mirror shape.

Some constant resolution designs have been presented in the literature, [12, 46, 15, 37] with a *different derivation* for each case. In this section, we present a *unified* approach that encompasses all the previous designs and allows for new ones. The key idea is to separate the equations for the reflection of light rays at the mirror surface and the mirror *Shaping Function*, which explicitly represents the linear projection properties to meet.

The Mirror Shaping Function

Combining the equations that describe the non-single projection centre model (Eqs. (2) and (3)) and expanding the trigonometric functions, one obtains an equation of the variables t, r, z encompassing the mirror shape, F and slope, F':

$$\frac{\frac{t}{F} + 2\frac{F'}{1-F'^2}}{1 - 2\frac{tF'}{F(1-F'^2)}} = -\frac{r-t}{z-F} \tag{7}$$

This is Hicks and Bajcsy's differential equation relating 3D points, (r, z) to the reflection points, $(t, F(t))$ which directly imply the image points, $(t/F, 1)$ [46]. We assume without loss of generality that the focal length, f is 1, since it is easy to account for a different (desired) value at a later stage.

Equation 7 allows to design a mirror shape, $F(t)$ given a desired relationship between 3D points, (r, z) and the corresponding images, $(t/F, 1)$. In order to compute $F(t)$, it is convenient to have the equation in the form of an explicit expression for F' [4]. Re-arranging Eq. (7) results in the following second order polynomial equation:

$$F'^2 + 2\alpha F' - 1 = 0 \tag{8}$$

where α is a function of the mirror shape, (t, F) and of an arbitrary 3D point, (r, z):

$$\alpha = \frac{-(z-F)\,F + (r-t)\,t}{(z-F)\,t + (r-t)\,F} \tag{9}$$

We call α the mirror *Shaping Function*, since it ultimately determines the mirror shape by expressing the relationship that should be observed between 3D coordinates, (r, z) and those on the image plane, determined by t/F. In the next section we will show that the mirror shaping functions allow us to bring the desired linear projection properties into the design procedure.

Concluding, to obtain the mirror profile first we specify the shaping function, Eq. (9) and then solve Eq. (8), or simply integrate:

$$F' = -\alpha \pm \sqrt{\alpha^2 + 1} \tag{10}$$

where we choose the + in order to have positive slopes for the mirror shape, F.

Setting Constant Resolution Properties

Our goal is to design a mirror profile to match the sensor's resolution in order to meet, in terms of desired image properties, the application constraints. As shown in the previous section, the shaping function defines the mirror profile, and here we show how to set it accordingly to the design goal.

For constant resolution mirrors, we want some world distances, D, to be *linearly* mapped to (pixel) distances, p, measured in the image sensor, i.e. $D = a_0.p + b_0$ for some values of a_0 and b_0 which mainly determine the visual field.

When considering conventional cameras, pixel distances are obtained by scaling metric distances in the image plane, ρ. In addition, knowing that those distances relate to the slope t/F of the ray of light intersecting the image plane as $\rho = f.\frac{t}{F}$. The linear constraint may be conveniently rewritten in terms of the mirror shape as:

$$D = a.t/F + b \tag{11}$$

Notice that the parameters a and b can easily be scaled to account for a desired focal length, thus justifying the choice $f = 1$.

We now specify which 3D distances, $D(t/F)$, should be mapped linearly to pixel coordinates, in order to preserve different image invariants (e.g. ratios of distances or angles in certain directions).

Constant Vertical Resolution - The aim of the first design procedure is to preserve the relative vertical distances of points located at a fixed distance,

[4] Having an explicit formula for F' allows to directly use matlab's ode45 function

Linear Property	Mirror Shaping Function	
$z = a.t/F + b$ $r = C$	$\alpha = \dfrac{-\left(a\frac{t}{F} + b - F\right) F + (C - t)\, t}{\left(a\frac{t}{F} + b - F\right) t + (C - t)\, F}$	(12)
$r = a.t/F + b$ $z = C$	$\alpha = \dfrac{-(C - F)\, F + \left(a\frac{t}{F} + b - t\right) t}{(C - F)\, t + \left(a\frac{t}{F} + b - t\right) F}$	(13)
$\varphi = a.t/F + b$ $r = C.cos(\varphi)$ $z = C.sin(\varphi)$	$\alpha = \dfrac{-\left(C\sin(a\frac{t}{F} + b) - F\right) F + \left(C\cos(a\frac{t}{F} + b) - t\right) t}{\left(C\sin(a\frac{t}{F} + b) - F\right) t + \left(C\cos(a\frac{t}{F} + b) - t\right) F}$	(14)

Table 1. Mirror Shaping Functions for constant vertical, horizontal and angular resolutions

C, from the camera's optical axis. In other words, if we consider a cylinder of radius, C, around the camera optical axis, we want to ensure that ratios of distances, measured in the vertical direction along the surface of the cylinder, remain unchanged when measured in the image. Such invariance should be obtained by adequately designing the mirror profile - yielding a constant vertical resolution mirror.

The derivation described here follows closely that presented by Gaechter and Pajdla in [30]. The main difference consist of a simpler setting for the equations describing the mirror profile. We start by specialising the linear constraint in Eq. (11) to relate 3D points of a vertical line l with pixel coordinates (see Fig. 4). Inserting this constraint into Eq. (9) we obtain the specialised shaping function of Eq. (12) in Table 1.

Hence, the procedure to determine the mirror profile consists of integrating Eq. (10) using the shaping function of Eq. (12), while t varies from 0 to the mirror radius. The initialization of the integration process is done by computing the value of $F(0)$ that would allow the mirror rim to occupy the entire field of view of the sensor.

Constant Horizontal Resolution (Bird's Eye View) - Another interesting design possibility for some applications is that of preserving ratios of distances measured on the ground plane. In such a case, one can directly use image measurements to obtain ratios of distances or angles on the pavement (which can greatly facilitate navigation problems or visual tracking). Such images are also termed *Bird's eye views.*

Figure 4 shows how the ground plane, l, is projected onto the image plane. The camera-to-ground distance is represented by $-C$ (C is negative because the ground plane is lower than the camera centre) and r represents radial distances on the ground plane. The linear constraint inserted into Eq. (9)

yields a new shaping function (as in Eq. (13)), which after integrating Eq. (10) results in the mirror profile proposed by Hicks and Bajcsy [46].

Constant Angular Resolution - One last case of practical interest is that of obtaining a linear mapping from 3D points spaced by equal angles to equally distant image pixels, i.e. designing a constant angular resolution mirror. Figure 4 shows how the spherical surface with radius C surrounding the sensor is projected onto the image plane. In this case the desired linear property relates angles with image points. Then, placing the constraints into Eq. (9) we finally obtain Eq. (14).

Integrating Eq. (10), using the shaping function just obtained (Eq. (14)), would result in a mirror shape such as the one of Chahl and Srinivasan [12]. The difference is that in our case we are imposing the linear relationship from 3D vertical angles, φ directly to image points, $(t/F, 1)$ instead of angles relative to the camera axis, $atan(t/F)$.

Shaping functions for Log-polar Sensors - Log-polar cameras are imaging devices that have a spatial resolution inspired by the human-retina. Unlike standard cameras, the resolution is not constant on the sensing area. More precisely, the density of the pixels is higher in the centre and decays logarithmically towards the image periphery. The organisation of the pixels also differs from the standard cameras, as a log-polar camera consists of a set of concentric circular rings, each one with a constant number of pixels. Advantageously, combining a log-polar camera with a convex mirror results in an omnidirectional imaging device where the panoramic views are extracted directly due to the polar arrangement of the sensor.

In a log-polar camera, the relationship of the linear distance, ρ, measured on the sensor's surface and the corresponding pixel coordinate, p, is specified by $p = \log_k(\rho/\rho_0)$, where ρ_0 and k stand for the fovea radius and the rate of increase of pixel size towards the periphery.

As previously stated, our goal consists of setting a linear relationship between world distances (or angles), D and corresponding (pixel) distances, p. Combining into the linear relationship the perspective projection, $\rho = t/F$ and the logarithmic law of the log-polar camera, results in the following constraint:

$$D = a.\log(t/F) + b \tag{15}$$

The only difference in the form of the linear constraint when using conventional or log-polar cameras, Eqs. (11) and (15), is that the slope t/F is replaced by its logarithm. Hence, replacing the slope by its log directly in Eqs. (12), (13) and (14), results in the desired shaping functions for the log-polar camera.

Concluding, we obtained a design methodology of constant resolution omnidirectional cameras, that is based on a shaping function whose specification allows us to choose a particular linear property. This methodology generalises a number of published design methods for specific linear properties. For example the constant vertical resolution design results in a sensor

Fig. 5. Svavisca camera equipped with the combined mirror (left) and world scene with regular patterns distributed vertically and over the floor (middle). Panoramic and bird's eye views (right). The bird's eye views have a transformation from cartesian to polar coordinates. The bird's eye view at right originated from the fovea area

equivalent to that of Gaechter et al [30]. Of particular interest is a constant angular resolution sensor, that is an implementation of a spherical sensor providing a constant number of pixels per solid angle. This is similar to Conroy and Moore's design [15], but with the difference that, due to the nature of the log-polar camera, we do not need to compensate for lesser pixels when moving closer to the camera axis.

Figure 5 shows an omnidirectional based on the prototype log-polar camera Svavisca [55]. The mirror is a combined design, encompassing constant vertical and horizontal resolutions, respectively, in the outer and in the two inner annular regions. Vertical and ground patterns in the real world are use to test for linear properties. The panoramic image results from a direct read out of the sensor and the bird's eye views are obtained after a change from polar to cartesian coordinates. In the panoramic image, the vertical sizes of black squares are equal to those of the white squares, thus showing linearity from 3D measures to image pixel coordinates. In the bird's eye views the rectilinear pattern of the ground was successfully recovered.

2.5 The Single Centre of Projection Revisited

A question related to the use of non-single centre of projection sensors is how different they are from single projection centre ones? What is the degree of error induced by a locus of viewpoints? We have studied this problem using the catadioptric sensor with a spherical mirror [33]. As outlined in Sect. 2.1, the Unifying Theory covers all catadioptric sensors with a single centre of projection. A projection model governing a catadioptric sensor with a generic mirror profile is given in Sect. 2.2. If the Unifying Theory can approximate a non-single centre of projection camera, one would expect that - using both models - the error between projecting 3D points to the image plane would be small. It turns out that for real-world points further than 2m away from the catadioptric sensor the error in the image plane is less than *1 pixel*.

Derrin and Konolige [23] also approximated a single centre of projection but used a concept they termed *iso-angle mapping*. They constructed a virtual system by displacing all incoming rays, each having a unique Euler angle, so as they converged at a single point. Thus, their method produced a camera with a single centre of projection, imaging a distorted scene. Since they did not derive an analytical expression for the distortion, it was measured as a change in the height of a small object, given a change in its elevation angle and remained less than 2.5%.

Concluding, many omnidirectional vision systems, despite not having a single projection centre, are well approximated by a single projection centre model. In this way models based on the single projection centre property may become the most common, in the same way as the pin-hole model is used for standard cameras even when it is just an approximation valid for the tasks at hand.

3 Environmental Perception for Navigation

Traditionally, localisation has been identified as a principal perceptual component of the navigation system of a mobile robot [53]. This has driven continuous research and development on sensors providing direct localisation measurements.

There is a large variety of self-localisation solutions available [5] in the literature. However, in general they are characterised by a hard and limiting tradeoff between robustness and cost. As paradigmatic and extreme examples we can refer to solutions based on artificial landmarks (beacons) and those based on odometry. Solutions based on beacons are robust but expensive in terms of the materials, installation, maintenance or configuration to fit a specific new purpose. The solutions based on odometry are inexpensive, but since they rely on the integration of the robot's internal measurements, i.e. not grounded to the world, errors accumulate over time.

We use vision to sense the environment as it allows navigation to be regulated by the world. In particular, we have noted the advantages of omnidirectional vision for navigation, including its flexibility for building environmental representations. Our robot combines two main navigation modalities: Visual Path Following and Topological Navigation. In Visual Path Following, the short-distance / high-accuracy navigation modality, the orthographic view of the ground plane is a convenient world model as it makes simple representing / tracking ground plane features and computing the pose of the robot. Panoramic views are a complementary representation, which are useful in the identification and extraction of vertical line features. These types of views are easily obtained from omnidirectional cameras using image dewarpings.

In Topological Navigation, the large-distance low-precision navigation modality, omnidirectional images are used in their raw format to characterise the environment by its appearance. Omnidirectional images are advantageous as they are more robust to occlusions created e.g. by humans. Visual servoing is included in topological navigation as the means of providing local control.

This eliminates the need to built highly detailed environment representations, thus saving computational (memory) resources.

In summary, both Visual Path Following and Topological Navigation rely upon environmental perception (self-localisation) for regulating movement. The main point here is that perception is linked to internal representations of the world which are chosen according to the tasks at hand. We will now detail *Geometrical Representations* for precise self-localisation, necessary for Visual Path Following, and *Topological Representations* for global positioning leading, necessary for Topological Navigation.

3.1 Geometric Representations for Precise Self-Localisation

Robot navigation in cluttered or narrow areas, such as when negotiating a door traversal, requires precise self-localisation in order to be successful. In other words, the robot has to be equipped with precise environmental perception capabilities.

Vision-based self-localisation derives robot poses from images. It encompasses two principal stages: image processing and pose-computation. Image processing provides the tracking of features in the scene. Pose-computation is the geometrical calculation to determine the robot pose from feature observations, given the scene model.

Designing the image processing level involves modelling the environment. One way to inform a robot of an environment is to give it a CAD model, as in the work of Kosaka and Kak [52], recently reviewed in [24]. The CAD model usually comprises metric values that need to be scaled to match the images acquired by the robot. In our case, we overcome this need by defining geometric models composed of features of the environment directly extracted from images.

Omnidirectional cameras based on standard mirror profiles, image the environment features with significant geometrical distortion. For instance, a corridor appears as an image band of variable width and vertical lines are imaged radially. Omnidirectional images must therefore be dewarped in order to maintain the linearity of the imaged 3D straight lines.

Pose-computation, as the robot moves in a plane, consists of estimating a 2D position and an orientation. Assuming that the robot knows fixed points in the environment (landmarks) there are two main methods of self-localisation relative to the environment: trilateration and triangulation [5]. Trilateration is the determination of a vehicle's position based on distance measurements to the landmarks. Triangulation has a similar purpose but is based on bearing measurements.

In general, a single image taken by a calibrated camera provides only bearing measurements. Thus, triangulation is a more "natural" way to calculate self-localisation. However, some camera poses / geometries provide more information. For example, a bird's eye view image (detailed in the following subsection) provides an orthographic view of the ground plane, providing simultaneous observation of bearings and distances to floor landmarks. Given distances and bearings, the pose-computation is simplified to the calculation of a 2D rigid transformation.

The fact that the pose-computation is based on feature locations, implies that they contain errors, propagated from the feature tracking process. To overcome this, we propose a complimentary pose-computation optimisation step, based on a photometric criterium. We term this optimisation fine pose adjustment, as opposed to the pose-computation based on the features which is termed coarse pose computation. It is important to note that the pose-estimation based on features is important for providing an initial guess for the fine pose adjustment step.

Image Dewarpings for Scene Modelling

Images acquired with an omni-directional camera, e.g. based on a spherical or hyperbolic mirror, are naturally distorted. Knowing the image formation model, we can correct some distortions to obtain Panoramic or Bird's Eye Views.

The panoramic view groups together, in each scan line, the projections of all visible points, at a constant angle of elevation. The bird's eye view is a scaled orthographic projection of the ground plane. These views are advantageous e.g. for extracting and tracking vertical and ground plane lines.

Panoramic and Bird's Eye Views are directly obtained by designing custom shaped mirrors. An alternative approach, as described next, is to simply dewarp the omnidirectional images to the new views.

Panoramic View: 3D points at the same elevation angle from the axis of the catadioptric omnidirectional vision sensor, project to a 2D circle in the image. Therefore, the image dewarping is defined simply as a cartesian to polar coordinates change:

$$I(\alpha, R) = I_0\left(R\cos(\alpha) + u_0, R\sin(\alpha) + v_0\right)$$

where (u_0, v_0) is the image centre, α and R are the angle and radial coordinates. The steps and range of α and R are chosen according to the resolution, and covering all the effective area, of the omnidirectional image. One rule for selecting the step of α is to make the number of columns of the panoramic image equal to the perimeter of the middle circle of the omnidirectional image. Hence inner circles are over-sampled and outer circles are sub-sampled. This rule gives a good tradeoff between data loss due to sub-sampling and memory consumption for storing the panoramic view.

Bird's Eye View: In general, 3D straight lines are imaged as curves in the omnidirectional image. For instance, the horizon line is imaged as a circle. Only 3D lines that belong to vertical planes containing camera and mirror axis project as straight (radial) lines.

In order to dewarp an omnidirectional image to a bird's eye view, notice that the azimuthal coordinate of a 3D point is not changed by the imaging geometry of the omnidirectional camera. Therefore, the dewarping of an omnidirectional image to a bird's eye view is a radial transformation. Hence, we can build a 1D look up table relating a number of points at different radial distances in the omnidirectional image and the respective real distances. The 1D look up table is the radial transformation to be performed for all directions on an omnidirectional image in order to obtain the bird's eye view.

However, the data for building the look up table is usually too sparse. In order to obtain a dense look up table we use the projection model of the omnidirectional camera. Firstly, we rewrite the projection operator, $\mathcal{P}_\rho$ in order to map radial distances, ρ_{ground} measured on the ground plane, to radial distances, ρ_{img}, measured in the image:

$$\rho_{img} = \mathcal{P}_\rho(\rho_{ground}, \vartheta) \tag{16}$$

Using this information, we build a look up table that maps densely sampled radial distances from the ground plane to the image coordinates. Since the inverse function cannot be expressed analytically, once we have an image point, we search the look up table to determine the corresponding radial distance on the ground plane.

Figure 6 illustrates the dewarpings of an omnidirectional image to obtain the Bird's Eye and Panoramic Views. Notice that the door frames are imaged as vertical lines in the Panoramic view and the corridor guidelines are imaged as straight lines in the Bird's Eye view, as desired.

As a final remark, notice that our process to obtain the look up table encoding the Bird's Eye View, is equivalent to performing calibration. However, for

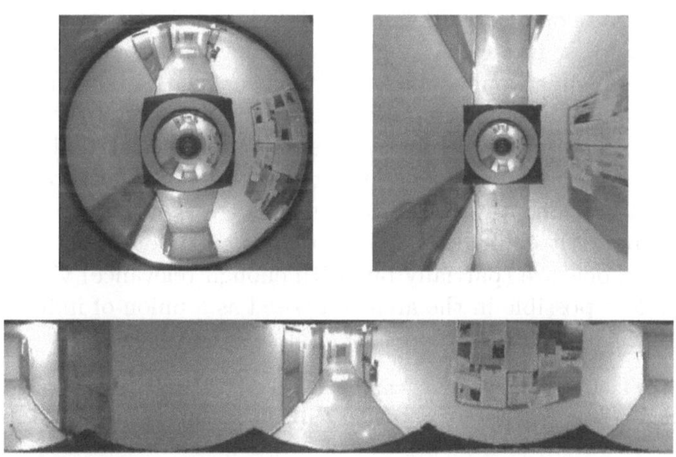

Fig. 6. Image dewarping for bird's eye and panoramic views. (Top-left) original omnidirectional image, (top-right) bird's eye view and (bottom) panoramic view

our purposes a good dewarping is simply the one that makes straight lines on the ground plane appear straight in the Bird's Eye View.

As long as the mirror, camera and support (mobile platform) remain fixed to each other, the dewarpings for panoramic and bird's eye views are time invariant and can be programmed with 2D lookup tables. The dewarpings are done efficiently in this way.

Doing fixed image dewarpings is actually a way to do (or help) *Scene Modelling*. The image dewarpings make geometrical properties of the scene clearly evident and as such simplify scene modelling to collecting a number of features.

Geometric Scene Modelling and Model Tracking

Geometric models of the scene are collections of segments identified in Bird's Eye and Panoramic views[5]. Ground segments are rigidly interconnected in the Bird's Eye views while vertical segments will vary their locations according to the viewer location. Considering both types of segments, the models are "wire-frames" whose links change according to the viewpoint.

Each scene model must have a minimal number of features (line segments) in order to allow self-localisation. One line of the ground plane permits finding only the orientation of the robot and gives a single constraint on its localisation. Two concurrent ground lines, or one ground and one vertical, already determine robot position and orientation. Given three lines either all vertical, one on the ground, two on the ground (not parallel) or three on the ground (not all parallel), always permit us to compute the pose and therefore form valid models[6].

Figure 7 shows one example of modelling the scene using line segments observed directly in the scene. The model is composed of three ground lines, two of which are corridor guidelines, and eight vertical segments essentially defined by the door frames. A single door frame (i.e. two vertical lines) and one corridor guideline would suffice but it is beneficial to take more lines than minimally required in order to improve the robustness of self-localisation.

In order to represent a certain scene area, and to meet visibility[7] and quality criteria, a minimal number of segments are required. Models characterising different world regions are related by rigid 2D transformations. These transformations are firstly defined between every two neighbour models at locations where both models are (partially but with enough relevance) visible. Navigation is therefore possible in the area composed as a union of individual areas, provided by each individual model.

[5] Despite the fact that localisation can be based on tracked image corners [74], more robust and stable results are obtained with line segments as noted for example by Spetsakis and Aloimonos in [78].

[6] Assuming known the xy coordinates of the intersection of the vertical line(s) with the ground plane.

[7] see Talluri and Aggarwal in [83] for a geometrical definition of visibility regions

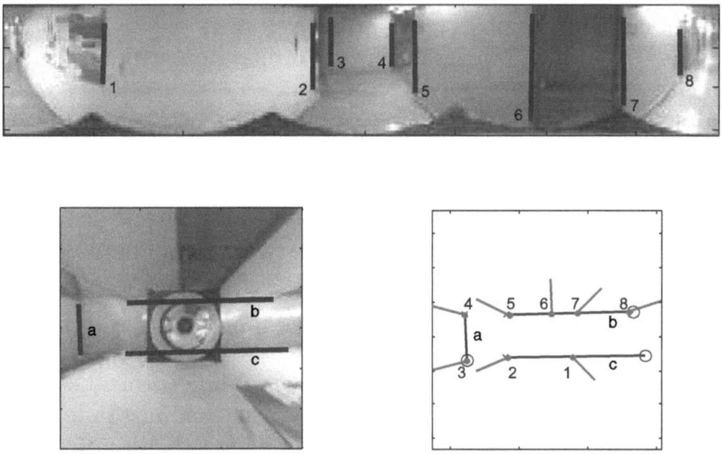

Fig. 7. Geometric models for a door crossing experiment. The segments composing the model (bottom right) are illustrated in the panoramic and bird's eye view images, respectively (top and bottom left)

Assuming that the robot pose evolves smoothly over time, the model segments need to be detected only once – at the initialisation stage. From then on, we need only *track* them, which is much more efficient in computational terms. We track both edges lying on the ground plane and vertical edge segments, using respectively the bird's eye and panoramic views (details in [31]).

Pose Computation

The self-localisation procedure is based on the tracking of the geometric models. The tracking of the models requires rigidity of the world structure (but naturally not rigidity of the observed model segments themselves).

A simple method of calculating pose from the models arises when the segments of the model intersect at ground points (as in the model shown in Fig. 7). In this case, the model, despite encompassing ground and vertical segments, is simplified to the case of a set of ground points. This set of points moves rigidly in the Bird's Eye View, and therefore self-localisation is in essence the computation of the 2D transformation tracking the movement of the points. This method requires intersecting segments, which is similar to tracking corners but in a much more stable manner. This is especially true when dealing with long segments, as the noise in the orientation of small segments may become significant, affecting the computation of the intersections and the quality of corner estimates.

Alternatively, localisation is achieved through an optimisation procedure, namely minimizing the distance between model and observed line segments, directly at the pose parameters. Intuitively, the best pose estimate should align the scene model and the observed lines as well as possible. This is computationally more expensive, but more robust to direction errors on the observed line segments [34].

Defining pose as $\mathbf{x} = [x\ y\ \theta]$ and the distance between the segments ab and cd as $d\,(cd, ab) = f\,(c-a, b-a) + f\,(d-a, b-a)$ where a, b, c, d are the segment extremal points and f is the normalised internal product, $f(\mathbf{v}, \mathbf{v_0}) = \left|\mathbf{v}^T.\mathbf{v}_0^{\perp}\right| / \left\|\mathbf{v}_0^{\perp}\right\|$, the problem of pose estimation based on the distance between model and observed segments can be expressed by the minimization of a cost functional:

$$\mathbf{x}^* = \arg_{\mathbf{x}} \min \sum_i d\,(s_i, s_{0i}(\mathbf{x})) \tag{17}$$

where s_i stands for observed vertical and ground line segments, and s_{0i} indicates the model segments (known a priori). The minimization is performed with a generic gradient descent algorithm provided that the initialisation is close enough. For the initial guess of the pose there are also simple solutions such as using the pose at the previous time instant or, when available, an estimate provided by e.g. a 2D rigid transformation of ground points or by a triangulation method [4].

The self-localisation process as described by Eq. (17), relies exclusively on the observed segments, and looks for the best robot pose justifying those observations on the image plane. Despite the optimization performed for pose-computation, there are residual errors that result from the low-level image processing, segment tracking, and from the method itself. Some of these errors may be recovered through the global interpretation of the current image with the a priori geometric model. Since the model is composed of segments associated with image edges, we want to maximize the sum of gradients, ∇I at every point of the model wire-frame, $\{P_i\}$. Denoting the pose by $\mathbf{x}$ then the optimal pose $\mathbf{x}^*$ is obtained as:

$$\mathbf{x}^* = \arg_{\mathbf{x}} \max \mu(\mathbf{x}) = \arg_{\mathbf{x}} \max \sum_i \left|\nabla I\,(\mathcal{P}(P_i; \mathbf{x}))\right| \tag{18}$$

where $\mathcal{P}$ is the projection operator and $\mu(\mathbf{x})$ represents the (matching) merit function. Given an initial solution to Eq. (17), the final solution can be found by a local search on the components of $\mathbf{x}$.

Usually, there are model points that are non-visible during some time intervals while the robot moves. This is due, for example, to camera (platform) self-occlusion or to the finite dimensions of the image. In these cases, the merit matching merit function does not smoothly evolve with pose changes: it is maximized by considering the maximum number of points possible, instead of the true segment pose. Therefore, we include a smoothness prior to the function. One solution is to maintain the gradient values at control points of the model for the images when they are not visible.

Visual Path Following

Visual Path Following can be described in a simple manner as a trajectory following problem, without having the trajectory explicitly represented in the scene. The trajectory is only a data structure learnt from example / experience or specified through a visual interface.

Visual Path Following combines the precise self-localisation (detailed in the preceding sections) with a controller that generates the control signals for moving the robot, such as that proposed by de Wit et al [21].

Experiments were conducted using an omnidirectional camera with a spherical mirror profile (shown in Fig. 3), mounted on a TRC labmate mobile robot. Figure 8 illustrates tracking and self-localization while traversing a door from the corridor into a room. The tracked features (shown as black circles) are defined by vertical and ground-plane segments, tracked in bird's eye view images.

Currently, the user initializes the relevant features to track. To detect the loss of tracking during operation, the process is continuously self-evaluated by the robot, based on gradient intensities obtained within specified areas around the landmark edges (Eq. 18). If these gradients decrease significantly compared to those expected, a recovery mechanism is launched.

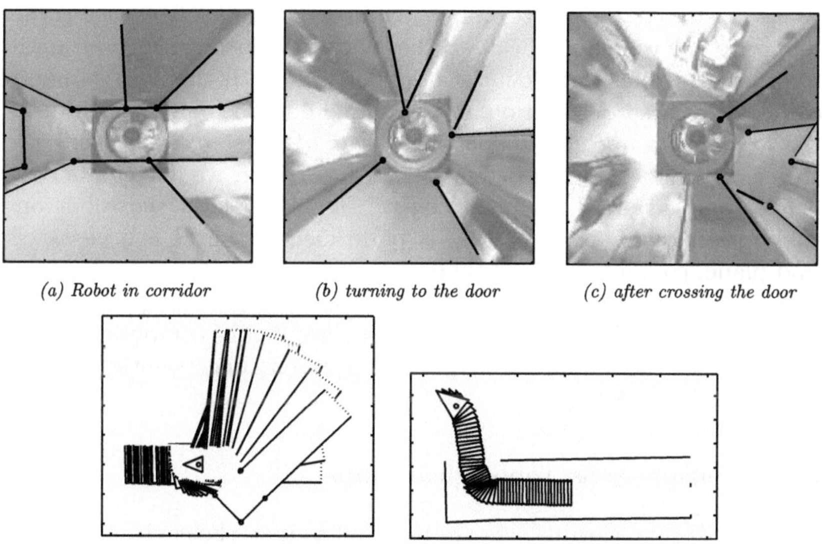

(a) Robot in corridor *(b) turning to the door* *(c) after crossing the door*

(d) Scene-model tracking *(e) Self-localisation*

Fig. 8. Feature tracking at three instants (a,b,c), scene-model tracking in the robot coordinate system (d) and the self-localisation result obtained by fixing the tracked scene-model (e)

The appropriate choice of the sensor and environmental representations, taking into account the task at hand, results in an efficient methodology that hardwires some tasks requiring precise navigation.

3.2 Topological Representations

A topological map is used to describe the robot's global environment and obtain its qualitative position when travelling long distances. It is represented as a graph: *nodes* in the graph correspond to landmarks, i.e. distinctive places such as corners. *Links* connect nodes and correspond to environmental structures that can be used to control the pose of the robot. In order to effectively use this graph the robot must be able to travel along a corridor, recognize the ends of a corridor, make turns, identify and count door frames. These behaviours are implemented through an appearance based system and a visual servoing strategy.

An appearance based system [62] is one in which a run–time image is compared to a database set for matching purposes. For example, in our corridor scene, the appearance based system provides qualitative estimates of robot position and recognizes distinctive places such as corner or door entrances.

Therefore, the topological map is simply a collection of inter-connected images. To go from one particular locale to another, we do not have to think in precise metric terms. For example, to move the robot from one corner to the opposite one we may indicate to the robot to follow one corridor until the first corner and then to follow the next corridor until the next corner, thus reaching the desired destination, or to complete more complex missions such as "*go to the third office on the left-hand side of the second corridor*".

To control the robot's trajectory along a corridor, we detect the corridor guidelines and generate adequate control signals to keep the robot on the desired trajectory. This processing is performed on bird's eye views of the ground plane, computed in real-time.

When compared to geometric approaches, topological maps offer a parsimonious representation of the environment, are highly computationally efficient [85], scale easily and can explicitly represent uncertainties in the real world [7].

Image Eigenspaces as Topological Maps

In general, sizeable learning sets are required to map the environment and so matching using traditional techniques, such as correlation, would incur a very high computational cost. If one considers the images as points in space, it follows that they shall be scattered throughout this space, *only* if they differ significantly from one other. However, many real-world environments (offices, highways etc.) exhibit homogeneity of structure, leading to a large amount of redundant information within the image set. Consequently, the images are not scattered throughout a high dimensional space but – due to their similarity – lie in a lower dimensional subspace.

We implement dimensionality reduction using the classical procedure of *Principal Component Analysis* (PCA)[8], as described by Murase and Nayar in [62], and detailed by Winters in [88] or Gaspar, Winters and Santos-Victor in [35]. Simply put, Principal Component Analysis **reduces the dimensionality** of a set of linearly independent input variables, while still accurately representing most of the original data. The reconstruction of this original data is optimal in the sense that the mean square error between it and the original data is minimized.

Imagine that we represent images as L-dimensional vectors in $\mathbb{R}^L$. Due to the similarity between images (data redundancy) these vectors will not span the entire space of $\mathbb{R}^L$ but rather, they will be confined (or close, to be more precise) to a lower-dimensional subspace, $\mathbb{R}^M$ where $M << L$. Hence, to save on computation, we can represent our images by their co-ordinates in such a lower-dimensional subspace, rather than using all of the pixel information. Each time a new image is acquired, its capture position can easily be determined by projecting it into the lower-dimensional subspace and finding its closest match from the *a priori* set of points (images).

A basis for such a linear subspace can be found through PCA, where the basis vectors are denominated **Principal Components**. They can be computed as the **eigenvectors** of the **covariance matrix** of the normalised set of images acquired by the robot. The number of eigenvectors that can be computed in such a way is the same as the number of images in the input data, and the eigenvectors are the same size as the images.

Each reference image is associated with a *qualitative* robot position (e.g. half way along the corridor). To find the robot position in the topological map, we have to determine the reference image that best matches the current view. The distance between the current view and the reference images can be computed directly using their projections (vectors) on the lower dimensional eigenspace. The distance is computed between M-dimensional coefficient vectors (typically 10 to 12), as opposed to image size vectors (128×128). The position of the robot is that associated with the reference image having the lowest distance.

When using intensity images, comparison of images is essentially a sum of squared differences of brightness (radiance) values and because of this the robot is prone to miscalculating its location where *large non-uniform* deviations in illumination occur. This can be overcome by using edge images, although these are not robust to edge-point position errors. The solution therefore, is to compare shapes instead of edge-points, or more specifically the *distance between shapes* present in both images. There are several possible

[8] It is sometimes known as the application of the *Karhunen-Loève transform* [67, 86].

definitions of the distance between shapes. Two very well known are chamfer distance and the the Hausdorff distance.

Localisation Based on the Chamfer Distance

The chamfer distance is based on the correlation of a template edge-image with a *distance transformed image*. The distance transform of an edge-image is an image of the same size as the original, that indicates at each point the distance to the closest edge point [6, 36, 16].

The *chamfer distance transform*[9] is computed from an edge-image using the forward and backward masks shown in Fig. 9 [6, 36]. There are various possible values for the constants in the masks. We use the values according to Montanari's metric [16].

The constants shown in the masks are added to each of the local values and the resulting value of the mask computation is the minimum of the set. Both masks are applied along the rows of the initialised image.

Figure 10 shows the distance transform of the edges of an omnidirectional image. We remove the inner and outer parts of the omnidirectional image as they contain artifact edges, i.e. edges not related to the scene itself, created by the mirror rim and the robot plus camera self-occlusion.

Finally, given the distance transform, the chamfer distance of two shapes is then computed as the correlation:

$+\sqrt{2}$	+1	$+\sqrt{2}$
+1	**+0**	

	+0	+1
$+\sqrt{2}$	+1	$+\sqrt{2}$

Fig. 9. Forward and backward masks for computing the distance transform. The element in bold face indicates the centre of the mask

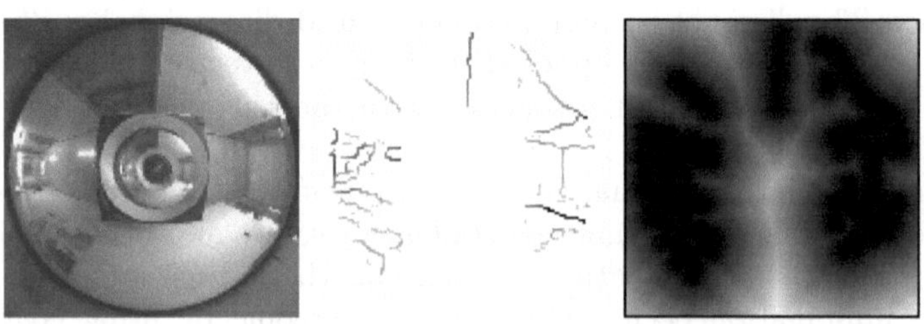

(a) (b) (c)

Fig. 10. Distance transform: (a) original omnidirectional image, (b) edges found in an annular region of the omnidirectional image and (c) the distance transform of the edge-image

[9] Not to be confused with the chamfer distance between two shapes. The chamfer distance transform is an image processing operation useful for computing the chamfer distance of two shapes.

$$d(D,T) = \frac{\sum_{i,j} D_{ij}T_{ij}}{\sum_{i,j} T_{ij}} \tag{19}$$

where T is a template edge-image and D is the distance transform of the edges of the current run-time image. As weaker edges (small gradient magnitudes) are more susceptible to noise, we set T_{ij} to the gradient magnitudes of the template images, instead of binary edges. Hence, we give more weight to the strongest edges.

Equation 19 says that the *chamfer distance* is an average of the distances between corresponding points of the shapes. In a strict sense, it is an approximation as the underlying *chamfer distance transform* is itself an approximation to the Euclidean distance. In practice this difference is not relevant as typically the shapes to compare are at similar poses and the distances between the points are small enough to make negligible the difference of the chamfer and the Euclidean distances.

In the topological localisation application, we want to find the database image corresponding to the current run-time image. In order to find the best matching we search the database using the chamfer distance as the comparison measure. The comparison of images is done from an edge-image to a distance transformed edge-image. The distance transformation may be applied either to the run time or to the database images [36]. We apply the distance transform to the run time edge-images and leave to the template edge-images the role of selecting the relevant edge locations.

The distance as defined by Eq. (19) is zero for perfectly matching images. Therefore we search for the image matching the current image I_m in a set $T_1 \ldots T_n$ by minimizing Eq. (19),

$$\hat{n} = \arg_n \min d\left(D(I_m), T_n\right). \tag{20}$$

Notice that, unlike recognition applications such as pedestrian and sign detection in an image [36], in the localisation application the template and run-time images have equal sizes. The search parameter is an image index instead of translation, rotation and scaling. The range of the index is the size of the database.

Usually there are a large number of database images, and thus undertaking localisation, as in Eq. (20), is computationally expensive. However it only needs to be performed once, when the robot is dropped-in-scene. During normal operation there is a causality constraint along the consecutive locations. We reduce the search range to a window around the last location, typically ±5 images.

Eigenspace Approximation to the Hausdorff Fraction

The Hausdorff distance [73] (of which the Hausdorff fraction is a subset) is a technique whereby one can measure the distance between two sets of points, in our case edge images. A number of Hausdorff distance measures are defined by the following equations:

$$H(A, B) = \max(h(A, B), h(B, A)) \tag{21}$$

where

$$h(A, B) = \max_{a \in A} \min_{b \in B} \parallel a - b \parallel \tag{22}$$

Here A and B represent sets of points. $h(A, B)$ measures the distance from each point in A to the nearest point in B and the maximum distance is termed the *directed* distance from A to B, and is the normal choice for critical time dependent systems.

The Hausdorff distance is very sensitive to even a single outlying point in one of the shapes. The Generalised Hausdorff distance, defined by Huttenlocher et al in [48], is thus proposed as a similar measure but that is robust to partial occlusions. The generalised Hausdorff distance is an fth quantile of the distances between all the points of one shape to the corresponding points of the other shape. The quantile is chosen according to the expected noise and occlusion levels.

In recognition applications, the generalised Hausdorff distance is further specialised to save computational power. The Hausdorff fraction, the measure we are interested in, instead of measuring a distance between shapes evaluates the percentage of superposition when one of the shapes is dilated. Furthermore, for computational efficiency, principal components analysis is included resulting in an eigenspace approximation to the Hausdorff fraction [49].

The eigenspace approximation is built as follows: Let I_m be an observed edge image and I_n^d be an edge image from the topological map, arranged as column vectors. The Hausdorff fraction, $\hbar(I_m, I_n^d)$, which measures the similarity between these images, can be written as:

$$\hbar(I_m, I_n^d) = \frac{I_m^T I_n^d}{||I_m||^2} \tag{23}$$

An image, I_k can be represented in a low dimensional eigenspace [62, 92] by a coefficient vector, $\mathcal{C}_k = [c_1^k, \cdots, c_M^k]^T$, as follows:

$$c_j^k = e_j^T.(I_k - \bar{I}).$$

Here, $\bar{I}$ represents the average of all the intensity images and can be also used with edge images. Thus, the eigenspace approximation to the Hausdorff fraction can be efficiently computed as:

$$\widehat{\hbar}(I_m, I_n^d) = \frac{C_m^T C_n^d + I_m^T \bar{I} + I_n^{dT} \bar{I} - ||\bar{I}||^2}{||I_m||^2}. \tag{24}$$

One important issue, when approximating the Hausdorff fraction, is to include some tolerance in matching step. Huttenlocher et al. [49] build the

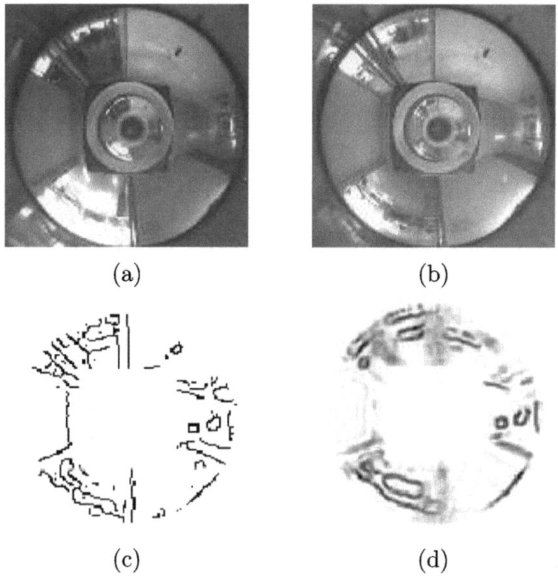

Fig. 11. (a) An omnidirectional image obtained at 11:00, (b) one obtained at 17:00; (c) An edge-detected image and (d) its retrieved image

eigenspace using both dilated and un-dilated model views and pre-process the run time edge images to dilate the edges. In our pre-processing we use low pass filtering instead of edge dilation. The purpose here is to maintain the local maxima of gradient magnitude at edge points while enlarging the matching area. We found this to be a good tradeoff between matching robustness and accuracy.

To test this view-based approximation we collected a sequence of images, acquired at different times, 11am and 5pm, near a large window. Figure 11 shows the significant changes in illumination, especially near the large window at the bottom left hand side of each omnidirectional image. Even so, the view based approximation can correctly determine that the unknown image shown in Fig. 11(a) was closest to the database image shown in Fig. 11(b), while PCA based on brightness distributions would fail. For completeness, Fig. 11 (c) and (d) shows a run-time edge image and its corresponding retrieved image using the eigenspace approximation to the Hausdorff fraction.

Integrating Topological Navigation and Visual Path Following

When continuously operating, the mobile robot is usually performing topological navigation. At some points of the mission the navigation modality is required to change to the visual path following. Thus, the robot needs to retrieve the scene features (straight lines in our case) chosen at the time of learning to specific this particular visual path following task.

The search for the features can be approached as a general pattern matching problem using e.g. a generalised Hough transform as in [93, 26]. We approach the problem by coordinating the two navigation modalities. To find the features, the uncertainty of the location of the robot is controlled by using more detailed topological maps and by increasing the searching regions of the features otherwise bounded according to the maximum speed of the robot.

During system initialisation, the robot will normally begin at a known docking place and the undocking visual path following task may be immediately elicited. Of course, the robot may have to start at an unknown (within the topological map) position, i.e. a drop-in-scene case. Should this occur, then self-localisation is performed using the topological localisation module.

The combination of omnidirectional images and the Topological and Visual Path Following navigation strategies are illustrated by the complete experiments described in this section. We believe that the complementary nature of these approaches and the use of omnidirectional imaging geometries result in a very powerful solution to build efficient and robust navigation systems.

Topological Localisation Results

We perform two experiments to test the three topological localisation methods, presented above. In the first experiment we test that the images after compression by the various methods are still sufficiently different to yield correct localisation results, and in the second experiment we test the robustness of the methods against illumination changes.

The experiments are based on three sequences of images: one database sequence describing the environment and two run-time sequences acquired along a fraction of the represented environment. One of the run time sequences was acquired at a time of the day different to the database set, resulting therefore in very different lighting conditions.

Experiment 1: the run time sequence, as compared to the database, is acquired under similar illumination conditions, the length of the traversed path is about 50% of the original and the images are acquired at a different sampling frequency (distance between consecutive images). Figure 12 shows that the three methods give similar localisation results, as desired. The small differences among the methods are due to the distinct image database (appearance set) construction techniques. The figure shows that in this experiment the three methods, despite compressing information, preserve enough detail to distinguish each image relative to all the others.

Experiment 2: Fig. 13 shows topological localisation performed by each of the methods for two sequences taken in the same path but at different times of the day, resulting in very different lighting conditions. As expected, the PCA-based method, i.e. the one using brightness values directly, fails to obtain correct locations in areas of large non-uniform illumination change (i.e. the last part of the test). The other two methods, which are based on edges, obtain better results.

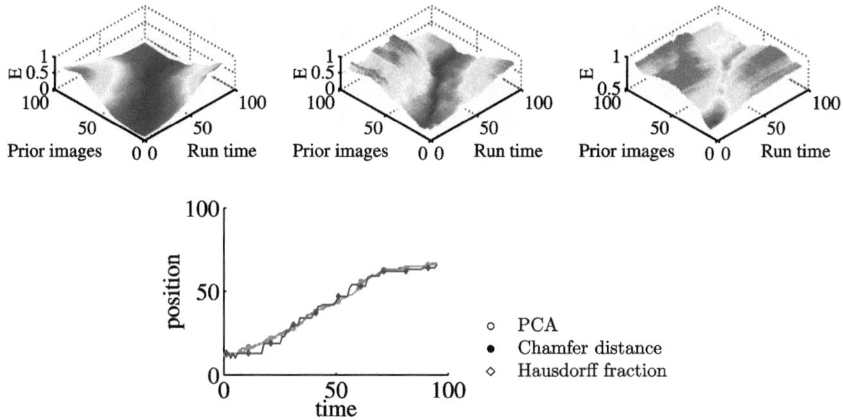

Fig. 12. Three methods of topological localisation, (top, from left to right): localisation based on PCA, Chamfer distance and Hausdorff distance. The clear valleys show that there is enough information to distinguish the robot locations. (Bottom) localisation as found by each of the methods i.e. ordinates corresponding to minimum values found at each time instant on the 3D plots

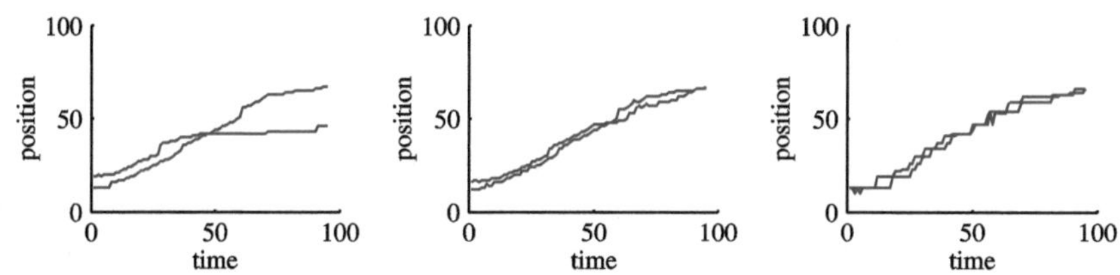

Fig. 13. Topological localisation experiments using the three methods over two sequences acquired under different lighting conditions. From left to right: localisation based on PCA, Distance transform and Hausdorff distance

As expected, the edges based methods are more suited to dealing with very different illuminations. In our navigation experiments we use mainly the PCA over brightness values, as most of our scenario is not subject to large illumination changes, and using brightness values is more informative than using only edges. For the parts of the scene where illumination can change significantly we use the Hausdorff based method. The reason of its choice when compared to the Distance transform, is that it is faster for the first localisation at dropped-in-scene situations.

Integrated Navigation Experiments

The concluding experiment integrates global and local navigational tasks, by combining the *Topological Navigation* and *Visual Path Following* paradigms.

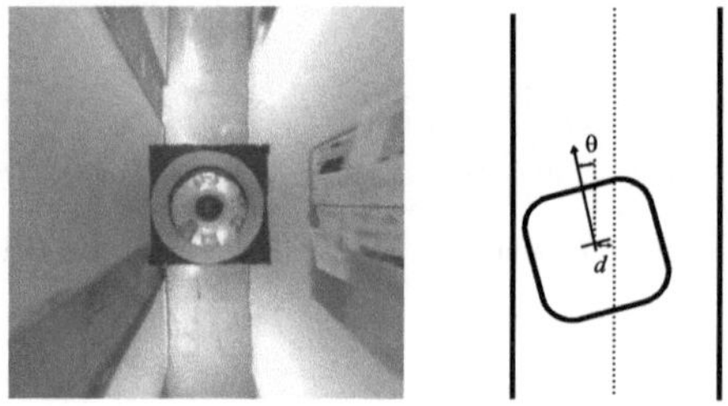

Fig. 14. (Left) Bird's eye view of the corridor. (Right) Measurements used in the control law: the robot heading θ and distance d relative to the corridor centre. The controller is designed to regulate to zero the (error) measurements actuating on the angular and linear speeds of the robot

To navigate along the topological graph, we still have to define a suitable vision-based behaviour for corridor following (*links* in the map). In different environments, one can always use simple knowledge about the scene geometry to define other behaviours. We exploit the fact that most corridors have parallel guidelines to control the robot heading direction, aiming to keep the robot centred in the corridor.

The visual feedback is provided by the omnidirectional camera. We use *bird's eye views* of the floor, which simplifies the servoing task, as these images are a scaled orthographic projection of the ground plane (i.e. no perspective effects). Figure 14 shows a top view of the corridor guidelines, the robot and the trajectory to follow in the centre of the corridor.

From the images we can measure the robot heading with respect to the corridor guidelines and the distance to the central reference trajectory. We use a simple kinematic planner to control the robot's position and orientation in the corridor, using the angular velocity as the single degree of freedom.

Notice that the use of bird's eye views of the ground plane simplifies both the extraction of the corridor guidelines (e.g. the corridor has a constant width) and the computation of the robot position and orientation errors, with respect to the corridor's central path.

Hence, the robot is equipped to perform Topological Navigation relying on appearance based methods and on its corridor following behaviour. This is a methodology for traversing long paths. For local and precise navigation the robot uses Visual Path Following as detailed in Sect. 3.1. Combining these behaviours the robot can perform missions covering extensive areas while achieving local precise missions. In the following we describe one such mission.

The mission starts in the Computer Vision Lab. Visual Path Following is used to navigate inside the Lab, traverse the Lab's door and drive the robot out into the corridor. Once in the corridor, control is transferred to the Topological Navigation module, which drives the robot all the way to the end of the corridor. At this position a new behaviour is launched, consisting of the robot executing a 180° turn, after which the topological navigation mode drives the robot back to the Lab entry point.

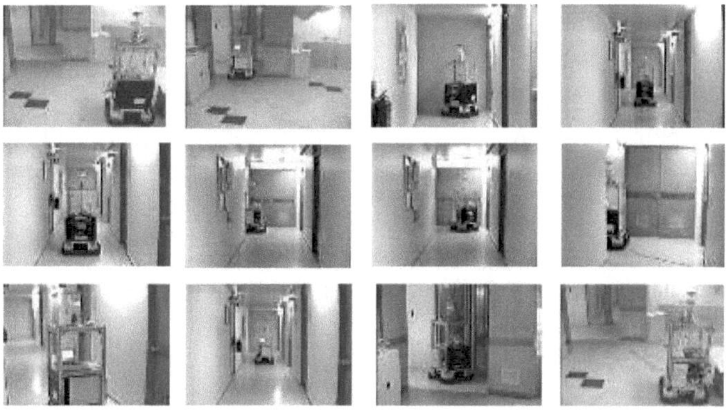

Fig. 15. Experiment combining visual path following for door traversal and topological navigation for corridor following

During this backward trajectory we use the same image eigenspaces as were utilised during the forward motion by simply rotating, in real-time, the acquired omnidirectional images by 180°. Alternatively, we could use the image's power spectrum or the Zero Phase Representation [69]. Finally, once the robot is approximately located at the lab entrance, control is passed to the Visual Path Following module. Immediately it locates the visual landmarks and drives the robot through the door. It follows a pre-specified path until the final goal position, well inside the lab, is reached. Figure 15 shows an image sequence to relate the robot's motion during this experiment.

In Fig. 16(a) we used odometric readings from the best experiment to plot the robot trajectory. When returning to the laboratory, the uncertainty in odometry was approximately 0.5m. Thus, door traversal would not be possible without the use of visual control. Figure 16(b), shows the actual robot trajectory, after using ground truth measurements to correct the odometric estimates. The mission was successfully accomplished.

This integrated experiment shows that omnidirectional images are advantageous for navigation and support different representations suitable both for Topological Maps, when navigating between distant environmental points, and Visual Path Following for accurate path traversal. Additionally, we have described how they can help in coping with occlusions, and with methods of achieving robustness against illumination changes.

4 Complementing Human and Robot Perceptions for HR Interaction

Each omnidirectional image provides a rich description and understanding of the scene. Visualization methods based on panoramic or bird's eye views provide a simple and effective way to control the robot. For instance, the

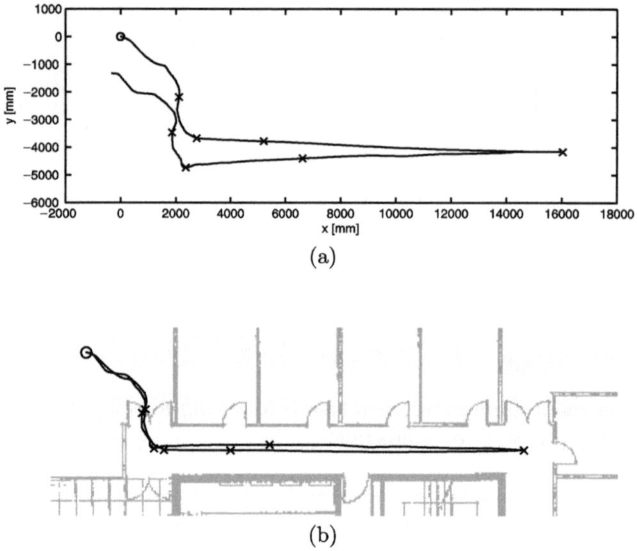

Fig. 16. A real world experiment combining Visual Path Following for door traversal and Topological Navigation for long-distance goals. Odometry results before (a) and after (b) the addition of ground truth measurements

robot heading is easily specified by clicking on the desired direction of travel in the panoramic image, and the desired (x, y) locations are specified by clicking in the bird's-eye view.

Using 3D models further improves the visualization of the scene. A unique feature of such a representation is that the user can tell the robot to arrive to a given destination at a certain orientation simply by rotating the 3D model. Beyond the benefits of immersion, it allows to group the information of many views and get a global view of the environment.

In order to build the 3D scene models, we propose *Interactive Scene Reconstruction*, a method based on the complimentary nature of Human and Robot perceptions. While Humans have an immediate qualitative understanding of the scene encompassing co-planarity and co-linearity properties of a number of points of the scene, Robots equipped with omnidirectional cameras can take precise azimuthal and elevation measurements.

Interactive scene reconstruction has recently drawn lots of attention. Debevec et al in [22], propose an interactive scene reconstruction approach for modelling and rendering architectural scenes. They derive a geometric model combining edge lines observed in the images with geometrical properties known a priori. This approach is advantageous relative to building a CAD model from scratch, as some information comes directly from the images. In addition, it is simpler than a conventional structure from motion problem because, instead of reconstructing points, it deals with reconstructing scene parameters, which is a much lower dimension and better conditioned problem.

In [79] Sturm uses an omnidirectional camera based on a parabolic mirror and a telecentric lens for reconstructing a 3D scene. The user specifies relevant points and planes grouping those points. The directions of the planes are computed e.g. from vanishing points, and the image points are back-projected to obtain parametric representations where the points move on the 3D projection rays. The points and the planes, i.e. their distances to the viewer, are simultaneously reconstructed by minimizing a cost functional based on the distances from the points to the planes.

We build 3D models using omnidirectional images and some limited user input, as in Sturm's work. However our approach is based on a different reconstruction method and the omnidirectional camera is a generalised single projection centre camera modelled by the Unified Projection Model [37]. The reconstruction method is that proposed by Grossmann for conventional cameras [43], applied to single projection centre omnidirectional cameras for which a back-projection model was obtained.

The back-projection transforms the omnidirectional camera to a (very wide field of view) pin-hole camera. The user input is of geometrical nature, namely alignment and coplanarity properties of points and lines. After back-projection, the data is arranged according to the geometrical constraints, resulting in a linear problem whose solution can be found in a single step.

4.1 Interactive Scene Reconstruction

We now present the method for interactively building a 3D model of the environment. The 3D information is obtained from co-linearity and co-planarity properties of the scene. The texture is then extracted from the images to obtain a realistic virtual environment.

The 3D model is a Euclidean reconstruction of the scene. As such, it may be translated and rotated for visualization and many models can be joined into a single representation of the environment.

As in other methods [50, 79], the reconstruction algorithm presented here works in structured environments, in which three orthogonal directions, "x", "y" and "z" shape the scene. The operator specifies in an image the location of 3D points of interest and indicates properties of alignment and planarity. In this section, we present a method based on [42].

In all, the information specified by the operator consists of:

- Image points corresponding to 3D points that will be reconstructed, usually on edges of the floor and of walls.
- Indications of "$x-$", "$y-$" and "$z=$constant" planes as and of alignments of points along the x, y and z directions. This typically includes the floor and vertical walls.
- Indications of points that form 3D surfaces that should be visualized as such.

The remainder of this section shows how to obtain a 3D reconstruction from this information.

Using Back-projection to form Perspective Images

In this section, we derive a transformation, applicable to single projection centre omnidirectional cameras that obtain images as if acquired by perspective projection cameras. This is interesting as it provides a way to utilize methodologies for perspective cameras directly with omnidirectional cameras. In particular, the interactive scene reconstruction method (described in the following sections) follows this approach of using omnidirectional cameras transformed to perspective cameras.

The acquisition of correct perspective images, independent of the scenario, requires that the vision sensor be characterised by a single projection centre [2]. The unified projection model has, by definition, this property but, due to the intermediate mapping over the sphere, the obtained images are in general not perspective.

In order to obtain correct perspective images, the spherical projection must be first reversed from the image plane to the sphere surface and then, re-projected to the desired plane from the sphere centre. We term this reverse projection *back-projection.*

The back-projection of an image pixel (u, v), obtained through spherical projection, yields a 3D direction $k \cdot (x, y, z)$ given by the following equations derived from Eq. (1):

$$
\begin{gathered}
a = (l + m), b = (u^2 + v^2) \\
\begin{bmatrix} x \\ y \end{bmatrix} = \frac{la - sign(a)\sqrt{a^2 + (1 - l^2)b}}{a^2 + b} \begin{bmatrix} u \\ v \end{bmatrix} \\
z = \pm\sqrt{1 - x^2 - y^2}
\end{gathered}
\tag{25}
$$

where z is negative if $|a|/l > \sqrt{b}$, and positive otherwise. It is assumed, without loss of generality, that (x, y, z) is lying on the surface of the unit sphere. Figure 17 illustrates the back-projection. Given an omnidirectional image we use back-projection to map image points to the surface of a sphere centred at the camera viewpoint[10].

At this point, it is worth noting that the set $M = \{P : P = (x, y, z)\}$ interpreted as points of the projective plane, already define a perspective image. By rotating and scaling the set M one obtains specific viewing directions and

---

[10] The omnidirectional camera utilized here is based on a spherical mirror and therefore does not have a single projection centre. However, as the scene depth is large as compared to the sensor size, the sensor approximates a single projection centre system (details in [33]). Hence it is possible to find the parameters of the corresponding unified projection model system and use Eq. (25).

Fig. 17. (Top) original omnidirectional image and back-projection to a spherical surface centred at the camera viewpoint. (Below) Examples of perspective images obtained from the omnidirectional image

focal lengths. Denoting the transformation of coordinates from the omnidirectional camera to a desired (rotated) perspective camera by R then the new perspective image $\{p : p = (u, v, 1)\}$ becomes:

$$p = \lambda KRP \tag{26}$$

where K contains intrinsic parameters and λ is a scaling factor. This is the pin-hole camera projection model [25], when the origin of the coordinates is the camera centre.

Figure 17 shows some examples of perspective images obtained from the omnidirectional image. The perspective images illustrate the selection of the viewing direction.

Aligning the Data with the Reference Frame

In the reconstruction algorithm we use the normalised perspective projection model [25], by choosing $K = I_{3\times3}$ in Eqs. (25) and (26):

$$p = \lambda RP \tag{27}$$

in which $p = [u\ v\ 1]^T$ is the image point, in homogeneous coordinates and $P = [x\ y\ z]^T$ is the 3D point. The rotation matrix R is chosen to align the camera frame with the reference (world) frame. Since the z axis is vertical, the matrix R takes the form:

$$R = \begin{bmatrix} \cos(\theta) & \sin(\theta) & 0 \\ -\sin(\theta) & \cos(\theta) & 0 \\ 0 & 0 & 1 \end{bmatrix}, \tag{28}$$

where θ is the angle formed by the x axis of the camera and that of the world coordinate system. This angle will be determined from the vanishing points [14] of these directions.

A vanishing point is the intersection in the image of the projections of parallel 3D lines. If one has the images of two or more lines parallel to a given 3D direction, it is possible to determine its vanishing point [79].

In our case, information provided by the operator allows for the determination of alignments of points along the x and y directions. It is thus possible to compute the vanishing points of these directions and, from there, the angle θ between the camera and world coordinate systems.

Reconstruction Algorithm

Having determined the projection matrix R in Eq. (27), we proceed to estimate the position of the 3D points P. This will be done by using the image points p to linearly constrain the unknown quantities.

From the projection equation, one has $p \times RP = 0_3$, which is equivalently written

$$S_p RP = 0_3, \tag{29}$$

where S_p is the Rodrigues matrix associated with the cross product with vector p.

Writing this equation for each of the N unknown 3D points gives the linear system:

$$\begin{bmatrix} S_{p_1}R & & & \\ & S_{p_2}R & & \\ & & \ddots & \\ & & & S_{p_N}R \end{bmatrix} \begin{bmatrix} P_1 \\ P_2 \\ \vdots \\ P_N \end{bmatrix} = A.\mathcal{P} = 0_{3N}. \tag{30}$$

where A is block diagonal and $\mathcal{P}$ contains the $3N$ coordinates that we wish to estimate:

Since only two equations from the set defined by Eq. (29) are independent, the co-rank of A is equal to the number of points N. The indeterminacy in this system of equations corresponds to the unknown depth at which each points lies, relatively to the camera.

This indeterminacy is removed by the planarity and alignment information given by the operator. For example, when two points belong to a $z =$ constant plane, their z coordinates are necessarily equal and there is thus a *single* unknown quantity, rather than *two*. Equation (30) is modified to take this information into account by replacing the columns of A (resp. rows of $\mathcal{P}$) corresponding to the two unknown z coordinates by a single column (resp. row) that is the sum of the two. Alignment information likewise states the equality of two pairs of unknowns.

Each item of geometric information provided by the user is used to transform the linear system in Equation (30) into a smaller system involving only *distinct* quantities:

$$A'\mathcal{P}' = 0_{3N}. \tag{31}$$

This system is solved in the total least-squares [39] sense by assigning to $\mathcal{P}'$ the singular vector of A' corresponding to the smallest singular value. The original vector of coordinates $\mathcal{P}$ is obtained from $\mathcal{P}'$ by performing the inverse of the operations that led from Eq. (30) to Eq. (31).

The reconstruction algorithm is easily extended to the case of multiple cameras. The orientation of the cameras is estimated from vanishing points as above and the projection model becomes:

$$p = \lambda(RP - Rt) \tag{32}$$

where t is the position of the camera. It is zero for the first camera and is one of $t_1 \ldots t_j$ if j additional cameras are present.

Considering for example that there are two additional cameras and following the same procedure as for a single image, similar A and $\mathcal{P}$ are defined for each camera. The problem has six new degrees of freedom corresponding to the two unknown translations t_1 and t_2:

$$\left[\begin{array}{ccc|c|c} A_1 & & & & \\ & A_2 & & -A_2.\mathbf{1}_2 & \\ & & A_3 & & -A_3.\mathbf{1}_3 \end{array}\right] \begin{bmatrix} \mathcal{P}_1 \\ \mathcal{P}_2 \\ \mathcal{P}_3 \\ t_1 \\ t_2 \end{bmatrix} = 0 \tag{33}$$

where $\mathbf{1}_2$ and $\mathbf{1}_3$ are matrices to stack the blocks of A_2 and A_3.

As before co-linearity and co-planarity information is used to obtain a reduced system. Note that columns corresponding to different images may be combined, for example if a 3D point is tracked or if a line or plane spans multiple images. The reduced system is solved in the total least-squares sense and the 3D points P are retrieved as in the single-view case. The detailed reconstruction method is given in [42].

Results

Our reconstruction method provides estimates of 3D points in the scene. In order to visualise these estimates, facets are added to connect some of the 3D points, as indicated by the user. Texture is extracted from the omnidirectional images and a complete textured 3D model is obtained.

Figure 18 shows an omnidirectional image and the superposed user input. This input consists of the 16 points shown, knowledge that sets of points belong to constant x, y or z planes and that other sets belong to lines parallel to the x, y or z axes. The table on the side of the images shows all the user-defined data. Planes orthogonal to the x and y axes are in light gray and white respectively, and one horizontal plane is shown in dark gray (the topmost horizontal plane is not shown as it would occlude the other planes).

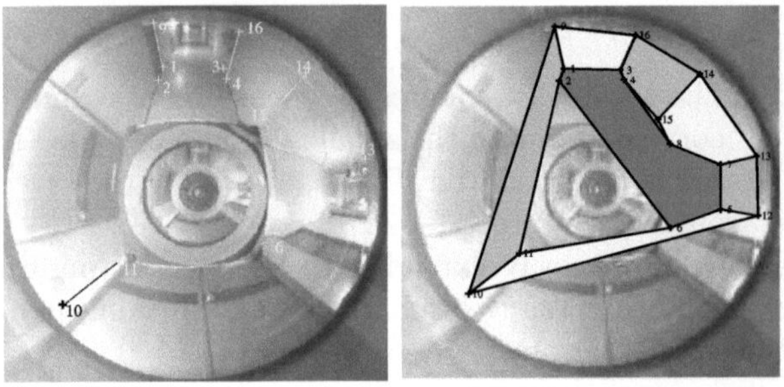

Planes orthogonal to the x axis:
(1, 2, 9, 10, 11),
(3, 4, 14, 15, 16),
(5, 7, 12, 13)
Planes orthogonal to the y axis:
(5, 6, 10, 11, 12),
(7, 8, 13, 14, 15),
(1, 3, 9, 16)
Planes orthogonal to the z axis:
(1, 2, 3, 4, 5, 6, 7, 8),
(9, 12, 13, 16)
Line parallel to the x axis:
(1, 2)

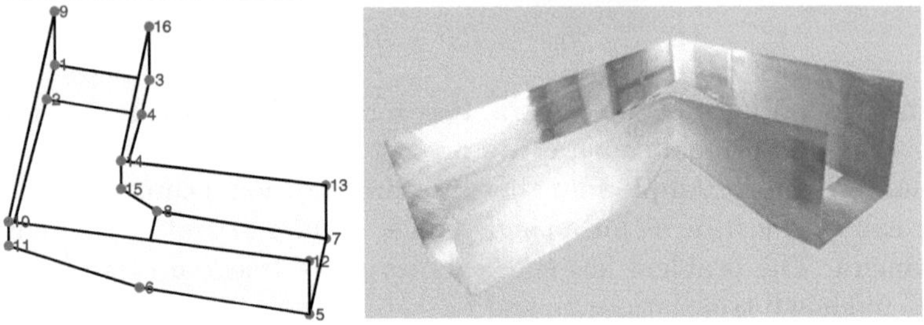

Fig. 18. Interactive modelling based on co-planarity and co-linearity properties using a single omnidirectional image. (Top) Original image with superposed points and lines localised by the user. Planes orthogonal to the x, y and z axis are shown in light gray, white, and dark gray respectively. (Table) The numbers are the indexes shown on the image. (Below) Reconstruction result and view of the textured mapped 3D model

Figure 18 shows the resulting texture-mapped reconstruction. This result shows the effectiveness of omnidirectional imaging to visualize the immediate vicinity of the sensor. It is interesting to note that just a few omnidirectional images are sufficient for building the 3D model (the example shown utilized a single image), as opposed to a larger number of "normal" images that would be required to reconstruct the same scene [50, 79].

4.2 Human Robot Interface based on 3D World Models

Now that we have the 3D scene model, we can build the Human Robot interface. In addition to the local headings or poses, the 3D model allows us to specify complete missions. The human operator selects the start and end locations in the model, and can indicate points of interest for the robot to undertake specific tasks. See Fig. 19.

Given that the targets are specified on interactive models, i.e. models built and used on the user side, they need to be translated as tasks that the robot understands. The translation depends on the local world models and navigation sequences the robot has in its database. Most of the world that the robot knows is in the form of a topological map. In this case the targets are images that the robot has in its image database. The images used to build

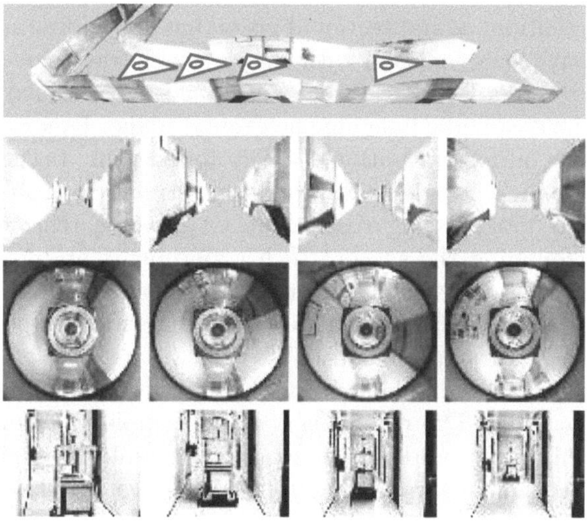

Fig. 19. Tele-operation interface based on 3D models: (top) tele-operator view, (middle) robot view and (bottom) world view

the interactive model are nodes of the topological map. Thus, a fraction of a distance on an interactive model is translated as the same fraction on a link of the topological map.

At some points there are precise navigation requirements. Many of these points are identified in the topological map and will be invoked automatically when travelling between nodes. Therefore, many of the Visual Path Following tasks performed do not need to be explicitly defined by the user. However, should the user desires, he may add new Visual Path Following tasks. In that case, the user chooses landmarks, navigates in the interactive model and then asks the robot to follow the same trajectory.

Interactive modelling offers a simple procedure for building a 3D model of the scene where a vehicle may operate. Even though the models do not contain very fine details, they can provide the remote user of the robot with a sufficiently rich description of the environment. The user can instruct the robot to move to desired position, simply by manipulating the model to reach the desired view point. Such simple scene models can be transmitted even with low bandwidth connections.

5 Conclusion

The challenge of developing perception as a key competence of vision-based mobile robots is of fundamental importance to their successful application in the real world. Vision provides information on world structure and compares favourably with other sensors due to the large amount of rich data available.

In terms of perception, omnidirectional vision has the additional advantage of providing output views (images) with simple geometries. Our sensors output Panoramic and Bird's Eye views that are images as obtained by cylindrical retinas or pin-hole cameras imaging the ground plane. Panoramic and Bird's Eye views are useful for navigation, namely for servoing tasks, as they make localisation a simple 2D rigid transformation estimation problem. Successful completion of the door crossing experiment, for example, relied on the tracking of features surrounding the sensor. Such experiments are not possible with limited field of view (conventional) cameras. Even cameras equipped with pan-and-tilt mounting would be unable to perform the many separate landmark trackings shown in our experiments.

Designing navigation modalities for the task at hand is easier and more effective when compared to designing a single complex navigation mode [8]. Therefore, in this work, emphasis was placed on building appropriate representations rather than always relying upon highly accurate information about the environment. The decision to use this representation was partly inspired by the way in which humans and animals model spatial knowledge. Our combined navigation modalities, Visual Path Following and Topological Navigation, constituted an effective approach to tasks containing both short paths to follow with high precision and long paths to follow qualitatively.

Interactive Scene Reconstruction was shown to be an effective method of obtaining 3D scene models, as compared to conventional reconstruction methods. For example, the model of the corridor corner, in Sect.4, was built from a single image. This constitutes a very difficult task for automatic reconstruction due to the low texture. These 3D models formed the basis for the human-robot interface. Unlike many other works, a unique feature of this representation was that the user could specify a given destination, at a certain orientation, simply by rotating the 3D model.

When considering the system as a whole, (i) our approach to visual perception was found to be useful and convenient because it provided world-structure information for navigation, tailored to the task at hand, (ii) the navigation modalities fulfilled the purpose of semi-autonomous navigation by providing autonomy while naturally combining with the human-robot interface, (iii) the human-robot interface provided intuitive way to set high level tasks, by combining limited user input with the simple output of the sensor (images).

In the future, omnidirectional vision will certainly have many developments. Many current catadioptric setups assume a rigid mounting of the mirror on the camera. Pan-tilt-zoom cameras have been demonstrated to be convenient for surveillance tasks, because of providing a large (virtual) field-of-view while having good resolution when zooming at regions of interest [75]. Adding convex mirrors will allow enlarging the field-of-view and achieving faster pan-tilt motions, obtaining the so termed active omnidirectional camera. Networking cameras poses new calibration challenges resulting from the mixture of various camera types, the overlapping (or not) of the fields-of-view, the different requirements of calibration quality (many times can be reduced just to a topological connection between cameras) and the type of calibration data used (as simple as static background or as dynamic as people moving) [76].

As suggested by the title, we believe there is a large amount of work still to be done before we have a full and true understanding of perception. We believe that key challenges can be addressed by building artificial vision systems. In the future our understanding of perception will allow for robots with visual perception systems, robust enough to cope with new and novel environments. Then, as happened with computers, almost every person will have their very own robot, or what we may term the *personal service robot.*

References

1. S. Baker and S. K. Nayar, *A theory of catadioptric image formation*, Proc. Int. Conf. Computer Vision (ICCV'97), January 1998, pp. 35–42.
2. *A theory of single-viewpoint catadioptric image formation*, International Journal of Computer Vision **35** (1999), no. 2, 175–196.
3. R. Benosman and S. B. Kang (eds.), *Panoramic vision*, Springer Verlag, 2001.
4. M. Betke and L. Gurvits, *Mobile robot localization using landmarks*, IEEE Trans. on Robotics and Automation **13** (1997), no. 2, 251–263.
5. J. Borenstein, H. R. Everett, and Liqiang Feng, *Navigating mobile robots: Sensors and techniques*, A. K. Peters, Ltd., Wellesley, MA, 1996 (also: Where am I? Systems and Methods for Mobile Robot Positioning, `ftp://ftp.eecs.umich.edu/people/johannb/pos96rep.pdf`).
6. G. Borgefors, *Hierarchical chamfer matching: A parametric edge matching algorithm*, IEEE Transactions on Pattern Analysis and Machine Intelligence **10** (1988), no. 6, 849–865.
7. R. Brooks, *Visual map making for a mobile robot*, Proc. IEEE Conf. on Robotics and Automation, 1985.
8. R. A. Brooks, *A robust layered control system for a mobile robot*, IEEE Transactions on Robotics and Automation **2** (1986), 14–23.
9. A. Bruckstein and T. Richardson, *Omniview cameras with curved surface mirrors*, Proceedings of the IEEE Workshop on Omnidirectional Vision at CVPR 2000, June 2000, First published in 1996 as a Bell Labs Technical Memo, pp. 79–86.
10. D. Burschka, J. Geiman, and G. Hager, *Optimal landmark configuration for vision-based control of mobile robots*, Proc. IEEE Int. Conf. on Robotics and Automation, 2003, pp. 3917–3922.
11. Z. L. Cao, S. J. Oh, and E.L. Hall, *Dynamic omni-directional vision for mobile robots*, Journal of Robotic Systems **3** (1986), no. 1, 5–17.
12. J. S. Chahl and M. V. Srinivasan, *Reflective surfaces for panoramic imaging*, Applied Optics **36** (1997), no. 31, 8275–8285.
13. P. Chang and M. Herbert, *Omni-directional structure from motion*, Proceedings of the 1st International IEEE Workshop on Omni-directional Vision (OMNIVIS'00) at CVPR 2000, June 2000.
14. R. Collins and R. Weiss, *Vanishing point calculation as a statistical inference on the unit sphere*, Int. Conf. on Computer Vision (ICCV), 1990, pp. 400–403.

15. T. Conroy and J. Moore, *Resolution invariant surfaces for panoramic vision systems*, IEEE ICCV'99, 1999, pp. 392–397.
16. Olivier Cuisenaire, *Distance transformations: Fast algorithms and applications to medical image processing*, Ph.D. thesis, U. Catholique de Louvain, October 1999.
17. K. Daniilidis (ed.), *1st international ieee workshop on omnidirectional vision at cvpr 2000*, June 2000.
18. ——, *Page of omnidirectional vision hosted by the grasp laboratory*, `http://www.cis.upenn.edu/~kostas/omni.html`, 2005.
19. P. David, D. DeMenthon, and R. Duraiswami, *Simultaneous pose and correspondence determination using line features*, Proc. IEEE Conf. Comp. Vision Patt. Recog., 2003.
20. A. Davison, *Real-time simultaneous localisation and mapping with a single camera*, IEEE Int. Conf. on Computer Vision, 2003, pp. 1403–1410 vol. 2.
21. C. Canudas de Wit, H. Khennouf, C. Samson, and O. J. Sordalen, *Chap.5: Nonlinear control design for mobile robots*, Nonlinear control for mobile robots (Yuan F. Zheng, ed.), World Scientific series in Robotics and Intelligent Systems, 1993.
22. P. E. Debevec, C. J. Taylor, and J. Malik, *Modeling and rendering architecture from photographs: a hybrid geometry and image-based approach*, SIGGRAPH, 1996.
23. S. Derrien and K. Konolige, *Approximating a single viewpoint in panoramic imaging devices*, Proceedings of the 1st International IEEE Workshop on Omnidirectional Vision at CVPR 2000, June 2000, pp. 85–90.
24. G. DeSouza and A. Kak, *Vision for mobile robot navigation: A survey*, IEEE Transactions on Pattern Analysis and Machine Intelligence **24** (2002), no. 2, 237–267.
25. O. Faugeras, *Three-dimensional computer vision - a geometric viewpoint*, MIT Press, 1993.
26. Mark Fiala, *Panoramic computer vision*, Ph.D. thesis, University of Alberta, 2002.
27. S. Fleck, F. Busch, P. Biber, H. Andreasson, and W. Straber, *Omnidirectional 3d modeling on a mobile robot using graph cuts*, Proc. IEEE Int. Conf. on Robotics and Automation, 2005, pp. 1760–1766.
28. J. Foote and D. Kimber, *Flycam: Practical panoramic video and automatic camera control*, Proc. of the IEEE Int. Conference on Multimedia and Expo, vol. III, August 2000, pp. 1419–1422.
29. S. Gaechter and T. Pajdla, *Mirror design for an omnidirectional camera with a uniform cylindrical projection when using svavisca sensor*, Tech. report, Czech Tech. Univ. - Faculty of Electrical Eng. ftp://cmp.felk.cvut.cz/pub/cmp/articles/pajdla/Gaechter-TR-2001-03.pdf, March 2001.
30. S. Gaechter, T. Pajdla, and B. Micusik, *Mirror design for an omnidirectional camera with a space variant imager*, IEEE Workshop on Omnidirectional Vision Applied to Robotic Orientation and Nondestructive Testing, August 2001, pp. 99–105.
31. J. Gaspar, *Omnidirectional vision for mobile robot navigation*, Ph.D. thesis, Instituto Superior Técnico, Dept. Electrical Engineering, Lisbon - Portugal, 2003.

32. J. Gaspar, C. Deccó, J. Okamoto Jr, and J. Santos-Victor, *Constant resolution omnidirectional cameras*, 3rd International IEEE Workshop on Omni-directional Vision at ECCV, 2002, pp. 27–34.
33. J. Gaspar, E. Grossmann, and J. Santos-Victor, *Interactive reconstruction from an omnidirectional image*, 9th International Symposium on Intelligent Robotic Systems (SIRS'01), July 2001.
34. J. Gaspar and J. Santos-Victor, *Visual path following with a catadioptric panoramic camera*, Int. Symp. Intelligent Robotic Systems, July 1999, pp. 139–147.
35. J. Gaspar, N. Winters, and J. Santos-Victor, *Vision-based navigation and environmental representations with an omni-directional camera*, IEEE Transactions on Robotics and Automation **16** (2000), no. 6, 890–898.
36. D. Gavrila and V. Philomin, *Real-time object detection for smart vehicles*, IEEE, Int. Conf. on Computer Vision (ICCV), 1999, pp. 87–93.
37. C. Geyer and K. Daniilidis, *A unifying theory for central panoramic systems and practical applications*, ECCV 2000, June 2000, pp. 445–461.
38. ——, *Catadioptric projective geometry*, International Journal of Computer Vision **43** (2001), 223–243.
39. Gene H. Golub and Charles F. Van Loan, *Matrix computations*, third ed., Johns Hopkins Studies in the Mathematical Sciences, The Johns Hopkins University Press, 1996. MR 1 417 720.
40. P. Greguss, *Panoramic imaging block for 3d space*, US patent 4,566,763, January 1986, Hungarian Patent granted in 1983.
41. P. Greguss (ed.), *Ieee icar 2001 workshop on omnidirectional vision applied to robotic orientation and non-destructive testing*, August 2001.
42. E. Grossmann, D. Ortin, and J. Santos-Victor, *Algebraic aspects of reconstruction of structured scenes from one or more views*, British Machine Vision Conference, BMVC2001, September 2001, pp. 633–642.
43. Etienne Grossmann, *Maximum likelihood 3d reconstruction from one or more uncalibrated views under geometric constraints*, Ph.D. thesis, Instituto Superior Técnico, Dept. Electrical Engineering, Lisbon–Portugal, 2002.
44. E. Hecht and A. Zajac, *Optics*, Addison Wesley, 1974.
45. R. Hicks, *The page of catadioptric sensor design*, http://www.math.drexel.edu/~ahicks/design/, 2004.
46. R. Hicks and R. Bajcsy, *Catadioptric sensors that approximate wide-angle perspective projections*, Proceedings of the Computer Vision and Pattern Recognition Conference (CVPR'00), June 2000, pp. 545–551.
47. A. Howard, M.J. Mataric, and G. Sukhatme, *Putting the 'i' in 'team': an egocentric approach to cooperative localization*, IEEE Int. Conf. on Robotics and Automation, 2003.
48. D. Huttenlocher, G. Klanderman, and W. Rucklidge, *Comparing images using the hausdorff distance*, IEEE Transactions on Pattern Analysis and Machine Intelligence **15** (1993), no. 9, 850–863.
49. D. Huttenlocher, R. Lilien, and C. Olsen, *View-based recognition using an eigenspace approximation to the hausdorff measure*, IEEE Transactions on Pattern Analysis and Machine Intelligence **21** (1999), no. 9, 951–956.
50. S. B. Kang and R. Szeliski, *3d scene data recovery using omnidirectional multibaseline stereo*, CVPR, 1996, pp. 364–370.

51. N. Karlsson, E. Di Bernardo, J. Ostrowski, L. Goncalves, P. Pirjanian, and M. Munich, *The vslam algorithm for robust localization and mapping*, Proc. IEEE Int. Conf. on Robotics and Automation, 2005, pp. 24–29.
52. A. Kosaka and A. Kak, *Fast vision-guided mobile robot navigation using model-based reasoning and prediction of uncertainties*, CVGIP: Image Understanding **56** (1992), no. 3, 271–329.
53. J. J. Leonard and H. F. Durrant-Whyte, *Mobile robot localization by tracking geometric beacons*, IEEE Trans. on Robotics and Automation **7** (1991), no. 3, 376–382.
54. R. Lerner, E. Rivlin, and I. Shimshoni, *Landmark selection for task-oriented navigation*, Proc. Int. Conf. on Intelligent Robots and Systems, 2006, pp. 2785–2791.
55. LIRA-Lab, *Document on specification*, Tech. report, Esprit Project n. 31951–SVAVISCA - available at `http://www.lira.dist.unige.it` - SVAVISCA–GIOTTO Home Page, May 1999.
56. A. Majumder, W. Seales, G. Meenakshisundaram, and H. Fuchs, *Immersive teleconferencing: A new algorithm to generate seamless panoramic video imagery*, Proceedings of the 7th ACM Conference on Multimedia, 1999.
57. D. Marr, *Vision*, W.H. Freeman, 1982.
58. B. McBride, *Panoramic cameras time line*, `http://panphoto.com/TimeLine.html`.
59. B. Micusik and T. Pajdla, *Structure from motion with wide circular field of view cameras*, IEEE Transactions on Pattern Analysis and Machine Intelligence (PAMI) **28** (2006), no. 7, 1135–1149.
60. K. Miyamoto, *Fish-eye lens*, Journal of the Optical Society of America **54** (1964), no. 8, 1060–1061.
61. L. Montesano, J. Gaspar, J. Santos-Victor, and L. Montano, *Cooperative localization by fusing vision-based bearing measurements and motion*, Int. Conf. on Intelligent Robotics and Systems, 2005, pp. 2333–2338.
62. H. Murase and S. K. Nayar, *Visual learning and recognition of 3d objects from appearance*, International Journal of Computer Vision **14** (1995), no. 1, 5–24.
63. V. Nalwa, *A true omni-directional viewer*, Technical report, Bell Laboratories, February 1996.
64. S. K. Nayar, *Catadioptric image formation*, Proc. of the DARPA Image Understanding Workshop, May 1997, pp. 1431–1437.
65. ——, *Catadioptric omnidirectional camera*, Proc. IEEE Conf. Computer Vision and Pattern Recognition, June 1997, pp. 482–488.
66. S. K. Nayar and V. Peri, *Folded catadioptric cameras*, Proceedings of the IEEE Computer Vision and Pattern Recognition Conference, June 1999.
67. E. Oja, *Subspace methods for pattern recognition*, Research Studies Press, 1983.
68. M. Ollis, H. Herman, and S. Singh, *Analysis and design of panoramic stereo using equi-angular pixel cameras*, Tech. report, Carnegie Mellon University Robotics Institute, TR CMU-RI-TR-99-04, 1999, comes from web.
69. T. Pajdla and V. Hlavac, *Zero phase representation of panoramic images for image based localization*, 8th Inter. Conf. on Computer Analysis of Images and Patterns CAIP'99, 1999.
70. V. Peri and S. K. Nayar, *Generation of perspective and panoramic video from omnidirectional video*, Proc. DARPA Image Understanding Workshop, 1997, pp. 243–246.

71. R. Pless, *Using many cameras as one*, Proc CVPR, 2003, pp. II: 587–593.
72. D. Rees, *Panoramic television viewing system, us patent 3 505 465*, postscript file, April 1970.
73. W. Rucklidge, *Efficient visual recognition using the hausdorff distance*, Lecture Notes in Computer Science, vol. 1173, Springer-Verlag, 1996.
74. J. Shi and C. Tomasi, *Good features to track*, Proc. of the IEEE Int. Conference on Computer Vision and Pattern Recognition, June 1994, pp. 593–600.
75. S. Sinha and M. Pollefeys, *Towards calibrating a pan-tilt-zoom camera network*, OMNIVIS'04, workshop on Omnidirectional Vision and Camera Networks (held with ECCV 2004), 2004.
76. S.N. Sinha and M. Pollefeys, *Synchronization and calibration of camera networks from silhouettes*, International Conference on Pattern Recognition (ICPR'04), vol. 1, 23–26 Aug. 2004, pp. 116–119 Vol. 1.
77. T. Sogo, H. Ishiguro, and M. Treivedi, *Real-time target localization and tracking by n-ocular stereo*, Proceedings of the 1st International IEEE Workshop on Omni-directional Vision (OMNIVIS'00) at CVPR 2000, June 2000.
78. M. Spetsakis and J. Aloimonos, *Structure from motion using line correspondences*, International Journal of Computer Vision **4** (1990), no. 3, 171–183.
79. P. Sturm, *A method for 3d reconstruction of piecewise planar objects from single panoramic images*, 1st International IEEE Workshop on Omnidirectional Vision at CVPR, 2000, pp. 119–126.
80. P. Sturm and S. Ramalingam, *A generic concept for camera calibration*, Proceedings of the European Conference on Computer Vision, Prague, Czech Republic, vol. 2, Springer, May 2004, pp. 1–13.
81. W. Sturzl, H. Dahmen, and H. Mallot, *The quality of catadioptric imaging - application to omnidirectional stereo*, European Conference on Computer Vision, 2004, pp. LNCS 3021:614–627.
82. T. Svoboda, T. Pajdla, and V. Hlaváč, *Epipolar geometry for panoramic cameras*, Proc. European Conf. Computer Vision, July 1998, pp. 218–231.
83. R. Talluri and J. K. Aggarwal, *Mobile robot self-location using model-image feature correspondence*, IEEE Transactions on Robotics and Automation **12** (1996), no. 1, 63–77.
84. G. Thomas, *Real-time panospheric image dewarping and presentation for remote mobile robot control*, Journal of Advanced Robotics **17** (2003), no. 4, 359–368.
85. S. Thrun and A. Bucken, *Integrating grid-based and topological maps for mobile robot navigation*, Proceedings of the 13th National Conference on Artifical Intelligence (AAAI'96), 1996.
86. S. Watanabe, *Karhunen-loève expansion and factor analysis*, Transactions of the 4th Prague Conference on Information Theory, Statistical Decision Functions and Random Processes, 1965, pp. 635–660.
87. R. Wehner and S. Wehner, *Insect navigation: use of maps or ariadne's thread?*, Ethology, Ecology, Evolution **2** (1990), 27–48.
88. N. Winters, *A holistic approach to mobile robot navigation using omnidirectional vision*, Ph.D. thesis, University of Dublin, Trinity College, 2002.
89. N. Winters, J. Gaspar, G. Lacey, and J. Santos-Victor, *Omni-directional vision for robot navigation*, 1st International IEEE Workshop on Omni-directional Vision at CVPR, 2000, pp. 21–28.
90. N. Winters and J. Santos-Victor, *Omni-directional visual navigation*, 7th International Symposium on Intelligent Robotics Systems (SIRS'99), July 1999, pp. 109–118.

91. N. Winters and G. Lacey, *Overview of tele-operation for a mobile robot*, TMR Workshop on Computer Vision and Mobile Robots. (CVMR'98), September 1999.
92. N. Winters and J. Santos-Victor, *Omni-directional visual navigation*, Proc. Int. Symp. on Intelligent Robotic Systems, July 1999, pp. 109–118.
93. P. Wunsch and G. Hirzinger, *Real-time visual tracking of 3-d objects with dynamic handling of occlusion*, IEEE Int. Conf. on Robotics and Automation, April 1997, pp. 2868–2873.
94. Y. Yagi, *Omnidirectional sensing and its applications*, IEICE Transactions on Information and Systems (1999), no. E82-D-3, 568–579.
95. Y. Yagi, Y. Nishizawa, and M. Yachida, *Map-based navigation for mobile robot with omnidirectional image sensor COPIS*, IEEE Trans. Robotics and Automation **11** (1995), no. 5, 634–648.
96. K. Yamazawa, Y. Yagi, and M. Yachida, *Obstacle detection with omnidirectional image sensor hyperomni vision*, IEEE ICRA, 1995, pp. 1062–1067.
97. J. Zheng and S. Tsuji, *Panoramic representation for route recognition by a mobile robot*, International Journal of Computer Vision **9** (1992), no. 1, 55–76.